FITS, PASSIONS, AND PAROXYSMS

Building upon his pioneering investigation of the colors of thin films, Isaac Newton developed two influential theories, one on the structure of matter, explaining the colors of bodies, and the other on fits, describing the periodicity of light. Alan Shapiro, editor of *The Optical Papers of Isaac Newton*, on the basis of his study of Newton's unpublished manuscripts, recounts the development of these theories and analyzes their experimental foundation. He also shows the essential role that Newton's philosophy of science played in the formulation and reception of these theories.

The second part of the book describes a vigorous dispute over Newton's theory of colored bodies waged by physicists and chemists for nearly fifty years from the late eighteenth to the early nineteenth century. Professor Shapiro's analysis of this previously unknown dispute and of the reasons for the chemists' attack on Newton's theory illuminates the nature and relation of physics and chemistry during this seminal period of their development.

FITS, PASSIONS, AND PAROXYSMS

Physics, method, and chemistry and Newton's theories of colored bodies and fits of easy reflection

ALAN E. SHAPIRO
University of Minnesota

CAMBRIDGE UNIVERSITY PRESS
Cambridge, New York, Melbourne, Madrid, Cape Town, Singapore, São Paulo, Delhi

Cambridge University Press
The Edinburgh Building, Cambridge CB2 8RU, UK

Published in the United States of America by Cambridge University Press, New York

www.cambridge.org
Information on this title: www.cambridge.org/9780521117555

First published 1993
This digitally printed version 2009

A catalogue record for this publication is available from the British Library

Library of Congress Cataloguing in Publication data
Shapiro, Alan E. (Alan Elihu), 1942–
Fits, passions, and paroxysms : physics, method, and chemistry and
Newton's theories of colored bodies and fits of easy reflection /
Alan E. Shapiro.
p. cm.
Includes bibliographical references and index.
ISBN 0-521-40507-6
1. Light, Corpuscular theory of – History. 2. Color – History.
3. Interference (Light) – History. 4. Chemistry, Physical and
theoretical. 5. Newton, Isaac, Sir, 1642–1727. Opticks.
I. Title.
QC402.S53 1993
535′.12 – dc20 92–14762
CIP

ISBN 978-0-521-40507-2 hardback
ISBN 978-0-521-11755-5 paperback

Additional resources for this publication at www.cambridge.org/9780521117555

To Asger Aaboe, Martin Klein,
and the memory of Derek de Solla Price,
who nurtured a community of scholars

CONTENTS

Preface	*page* ix
Abbreviated references	xiv
Frequently cited writings by Newton	xvi

PART I. PHYSICS AND METHOD: NEWTON'S THEORIES OF COLORED BODIES AND FITS

1 Historical and philosophical background	3
1.1 Introduction, historiographical and historical	3
1.2 Hypotheses and the quest for certainty	12
1.3 Transduction	40
2 Newton's rings	49
2.1 Newton, Hooke, and the colors of thin films	50
2.2 Observations of Newton's rings	59
2.3 The hypothesis of aethereal vibrations	72
2.4 Describing Newton's rings: The nomograph	89
3 The colors of natural bodies	98
3.1 Boyle's considerations and Hooke's hypothesis	99
3.2 Newton's phenomenological theory	106
3.3 Newton's analogical theory	110
3.4 No rational doubts	129
4 The theory of fits	136
4.1 Completing the *Opticks*	138
4.2 The colors of thick plates	150
4.3 From vibrations to fits	171

4.4 The formal theory 179
4.5 Conclusion and historical postscript 199

PART II. PHYSICS AND CHEMISTRY: THE THEORY OF COLORED BODIES, THE CHEMISTS' REVOLT, AND ABSORPTION SPECTROSCOPY

5 The glory years: 1704–1777 211
5.1 Background: Chemistry and physics in the eighteenth century 216
5.2 Newton's theory and natural philosophy 225

6 The chemistry of light in France: 1776–1790 242
6.1 The chemists' revolt 242
6.2 The chemistry of color 253

7 The chemical philosophy in Britain: 1785–1815 268
7.1 The chemistry of light and color 269
7.2 The physicists' defection 281

8 Debate and absorption spectroscopy in France: 1791–1816 291
8.1 An Italian interlude: Venturi 292
8.2 The physicists' defense of law and order 302

9 Absorption spectroscopy in Britain: 1822–1833 330
9.1 Brewster and Herschel 331
9.2 Conclusion 354

Appendix 1 Jordan's criticism of transduction in Newton's theory of colored bodies 363
Appendix 2 A derivation of the diameters of the colored rings in thick plates according to the theory of fits 367
Bibliography 370
Index 391

PREFACE

Many years ago when I was a graduate student, I first read Newton's *Opticks* and encountered his theory of the colors of natural bodies. With the self-confidence typical of a graduate student, I found the central idea of his theory, namely, that the colors of bodies are produced by tiny transparent corpuscles in exactly the same way as in thin films, too clever by far, if not downright silly. I tacitly assumed that such a strange theory could not have been taken seriously by anyone, and so could not have a history. Inasmuch as the theory of colored bodies still remains without a history, other scholars had evidently arrived at the same conclusion and confirmed my earlier hasty assumption. About six or seven years ago, for a project on the history of color theory, I set out to examine as many manuals and handbooks on painting and dyeing from approximately 1600 to 1850 as I could find. I was already convinced that the history of optics and color theory could not be confined to physics, or even to the scientific literature, and that to comprehend that history properly this broader literature must be searched. At some point I came upon Edward Bancroft's and Claude-Louis Berthollet's treatises on dyeing from the 1790s and was puzzled, and astonished, to find sustained, aggressive attacks on Newton's theory of colored bodies. These soon led me to the work of Edward Hussey Delaval and Jean-Henri Hassenfratz, which refuted Newton's theory, and to that of René-Just Haüy and Jean-Baptiste Biot, which defended it. My net rapidly spread wider and snared scores of additional participants. It soon became evident that for about fifty years surrounding the turn of the nineteenth century a heated battle between chemists and physicists was waged over Newton's theory, until it was finally refuted by John Herschel and

David Brewster around 1830. For a theory without a history, Newton's theory of colored bodies turned out to have had a long and turbulent one involving many major scientists. I also came upon another, equally puzzling aspect to this history: the repeated independent invention of absorption spectroscopy around 1800 as a means to test Newton's theory, and its subsequent total neglect until it was once again independently developed by Herschel and Brewster in 1822. Only with the work of these latter two does absorption spectroscopy enter the historical literature.

I found that the long and acrimonious dispute over Newton's theory of colored bodies and the lost history of absorption spectroscopy could best be explained by viewing it within the broader context of the varying relation and mutual influence of physics and chemistry within this period. My initial interpretation of the conflict as simply a clash between chemists and physicists was too crude. It could not, for example, explain why the battle erupted when it did and continued so long in France, where, through the alliance of Laplace with Lavoisier and Berthollet, physics and chemistry were drawn so close together; nor could it explain the stillborn creation of absorption spectroscopy. A more nuanced interpretation evolved that took into account such differences between the two sciences as their epistemological foundations and experimental procedures, as well as the different historical development of these disciplines in France and in Britain, where no physicist ever stepped forward to defend Newton's theory against the chemists' attack. My focus upon a second-level theory, colored bodies (as opposed to, say, the wave theory of light or phlogiston theory), provides a unique and somewhat unusual perspective on the nature of the physical sciences in this period and offers some unexpected insights. For instance, a number of historians have touched on the chemistry of light – the idea that light interacts with matter just as any chemical substance – but I was surprised to see how broadly accepted it was at the turn of the eighteenth century. In fact, I was surprised to see the extent to which qualities had been readmitted into the physical sciences around the turn of the century. This should cause us to reevaluate our traditional emphasis on the mathematical and experimental aspects of physical optics in this period and somehow relate it to the apparently more widespread chemical view of light.

It was only after I had uncovered the broad outline of the history of the theory of colored bodies that, in the natural course of preparing my edition of Newton's optical papers, I arrived at Book II of the

Opticks and related papers. This contains the theory of colored bodies as well as the theory of fits of easy reflection and transmission (Newton's account of the periodicity of light). The theory of fits has long been dismissed by most scholars as hypothetical, opaque, and, worst of all, wrong. These were obviously not judgments arrived at after careful deliberation, as there is scarcely any serious historical literature on that theory. The more I studied the development of the theory of fits through various published and unpublished papers of Newton the more I came to admire it. The theory is certainly hypothetical, but its obscurity has been exaggerated, and it is "wrong" only insofar as it was replaced by a better theory, the wave theory of light.

In order to explain the abstract and rather obscure formulation of the theory of fits and its subsequent poor reception, I gradually recognized that it was necessary to take into account Newton's scientific methodology, in particular, his views on hypotheses. Likewise, I found that in order to explain the development of the theory of colored bodies it was necessary to understand Newton's use of transduction, which is a method for extending knowledge of macroscopic bodies to their unobservable, microscopic components.

I am not a philosopher of science, but as a historian of science I was compelled to study Newton's philosophy because it penetrated into the structure and formulation of his theories. For very different reasons, many contemporary scientists and historians of science believe that scientists' methodology is of little consequence and of even less interest. Yet Newton's two theories and their reception cannot be properly understood without appreciating his methodology. Just as I have drawn upon the work of philosophers, so I hope they will be able to pursue in a more analytical way the philosophical issues that I have treated historically. Indeed, when I presented the earliest version of what are now the first three chapters as a paper at a conference in Jerusalem in 1987, I was encouraged by philosophers of science – Ian Hacking, Ernan McMullin, and Hillary Putnam – to develop it into a book.

This book consists of two distinct studies: Part I, on Newton's two theories, and Part II, on the history of his theory of colored bodies. Although I hope everyone will be enthralled by both parts, I have provided sufficient cross-references so that they may be read separately. The obvious link between the two parts is Newton's theory of colored bodies and its history over 150 years from its formulation through its ultimate refutation. Another, more subtle link is my continuing concern with the development of mathematical and quantitative physics. There are, however, major differences between the two parts, the most evident being the more than a century separating them. More fundamentally, in the first part I have been able to treat Newton's formulation of his theories almost in complete isolation from the scientific community, except for the initial stimulus from his readings and his priority dispute with Hooke, whereas in the second I treat scores of participants within the context of the chemical and physical communities. This difference arises from the different sorts of historical questions I treat in the two parts: the creation of a theory, which Newton happened to carry out in near isolation, and its reception, which necessarily involves the larger community.

By adopting a rather narrow focus for this study, I have been able to cover a broader sweep of time than is now generally adopted for the study of modern science. A fundamental task of the history of science is to treat the persistence of scientific problems and theories within different scientific milieus. The manageable scope and crucial position of the history of colored bodies allow us to see how both the problem and the theory were at the same time transmitted and transformed by a diversity of factors in different communities. Nonetheless, the principal justification for this book is that there is virtually no literature on Newton's two theories and their subsequent history, which, I believe, is an exciting story that raises fundamental issues about the development of science.

For readers curious about my title, "fits," "passions," and "paroxysms" are all legitimate Newtonian terms for the states of easy reflection and transmission of light. They can equally well describe the states that generations of scholars have been put into by the two theories of Book II. I must also confess that I was motivated by my own passion for mysteries and thrillers. Recognizing my evident lack of writing talent in this genre, I seized at the opportunity to choose at least a title that belonged to it. Archimedes' decision to title one of his works *On Floating Bodies* may serve as a venerable precedent for my choice.

Editorial note: Because I have worked with Newton's manuscripts for years as the editor of his *Optical Papers* and much of my story depends on manuscript evidence, I generally cite the manuscript and also, when there is one, the published version. Unless otherwise indicated all manuscripts are in Cambridge University Library. All quotations, whatever the source, follow the original, and any changes, such as in punctuation or italicization, are indicated in footnotes.

In the first edition of the *Opticks* the pagination begins anew with

Book II, and page references to this portion of the work are preceded by a subscript 2. This notation is extended to any multipaginated work. In queries to the *Opticks* that are indicated by two numbers such as 23/31, the first number is that of the Latin translation of 1706, when the query initially appeared, and the second one is its new number in the second English edition of 1716, when it was renumbered and often revised. Because the new queries for the Latin edition were written in English by Newton and then translated into Latin by Samuel Clarke, I cite Newton's English drafts of them rather than my own translations. Newton himself used these drafts as the basis for the queries in the second English edition and did not retranslate Clarke's Latin back into English. Significant differences from the 1716 edition or Clarke's translation are indicated in footnotes.

This research has been supported by the John Simon Guggenheim Memorial Foundation, the Rowland Foundation, and the National Science Foundation. I am truly grateful to them for their generous assistance. The Guggenheim Foundation has continued its munificence by providing a subvention for the two colored plates, which, the foundation recognized, are not a luxurious indulgence in a book on color. Peter Harman and Casper Hakfoort read parts of the book and provided useful comments and suggestions. Arthur Donovan's criticisms were particularly insightful and valuable. I thank Yehuda Elkana for inviting me to deliver a paper at a conference sponsored by the Van Leer Jerusalem Institute celebrating the three hundredth anniversary of the publication of the *Principia* in April 1987. This book grew from that paper. I also thank John Beatty for his many helpful discussions and suggestions. The staff of Cambridge University Library has been exceedingly helpful and accommodating, and I especially wish to acknowledge the cooperation of Godfrey Waller of the Manuscripts Room, where I have spent so much time poring over Newton's manuscripts. All quotations from those manuscripts are by permission of the Syndics of Cambridge University Library. Maurine Bielawski typed the various drafts of the book with efficiency and her usual good humor.

ABBREVIATED REFERENCES

Add.	Additional Manuscript, Cambridge University Library
Birch	Thomas Birch, *The History of the Royal Society of London*, 4 vols. (London, 1756–7)
Correspondence	*The Correspondence of Isaac Newton*, ed. H. W. Turnbull et al., 7 vols. (Cambridge, 1959–77)
Mathematical Principles	*Sir Isaac Newton's Mathematical Principles of Natural Philosophy*, trans. Andrew Motte and Florian Cajori (Berkeley, 1934)
Optical Papers	*The Optical Papers of Isaac Newton*, ed. Alan E. Shapiro, 1 vol. to date (Cambridge, 1984–)
Optice	Isaac Newton, *Optice*, trans. Samuel Clarke (London, 1704)
Opticks	Isaac Newton, *Opticks: Or, a Treatise of the Reflexions, Refractions, Inflexions and Colours of Light* (London, 1704)
Opticks (Dover)	Isaac Newton, *Opticks*. Based on the 4th ed. London, 1730 (New York, 1952)
Principia	*Isaac Newton's 'Philosophiae naturalis princi-*

pia mathematica.' The Third Edition (1726) with Variant Readings, eds. Alexandre Koyré and I. Bernard Cohen (Cambridge, Mass., 1972)

Questions J. E. McGuire and Martin Tamny, *Certain Philosophical Questions: Newton's Trinity Notebook* (Cambridge, 1983)

FREQUENTLY CITED WRITINGS BY NEWTON

Ca. 1664	"Of Colours"	Essay in the notebook "Questiones quaedam philosophicae," Add. 3996.
Ca. 1666	"Of Colours"	Essay in a chemical notebook, Add. 3975.
1670–2	*Optical Lectures*	Delivered at Cambridge University in Latin. Published posthumously.
Ca. 1671	"Of ye coloured circles twixt two contiguous glasses"	First systematic investigation of the colors of thin films.
1672	"New theory about light and colors"	Theory of compound nature of white light in *Philosophical Transactions*.
1672	"Observations"	First, unpublished paper on the colors of thin films and natural bodies.
1675	"Observations" and "Hypothesis explaining the properties of light"	A revision of the preceding plus a physical model of light and the aether. Read at Royal Society.

1687	*Principia*	*Mathematical Principles of Natural Philosophy.*
Ca. 1687–8	"Fundamentum opticae"	First draft of *Opticks*, Bk. I.
Ca. 1690–1	"The Fourth Book"	Sketch of a projected fourth book of *Opticks.*
Ca. 1691	"Book IV, Part I"	The colors of thick plates in the first completed state of *Opticks.* Revised to become Book II, Part IV, of published *Opticks.*
1704	*Opticks*	Contains Queries 1–16.
1706	*Optice*	Latin translation by Clarke. Queries 17–23 added.
1713	*Principia*, 2nd ed.	General Scholium and Rules of Reasoning added.
1716	*Opticks*, 2nd ed.	Queries 17–23 renumbered 25–31, and 17–24 added.

PART I

PHYSICS AND METHOD: NEWTON'S THEORIES OF COLORED BODIES AND FITS

1

HISTORICAL AND PHILOSOPHICAL BACKGROUND

1.1. INTRODUCTION, HISTORIOGRAPHICAL AND HISTORICAL

In Book II of the *Opticks,* Newton set forth his observations on the colors of thin films and thick plates. He also expounded two theories there, one on the colors of natural bodies and the other on fits of easy reflection and transmission, which was his explanation of the periodicity of light. Each of these theories has its roots in his *anni mirabiles* nearly forty years earlier and was still in the mainstream of scientific thought over a century later. Yet both are entirely bereft of a history. The most obvious reason for this neglect is that the theories are "wrong" and therefore, it is implicitly assumed, unfruitful and unworthy of the great man who gave us the theories of gravitation and of white light and color. Historians need not worry about embarrassing Newton. Both were fine theories capable of explaining a broad range of phenomena known in the seventeenth century and some yet to be discovered.

There are, nonetheless, understandable reasons for the unease of historians, for these theories are formulated in ways that are foreign to the modern sensibility. To appreciate them, we have to enter into seventeenth-century scientific methodology and Newton's philosophy of science. Traditionally, studies of Newton's investigations of the colors of thin films have isolated his demonstration of the periodicity of light from the rest of Book II as an instance of his experimental skill, which it certainly is. They then turn only briefly, if at all, to his enigmatic theory of fits. At this point the studies generally either denounce the theory of fits as an arbitrary hypothesis or declare it to be incomprehensible, and sometimes even both. A more fruitful approach, I suggest, is provided by posing two novel questions: Why, in the *Opticks*, did Newton transform what had earlier been a *hypothesis* of aethereal vibrations, which he had used to explain the colors of thin films, into a *theory* of experimentally demonstrated propositions?[1] Why did he construct that theory in such a stark, nearly unintelligible form (though its difficulty has been exaggerated)? The answer to both questions depends on Newton's methodology, his insistence that scientific principles be founded on experiment and observation, and his consequent requirement that hypotheses not be mingled with demonstrated principles.

In the early 1690s, during the composition of the *Opticks*, Newton discovered a new phenomenon, the periodic colors of thick plates. He was able to explain them by his concept of vibrations, which as a consequence were "rendered general by induction" or confirmed by new experiments and observations. According to his methodology, the vibrations could be reformulated as a series of demonstrated principles or properties, provided they contained no hypothetical elements or conjectural causal explanations, such as the aether. Thus, in attempting to purge all hypotheses from his model of aethereal vibrations, Newton rendered the theory of fits so abstract that it suffered a loss of physical intelligibility. To appreciate the theory of fits on Newton's own terms, we will have to examine his philosophical views on hypotheses as well as his physical research on thick plates. The total neglect of this research by historians is ironic, for it was Newton's last major optical investigation and, appropriately enough, his most sophisticated in its combination of experiment, mathematical analysis, and physical explanation.

In his theory of the colors of natural bodies – that is, the quotidian colors of the sky, vegetables, paper, and linen – Newton demonstrates that the colors of all bodies are produced by the corpuscles composing them in exactly the same way as they are in thin films. Thus all bodies are composed of invisible, transparent corpuscles that are, in effect, tiny fragments of a thin film that produces the identical color in Newton's rings. Newton was not simply proposing a conjectured analogy but, rather, an experimentally demonstrated theory based on the method of transduction. Transduction is a method of making inferences about the unobservable, microscopic components of bodies from knowledge of the observed laws and properties of macroscopic

[1] Strictly speaking, as we shall see in the next section, Newton did not transform or elevate hypotheses to demonstrated principles. Rather, he would say that he used hypotheses to discover properties that become principles.

bodies. Intrinsic to the newly developed mechanical philosophy, transduction was widely utilized and, as Newton applied it, closely akin to induction.

Although transduction can be a fruitful method of reasoning, it is problematic. In the extreme case of quantum mechanics, for example, it breaks down completely, for the laws governing the smallest particles of matter are presumed to be fundamentally different from those governing the "classical," macroscopic world. Thus the method of transduction and its place within the mechanical philosophy will have to be scrutinized in order to appreciate Newton's theory of colored bodies and its broad acceptance. It is particularly surprising that the theory of colored bodies has been so neglected by historians: It occupies a unique position in Newton's optical thought, as it is actually a theory of matter, not a theory of light, that accounts for certain optical properties of bodies.

There is a deeper historiographical reason for the neglect of Newton's theory of colored bodies – indeed, of the entire history of colored bodies – than that it is "wrong." After all, most scientific theories are wrong. Since the early seventeenth century, the explanation of the colors of bodies has gradually disappeared as a fundamental problem in physics, and historians have taken their lead from physicists. Before that time, the colors of bodies held a certain conceptual primacy. Medieval natural philosophers, developing ideas of Aristotle, conceived of color as a *quality of a body* that was revealed by light. Thus they variously attributed the origin of different colors to the constituents of bodies, either to different mixtures of the four elements (Averroes, for example) or of the four primary qualities (Albertus Magnus and Nicole Oresme).[2] Other interpretations were proposed, but in all of them color was an intrinsic quality of bodies. Radiant colors such as those of the rainbow and prism were particularly troublesome, because colors appeared where there were evidently no colored bodies. These colors then either had to be assimilated to those of bodies (Theodoric of Freiberg), in which case the colors of the bow

[2] Averroes, *Epitome of "Parva naturalia,"* p. 10; Albertus Magnus, *De anima*, Bk. II, Tract. 3, Ch. 7, p. 109; and Stephen C. McCluskey, Jr., "Nicole Oresme on light," pp. 73, 268–71. In Newton's undergraduate textbook on Scholastic natural philosophy, *Physiologiae peripateticae* (1642), Johannes Magirus indifferently considers colors to derive from the four elements and the four qualities, pp. 166, 322. By identifying light with the element fire, the transparent with air and water, and the opaque with earth (or equivalently with the qualities that compose the elements), medieval scholars were able to arrive at various formulations of Aristotle's view that colors derive from a mixture of light and transparent.

were explained in the same terms as those of bodies, or reduced to a mere appearance (Roger Bacon and Oresme).[3] Whatever solution was adopted, it was the permanent colors of bodies that were accepted as "real" or "true," whereas the other sort were only "apparent." The very terms show which was considered more fundamental and that color was considered as a property of bodies. Robert Boyle, in his *Experiments and Considerations Touching Colours*, captured this feature of Scholastic color theory when he observed that "the Peripatetick Schools, though they dispute amongst themselves divers particulars concerning Colours, yet in this they seem Unanimously enough to Agree, that Colours are Inherent and Real Qualities, which the Light doth but Disclose, and not concurr to Produce."[4]

The mechanical philosophers of the seventeenth century rejected the distinction between real and apparent colors. Though various interpretations of the mechanical philosophy were advanced, there was broad agreement that natural phenomena were to be explained in terms of matter and its size, shape, arrangement, and, above all, its motion. In this view color is a sensation excited by light in the brain, and different colors arise from different *properties of light*. It was only of secondary consequence that light of a particular color arrived from a colored body or a rainbow. Both were equally light and caused the same sensation. All colors were now, in one sense (though in a very different one), real or true and, in another, only apparent. "And I cannot approve the distinction," René Descartes observed in the *Meteores*, "made by the philosophers when they say that there are some true colors, and others which are only false or apparent. For because the entire *true* nature of colors consists only in their *appearance*, it seems to me to be a contradiction to say that they are false, *and* that they appear."[5] The relation of the two sorts of color was now inverted, and radiant colors had attained conceptual primacy over those of bodies. In order to study color, whatever its origin, one need only study colored light. One of the more important consequences of this new approach was Newton's theory of white light and color. It allowed him to found a theory of all colors based almost exclusively on the study of light refracted by prisms, that is, on the former ap-

[3] For a particularly patent reduction of the colors of the bow to those of bodies, see Theodoric of Freiberg's use of the four elements in William A. Wallace, *The Scientific Methodology of Theodoric of Freiberg*, p. 193. For the contrary view that they are only an appearance, see Roger Bacon, *The Opus Majus*, Pt. VI, Chs. 8, 12, 2:605, 611; and McCluskey, "Nicole Oresme," pp. 73, 276–7.

[4] Boyle, *Experiments and Considerations Touching Colours* (1664), Pt. I, Ch. 5, p. 84.

[5] Descartes, *Discourse on Method, Optics, Geometry and Meteorology*, p. 338; italics added.

parent colors. As it turned out, Newton's discovery that bodies only return and do not alter the colors of the light falling on them diminished still more the role of bodies in producing color. Now they were limited either to selecting or absorbing, but not creating, the colors incident upon them.

So effective was the new philosophy's revolutionary concept of color that historians have devoted virtually all their attention to the study of radiant colors and have confined the study of colored bodies to an insignificant position. The rainbow remains at the center of accounts of medieval color theory. In some respects this is perfectly legitimate, for the rainbow exercised the imagination of ancient and medieval scientists; and there is a continuity between these earlier investigations, especially in the mathematical tradition, and those of the seventeenth century in the use of the prism to study radiant colors and in the concept of modification theories of color. In other respects, however, this approach distorts the past, for insufficient attention is paid to the colors of bodies, which were treated primarily in the philosophical tradition. This historical approach thus fails to acknowledge the fundamental discontinuity with the modern view introduced in the Scientific Revolution. It is history written with one eye on Descartes and the other on Newton.

Historians of the early modern era performed little better than the medievalists. It is true that I am arguing that after the mechanical philosophers introduced the idea that color was a property of light, the problem of the colors of bodies became a secondary one, simply to determine how bodies modified or analyzed incident light to produce this or that color. Nonetheless, because the colors of bodies are such a pervasive phenomenon and an integral part of the optical tradition, it remained necessary for the new philosophers to explain them if their theories were to have any pretension to a comprehensive explanation of nature. Boyle devoted an entire book, his *Experiments and Considerations Touching Colours,* to exploring the problem, and in the following year, Robert Hooke proposed a detailed explanation of the colors of bodies in one chapter of his *Micrographia.* Both of these works had a major influence on Newton's theory of colored bodies.

By about 1830, when Newton's theory was conclusively refuted and replaced by a theory of selective absorption, the colors of bodies was no longer a distinct subject in optics, as it had been for Newton and his contemporaries, let alone a privileged one, as it had been in medieval optics. Selective absorption was studied as a physical process, just one of whose applications was an explanation of the colors of bodies. Moreover, it was now recognized that bodies are colored in a number of physically distinct ways, and not by a universal cause, as had been believed for millennia. At this point physicists and historians are at one, and colored bodies is not considered a proper subject of study. This is unfortunate because historical study of explanations of the colors of bodies is particularly fascinating. As a theory of matter and its sensible qualities, it is simultaneously central to natural philosophy and at the boundary of physics and chemistry and thus reveals fundamental assumptions of each.

Before we embark on our study of Newton's investigations of the colors of thin films and natural bodies, an overview of his early optical work will be useful. We will return to much of it when the dating is justified and the contents are treated in greater detail. Newton's introduction to contemporary optics in 1664 (his last year as an undergraduate) and 1665 came primarily through the works of Descartes, Boyle, and Hooke and so was inextricably in a mechanical framework. His study of Descartes' various works – *Principles of Philosophy, Dioptrics, Geometry,* and *Meteorology* – in mid-1664 introduced him to the sine law of refraction (Snel's law), the principles of geometrical and applied optics, and the modification theory of color. Most importantly, it was in Descartes' work that Newton first encountered the power of a mathematical description of nature.

Descartes' refraction model, which was based solely on mechanical parameters (velocities and "impulses"), made a profound and enduring impression on Newton, both as a persuasive example of how all nature might be reduced to similar mechanical principles and in its own right as a mathematical model of a general optical law, which he quickly incorporated into his own research. Newton's natural philosophy was never strictly mechanical, for he always invoked some sort of "active principles" to supplement the passivity of "brute, inanimate matter." Yet his physical explanations of observed optical properties could always be interpreted mechanically (even when other sorts of interpretation were possible), if only because this approach alone offered the opportunity for a mathematical description.

In 1664, Newton began a notebook, which he entitled "Questiones quaedam philosophicae" ("Certain philosophical questions"), to record his gleanings from the new philosophy. It had numerous entries, such as "Of Attomes" and "Of heate & cold," but we are most concerned with the essay-length "Of Colours." Well over half of it consists of extracts, probably made in late 1664, from Boyle's *Touching*

Colours, which appeared early in that year. Boyle's principal aim was to establish that the colors of natural bodies arise from a modification of the incident light and, in particular, to explain those colors and their alterations in terms of the arrangements of the corpuscles composing them. In his cautious way he refused to propose any specific mechanism for the way in which light is modified to cause colors, but, as befit a chemist, he did propose a number of ways that changes in the size and arrangements of the corpuscles of bodies could alter their color. Although Newton drew many observations and his interpretation of chemical change from *Touching Colours*, its most important contribution to his development was in exciting his interest in the relation of the colors of bodies and their corpuscular structure. Newton's "Of Colours" reflected Boyle's *Touching Colours* in being almost exclusively devoted to the colors of bodies, but Newton had already set out on his own path and made his first fundamental discovery in physical optics. By viewing through a prism light reflected from bodies of various colors, he discovered that rays of different color are refracted unequally.

In January of the following year, the efflorescence of Restoration science continued with the publication of the *Micrographia* by Robert Hooke (Boyle's former assistant). Probably late that year, or early in 1666, Newton read it attentively, taking seven closely written pages of notes. Newton's antipathy to Hooke is already evident in the many critical notes (unlike those on Boyle), but he drew much from the *Micrographia*, more than he was ever able to admit. Hooke recognized that a complete account of the various ways in which colors are produced – by refraction, by thin films (interference), and by bodies – was required for any theory of color to be valid. In two chapters of the *Micrographia* he carried out this enterprise and, adopting a wave theory, provided the most comprehensive treatment of color yet to appear in the mechanical philosophy. His account of the colors of thin films was the first in the literature and was Newton's starting point. Unlike his predecessors, such as Descartes and Boyle, Hooke was not content to sketch in bare outline how the corpuscles of bodies may modify white light to produce colors. His theory of colored bodies was certainly not the first to appear, but it was the first detailed mechanical explanation. Many of Hooke's ideas on the colors of natural bodies were incorporated directly into Newton's theory. Seven years after the publication of the *Micrographia*, Hooke introduced another fundamental mode of the production of colors, Grimaldi's discovery of diffraction, to England and hence to Newton. Whatever failings Hooke's explanations of these phenomena had, we should recognize his grasp of the fundamental phenomena of optics and his pioneering attempt to unify them. Newton did, and Hooke's optical program served as a stimulus and a challenge to his own research.

By the beginning of 1666, Newton had a firm command of contemporary optics and had even advanced it somewhat. According to his own recollection, it was at this time that he made his most important optical discovery – that sunlight consists of a mixture of rays differing in degree of refrangibility and exhibited color. He found that at the same angle of incidence, rays of different color are refracted different amounts and that there is a constant correspondence between color and degree of refrangibility; that is, the red rays are always least refracted, the violet most, and the intermediate colors intermediate amounts. Rays of each color apart obey the sine law of refraction but with a different index of refraction for each.

Although Newton did not present a formal version of his theory of light and color until 1670, it will be convenient to summarize its main points here.[6] The color of a ray, he found, cannot be changed by reflection, refraction, transmission, or any other means. In order to develop further his new theory, he introduced his new concepts of simple and compound colors. Though these two sorts of color appear identical to the eye, simple or primary colors consist of rays of a single degree of refrangibility and compound ones are a mixture of rays of different refrangibility. Each species of color can be either simple or compound. Orange, for example, can consist of simple spectral orange rays or a compound of red and yellow rays. The two sorts can always be distinguished by refraction, which separates or decomposes the rays of different refrangibility that make up compound colors while leaving simple colors unchanged.

The colors of the spectrum – red, orange, yellow, green, blue, indigo, and violet – together with their intermediate gradations, are primary colors. "But," Newton announced, "the most surprising and wonderful composition was that of *Whiteness*. . . . 'Tis ever compounded."[7] This was the most revolutionary, and most resisted, part of the theory, for universally sunlight had been considered to be

[6] I am in general following the formulation of the "New theory"; Newton to Oldenburg, 6 February 1671/2, *Correspondence*, 1:92–102; which was published as "A letter of Mr. Isaac Newton . . . containing his new theory about light and colors," *Philosophical Transactions*. 6(1671/2):3075–87; reprinted in Newton, *Papers and Letters on Natural Philosophy*. See Richard S. Westfall, "The development of Newton's theory of color," and my "The evolving structure of Newton's theory of white light and color."

[7] *Correspondence*, 1:98.

simple, homogeneous, and pure, whereas colors were assumed to be some modification of sunlight. "Colours are not *Qualifications of Light*," Newton concluded, "derived from Refractions, or Reflections of natural Bodies (as 'tis generally believed,) but *Original* and *connate properties*."[8] Whenever colors appear, they are only separated from sunlight; they are never created. Thus bodies appear of a particular color because they reflect some rays in the incident light in greater quantity than others. The theory of color was the foundation for all of Newton's subsequent optical research.

Never in a hurry to publish, Newton, probably some time in 1666, wrote up the experiments establishing his theory of color in a notebook (Add. 3975) and once again entitled his account "Of Colours." However, he interrupted the record of his results on color to set forth his discovery of a method – afterward known as "Newton's rings" – for measuring the thickness at which colors appear in a thin film, a problem laid down in the *Micrographia*. Newton then set aside his research on the colors of thin films until about 1671. In the late 1660s, his efforts in optics were devoted to constructing the first successful reflecting telescopes, one in 1668 and two more in the early 1670s. His continuing interest in telescope mirrors led more than twenty years later to his discovery of the colors of thick plates, his last major optical research. Newton did not return to his theory of color until he was appointed to succeed Isaac Barrow, the first Lucasian Professor of mathematics at the University of Cambridge, in late 1669. Newton was not yet twenty-seven years old. He took the opportunity of his inaugural series of lectures, delivered between 1670 and 1672, to refine and further develop his theory of color. His *Optical Lectures* (which remained unpublished in his lifetime) contains the most comprehensive account of his theory of color and served as the source for his subsequent expositions. Having revised his *Optical Lectures* with his theory of color, he once again took up the colors of thin films in 1670 or, most likely, 1671 and carried out an extensive series of experiments that led to his command of the phenomenon. His experiments and tentative explanation of them are recorded in a working paper, "Of ye coloured circles twixt two contiguous glasses."

In the early spring of 1672, Newton wrote up his experiments on the colors of thin films and his theory of the colors of natural bodies in a paper I call the "Observations" of 1672. He had intended to send the report to the Royal Society, together with his reply to Hooke's

[8] Ibid., p. 97.

critique of his theory of white light and color. However, he became so distraught at the growing controversy over his theory of color that he abandoned the idea. A few months earlier, in December 1671, Newton had finally decided to make his theory of color public, and in January he sent his "New theory about light and colors" to be read to the Royal Society and published in *Philosophical Transactions*. The paper was not very well received, and Newton was compelled to defend it in an extensive correspondence over the next few years. This was the occasion for his first dispute with Hooke, and their relation ever after was, at best, tense. Late in 1675, Newton recovered his equanimity and revised his abandoned "Observations" from 1672 on the colors of thin films and natural bodies and composed a companion piece, "An Hypothesis explaining ye properties of Light." These were read at meetings of the Royal Society in December, January, and February, but Newton would not allow their publication. More than forty years after it was initially carried out, this research was finally published in the *Opticks*. It was now substantially enriched by the theory of fits and the analysis of the colors of thick plates, both of which were added during the composition of the *Opticks* in about 1691.

1.2. HYPOTHESES AND THE QUEST FOR CERTAINTY

One aim of my account of Newton's investigation of periodic colors is to study his use of hypotheses and how it affected the formulation of the theory of fits. This enterprise may not seem to square with Newton's many pronouncements that appear to exclude hypotheses altogether from natural science. However, his primary intention in these statements, I will argue, was not to exclude hypotheses but rather to demarcate the conjectural from the certain. To conflate hypotheses – conjectural causal explanations – with rigorously demonstrated principles would, he believed, grieviously compromise the certitude of the latter. Newton's lifelong attack on hypotheses was a consequence of his equally long quest for certainty.

Early in his career Newton set forth a program to reform the methods of contemporary science, which he judged to be rife with "conjectures and probabilities." In the seventeenth century the dual Aristotelian concepts of demonstrative knowledge of nature and belief in the possibility of attaining knowledge of the essence of things were gradually abandoned. Two not altogether independent alternatives emerged to fill this epistemological void, one mathematical and the other probabilistic. Some saw a mathematical description of nature as a new source of demonstrative knowledge that could approach the absolute certainty of pure mathematics. Many others adopted a probabilistic interpretation of scientific knowledge, according to which all explanations and theories were hypotheses with varying degrees of probability. Newton, whose outlook derived primarily from the mathematical tradition with its ideal of demonstrative knowledge, was reacting against such probabilistic interpretations. He believed that there were scientific theories – his theory of color, for instance – that were rigidly demonstrated, and not just probable conjectures. He was also attacking the proliferation of unrestrained conjectures to explain phenomena and their commingling with certain results. Amid such confusion no truth could emerge.

Part of the problem in understanding Newton's attitude toward hypotheses is that he uses the single word "hypothesis" to express two different concepts. It is what I call "imaginary hypotheses," which are based on no empirical evidence and are totally fictitious, that he banishes altogether from natural philosophy. Newton did believe that what I call "experimental hypotheses," which have some experimental foundation but not enough to be judged demonstrated principles, do have a legitimate role in natural philosophy, provided they are kept distinct from rigorously demonstrated principles: to suggest new experiments and to explain and render intelligible previously discovered properties. Besides using hypotheses in these ways, he also used them to develop theories and predict new properties, very much like the hypothetico-deductive method. When he used hypothetical causes such as light corpuscles and the aether in the latter way, he then purged them (or, at least, he attempted to do so) and reformulated his theory in terms of experimentally discovered "properties" – an important term in Newton's philosophy of science – such as unequal refrangibility and periodicity. These phenomenological theories were one way in which he attempted to achieve greater certainty. Newton appears never to have questioned the possibility of constructing a hypothesis-free science. To have denied such a possibility would have been tantamount to denying his conception of science.

Newton's quest for certainty is intimately connected to his commitment to the development of mathematical physics and the deductive model of scientific knowledge. Following an English tradition represented by Christopher Wren and Barrow, these sciences could approach the certitude of mathematics as their principles were established with greater certainty. In the early years of his career, in his zeal to reform natural philosophy and create a new mathematical science, Newton followed this mathematical tradition and stressed both the mathematical nature of his theories and the certitude of his deductions from experiment. Although he soon modified his excessive claims and more carefully distinguished mathematical demonstrations from less than certain experimental conclusions, he continued to prize mathematical theories and seek "truth" and "certainty," words that appear frequently in his writings. Only in the last decades of his life did he accept the probabilism of his contemporaries.

Newton's lifelong quest for certainty is, I claim, responsible for many of the characteristic features of his science: the rejection of hypotheses, the adoption of phenomenological theories, and the development of mathematical theories. All of these features, as we shall see, play a role in Newton's investigation of periodic colors and formulation of the theory of fits.

Although I am primarily concerned with Newton's views on hypotheses, I will perforce devote much attention to his concern with certainty because I believe that this was his motivation for his frequent attacks on hypotheses. It is difficult to define exactly his concept of certainty (or, for that matter, Wren's or Barrow's) at any particular time. For my purpose, however, it will suffice to show that he was claiming a greater degree of certainty than most of his contemporaries allowed, especially those gathered about the Royal Society, and that he subsequently modified those claims. Newton's science was rich, diverse, and ever-changing, so that one must be cautious in extending generalizations too broadly. What is true of the theory of color may not apply to the *Principia*, and what is valid in 1672 may no longer be so in 1687 or 1712. In particular, we must recognize that though the deductive mode of scientific knowledge remained Newton's ideal, it was not the only one that he utilized, as we shall see in his theory of colored bodies.

I shall begin with the obvious: Newton's declaration *hypotheses non fingo* (I feign no hypotheses), which appeared in the General Scholium appended to the second edition of the *Principia* (1713). In the penultimate paragraph of the *Principia*, he enumerates the various properties of gravity that he had succeeded in discovering and then declares:

> But hitherto I have not been able to deduce the cause of those properties of gravity from phenomena, and I feign no hypotheses; for whatever is not

deduced from the phenomena is to be called an *hypothesis;* and hypotheses, whether metaphysical or physical, whether of occult qualities or mechanical, have no place in *experimental philosophy*. In this philosophy propositions are deduced from the phenomena and rendered general by induction. . . . And to us it is enough that gravity does really exist and act according to the laws which we have expounded and suffices to account for all the motions of the celestial bodies, and of our sea.[9]

Newton is unequivocally rejecting hypotheses from experimental philosophy, but we must carefully examine what he means by "hypothesis."[10] He states that a hypothesis is "whatever is not deduced from phenomena." However, in the letter to Roger Cotes (the editor of this edition of the *Principia*) in which Newton forwarded this passage to be added to the General Scholium, he explained more fully that by hypothesis he here means a proposition that "is not a Phae-

[9] *Mathematical Principles*, p. 547; and *Principia*, 2:764. Here and elsewhere I have felt free to alter the Motte-Cajori translation. In changing "frame" to "feign," I have followed Koyré, *Newtonian Studies*, Ch. 2, "Concept and experience in Newton's scientific thought," pp. 35–6. However, Koyré's claim that Newton "nowhere used the word 'frame,' " which he supports by a reference to Query 28, is too strong. In his English draft of Query 20/28 for the Latin translation of the *Opticks* in 1706, Newton wrote: "Later Philosophers banish the consideration of the supreme cause out of natural Philosophy *framing* Hypotheses for expla[in]ing all things without it & referring it to Metaphysicks: whereas the main business of natural Philosophy is to argue from effects to causes" (Add. 3970, f. 249 *bis;* italics added). (Clarke, incidentally, translated "framing" as "confingentes"; *Optice*, p. 314.) In his initial revise of this passage for the second English edition Newton altered this to: "Later Philosophers banish the consideration of other causes then mechanism out of natural Philosophy, *framing* Hypotheses for expla[in]ing all things mechanically & referring other causes to Metaphysicks: whereas the main business of natural Philosophy is to argue from Phaenomena without *feigning* Hypotheses." (Add. 3970, f. 247; italics added). Finally, in the manuscript that was sent to the printer (and afterwards returned) he replaced the first "framing" by "feigning" (f. 270; *Opticks* [Dover], p. 369). Thus, after ten years Newton's considered choice was "feigning." See also Cohen, "The first English version of Newton's *hypotheses non fingo.*" For an account of the various stages of this paragraph of the General Scholium, see Cohen, *Introduction to Newton's 'Principia,'* pp. 240–5.

[10] I will be concerned only with Newton's use of hypothesis as a causal explanation. See Cohen, "Hypotheses in Newton's philosophy," and, for a discussion of the varieties of hypotheses in Newton's work, *Franklin and Newton*, Ch. 5, and Appendix 1. The literature on Newton's concept of hypothesis is immense. See, for example, Edwin Arthur Burtt, *The Metaphysical Foundations of Modern Physical Science;* A. C. Crombie, "Newton's conception of scientific method"; Maurice Mandelbaum, *Philosophy, Science and Sense Perception*, Ch. 2, "Newton and Boyle and the problem of 'Transdiction' "; J. E. McGuire, "Atoms and the 'analogy of nature' "; Norwood Russell Hanson, "Hypotheses fingo"; John F. McDonald, "Properties and causes"; Anita M. Pampusch, " 'Experimental,' 'metaphysical,' and 'hypothetical' philosophy in Newtonian methodology"; Koyré, *Newtonian Studies*, Ch. 2; and Peter Achinstein, "Newton's corpuscular query and experimental philosophy."

nomenon nor deduced from *any* Phaenomena but assumed or supposed wthout *any* experimental proof."[11] Thus Newton is altogether excluding hypotheses that have no experimental evidence; following Newton, I call these "imaginary hypotheses."[12] In a draft of his letter to Cotes, Newton was still more explicit about the kinds of hypotheses (and their authors) that he intended to reject:

> Experimental Philosophy reduces Phaenomena to general Rules & looks upon the Rules to be general when they hold generally in Phaenomena. . . . Hypothetical Philosophy consists in *imaginary* explications of things & *imaginary* arguments for or against such explications, or against the arguments of Experimental Philosophers founded upon Induction. The first sort of Philosophy is followed by me, the latter too much by Cartes, Leibnitz & some others.[13]

It is totally fictitious, a priori constructs such as Cartesian vortices that have no role at all in experimental philosophy. In the next place, Newton considers it sufficient for his theory of gravity that the discovered properties can account for all the phenomena. A causal explanation of those properties, however desirable, is not necessary.

Newton's aim in the General Scholium was also to demarcate and clearly distinguish hypothetical explanations of the cause of gravity from his mathematical derivation of the laws of gravitation from the phenomena. His rejection of hypotheses, as we shall encounter in numerous passages in his writings, always occurs in a contrast with rigorously demonstrated principles: As he put it in the "New theory," one must "not mingle conjectures with certainties."[14] In this same vein he began the *Opticks* by declaring: "My Design in this Book is not to explain the Properties of Light by Hypotheses, but to propose and prove them by Reason and Experiments."[15] That Newton never condemns hypotheses in themselves should not be surprising, as he himself freely proposed and employed them. For instance, in the next paragraph of the General Scholium he describes an electric spirit that may account for the cohesion of bodies, electricity, sensation, and the reflection, refraction, and diffraction of light. In addition, in the course of my study, I shall devote much attention to his 1675 paper, "An

[11] Newton to Cotes, 28 March 1713, *Correspondence*, 5:397; italics added.
[12] Mandelbaum, *Philosophy, Science and Sense Perception*, pp. 74–7, stresses the importance of Newton's requirement that explanatory propositions possess an empirical origin.
[13] *Correspondence*, 5:398–9; italics added.
[14] Ibid., 1:100.
[15] *Opticks*, p. 1.

Hypothesis explaining ye properties of Light discoursed of in my several papers."

All those hypotheses that Newton chose to make public (including the queries) were supported by some experimental evidence, though it was inadequate to raise them to the level of more certain principles; I call these "experimental hypotheses." For instance, Newton abruptly ended his discussion of the electric spirit in the General Scholium because "an insufficient number of experiments has been undertaken whereby the laws of the actions of this electric and elastic spirit must be accurately determined and demonstrated."[16] It is a hypothesis because there were insufficient experiments to determine laws or principles, but a legitimate one because founded on experiment. Although such experimental hypotheses were acceptable to Newton, they too had to be kept distinct from established principles, as we shall see in his investigation of the colors of thin films. By 1713, Newton had been stung by criticisms of his theory of gravity as hypothetical and was more sensitive to the use of the word. Nonetheless, although his views matured since his inaugural *Optical Lectures*, his attitude toward hypotheses did not fundamentally change. Throughout his career three points remained constant: His railing against hypotheses is always in the context of distinguishing them from rigorously established principles; imaginary hypotheses, which are supported by no experimental evidence, have no place in natural philosophy; and experimental hypotheses, which have some, but inadequate, experimental support, may be used provided they are clearly demarcated from certain principles.[17]

In the spring of 1672, Newton was offended when the French Jesuit Ignace Gaston Pardies called his theory of color a hypothesis. Newton took the opportunity to make his clearest and most complete statement on the proper role of hypotheses, but not before insisting on the distinction between his theory, which "consists only in some *properties* of light," and "*hypotheses*, by which those properties might be explained":

For the best and safest method of philosophizing seems to be, first that we should inquire diligently into the properties of things and establish them by

[16] *Mathematical Principles*, p. 547; *Principia*, p. 765.

[17] The distinction between two sorts of causal hypotheses has necessarily been introduced in one way or another by many commentators to resolve the apparent contradiction between Newton's simultaneous condemnation and use of hypotheses; but Pampusch, " 'Experimental,' 'metaphysical,' and 'hypothetical' philosophy," has clearly defined the two sorts in a way similar to that adopted here.

experiments, and then more slowly pursue hypotheses for the explanation of them; for hypotheses ought to be accommodated only to explaining the properties of things, and not unlawfully assumed [*usurpari*] for defining them, except insofar as they may furnish experiments. For if solely from the possibility of hypotheses one conjectures about the truth of things, I see not how anything certain can be obtained in any science; since numerous hypotheses may always be devised, which shall seem to overcome new difficulties. Wherefore I judge one must here refrain from the consideration of hypotheses as an improper point of argument and abstract the force of the objection so that it may receive a fuller and more general answer.[18]

Here Newton is making a legitimate distinction – fundamental to his entire natural philosophy – between experimentally discovered properties of light, such as the unequal refrangibility of rays of different color, and hypotheses, such as the diffusion of light proposed by Hooke and Pardies to explain them. Not only had Newton not proposed explanations of those discovered properties, but Hooke and Pardies had unlawfully invoked hypothetical properties that were contrary to fact and incompatible with his discoveries. More simply, Newton is suggesting that we must approach nature with an open mind and not try to force the phenomena into some a priori explanatory scheme.

In his second major point he proposes how experimental hypotheses may be legitimately used: to help understand properties already discovered and to suggest new experiments, ways in which he himself employed them. Newton, however, also used hypotheses for the deduction of new properties, rather as in the hypothetico-deductive method or a modern scientist's use of models. This would lead him into a practice that he deplored, the generation of hypothetical properties. We shall encounter this in his theory of fits, where physical hypotheses about the causes of periodic colors are embedded in his mathematical description of them. It is at this point that one might be tempted to accuse Newton of hypocrisy. That he failed to eliminate all hypothetical properties and entities from his theories, despite a genuine effort to do so, shows only that his program to construct a hypothesis-free science was a failure, not that it was hypocritical.

In Newton's final point to Pardies, he attacks the ease with which imaginary hypotheses – "conjectures solely from possibility" – may

[18] Newton to Oldenburg for Pardies, 10 June 1672, *Correspondence*, 1:164; I have added italics and revised the translation from *Philosophical Translation Abridged* reprinted in Newton, *Papers and Letters*, p. 106.

be arbitrarily multiplied when they are not constrained by experiment, and he rejects their use against his discoveries as an improper mode of arguing. This is a recurrent theme in Newton's thought, and we may recognize it as the origin of the fourth of the Rules of Reasoning that was added to the *Principia* more than fifty years later:

> In experimental philosophy we are to look upon propositions inferred by induction from phenomena as accurately or very nearly true, notwithstanding any contrary hypotheses, till such time as other phenomena may occur, by which they may either be made more accurate, or liable to exceptions.
>
> This rule must be followed lest the argument of induction be evaded by hypotheses.[19]

We should note that Newton is still so confident of the validity of principles deduced from experiment that he allows only that they may be refined or restricted, but not erroneous and rejected.[20]

The contrast that Newton lays down for Pardies between hypothetical explanations and "the truth of things" and "anything certain" serves to bring out the difference between his concept of science and that of his contemporaries.[21] As van Leeuwen, Hacking, and Shapiro have shown, by the 1660s, most natural philosophers, especially in England, had abandoned the demand for certain scientific knowledge and developed the alternative concept of probable knowledge.[22] They rejected equally the Aristotelian model of demonstrative knowledge and Cartesian a priorism. Though Descartes was convinced that by reason alone he had deduced new and certain general laws of nature (such as his laws of motion), at the same time he also encouraged probabilism. He held that his explanations of particular phenomena, such as magnetism and heat, were only probable conjectures that possessed moral certainty because there is no direct knowledge of the invisible particles of matter. In England, a moderate or Ciceronian skepticism of the Renaissance prevailed, and mathematical demonstrations, some intuitive truths (such as the whole is greater than a part), and immediate objects of sense were alone granted certainty.

[19] *Mathematical Principles*, p. 400; *Principia*, p. 555.

[20] Rule IV is actually a paraphrase of a passage that Newton added to Query 23/31 in 1716, which is quoted in note 73.

[21] Zev Bechler first showed how sharply Newton's dogmatic claims to certainty contrasted with the prevailing philosophy of science at the Royal Society; "Newton's 1672 optical controversies."

[22] Henry G. van Leeuwen, *The Problem of Certainty in English Thought, 1630–1690;* Ian Hacking, *The Emergence of Probability;* and Barbara J. Shapiro, *Probability and Certainty in Seventeenth-Century England.* See also Ernan McMullin, "Conceptions of science in the scientific revolution."

There were differences among philosophers as to the exact degree of certainty to assign to these categories, but it was generally agreed, especially among the experimental virtuosi at the Royal Society, that theories and causal explanations attained only various degrees of probability and were at best morally certain.

Joseph Glanvill, whose *Vanity of Dogmatizing* Newton read about 1665, clearly expressed this probabilistic outlook when he wrote in his 1676 essay "Of scepticism and certainty" that the new philosophers in their attempt "to seek Truth in the Great Book of Nature . . . proceed with wariness and circumspection without too much forwardness in establishing Maxims, and positive Doctrines: To propose their Opinions as *Hypotheseis*, that *may probably* be the true accounts, without peremptorily affirming that *they are*."[23] Boyle, whose works Newton read assiduously throughout his career, was perhaps the most skeptical about establishing scientific theories. Accordingly, he was reluctant to propose hypotheses and devoted most of his effort to experiments and observations (or natural histories). In his *Excellency of Theology, Compared with Natural Philosophy*, which was written in 1665 but published in 1674, he cautioned that "even in many things, that are looked upon as physical demonstrations, there is really but a moral certainty . . . as he will not scruple to acknowledge, that knows by experience, how much more difficult it is, than most men imagine, to make observations about such nice subjects, with the exactness, that is requisite for the building of an *undoubted theory* upon them."[24]

Let us turn to one more, immediately relevant illustration of contemporary skepticism from Hooke's *Micrographia*. Hooke published observations on a broad range of subjects, including optics and Boyle's law, but with regard to the theories he invoked to explain them, he advised:

> If therefore the Reader expects from me any infallible Deductions, or certainty of *Axioms*, I am to say for my self, that those stronger Works of Wit and Imagination are above my weak Abilities; or if they had not been so, I would not have made use of them in this present Subject before me: Whereever he finds that I have ventur'd at any small Conjectures, at the causes of the things that I have observed, I beseech him to look upon them only as *doubtful Problems*, and *uncertain ghesses*, and not as unquestionable Conclusions, or matters of unconfutable *Science*.[25]

[23] Joseph Glanvill, *Essays on Several Important Subjects in Philosophy and Religion* (1676), p. 44.
[24] Boyle, *Works*, 4:42; italics added.
[25] Hooke, *Micrographia* (1665), preface, sig. b[1^r].

In his first publication, in 1672, "A New theory about light and colors," Newton boldly announced a new, more certain science, which seemed almost the antithesis of Hooke's:

A naturalist would scearce expect to see ye science of [colors] become mathematicall, & yet I dare affirm that there is as much certainty in it as in any other part of Opticks. For what I shall tell concerning them is not an Hypothesis but most rigid consequence, not conjectured by barely inferring 'tis thus because not otherwise or because it satisfies all phaenomena (the Philosophers universall Topick,) but evinced by ye mediation of experiments concluding directly & wthout any suspicion of doubt.[26]

At the beginning of his career, before he had entered into any controversies, Newton was already using the word "hypothesis" pejoratively and (as is constantly the case) in contrast to demonstrated principles. The confident, even dogmatic, tone of this declaration was in such sharp contrast to the Royal Society's ideology that Henry Oldenburg, its secretary and editor of *Philosophical Transactions*, excised it from the published version. The most perplexing aspect of Newton's assertion is that, though he begins by announcing the creation of a mathematical science of color as certain as geometrical optics (then the only mathematical part of optics), he does not describe that new science anywhere in the paper. It is devoted entirely to his experimental theory. And despite his claim that the theory is derived from experiment "wthout any suspicion of doubt," he leaves all but one proposition undemonstrated. In other passages from this period in which Newton insists on the certainty of his experimental theory of color – for instance, that it is "infallibly true & genuine"[27] – he also appeals to the existence of his mathematical theory (which remained unknown in his unpublished *Optical Lectures*), as if the certainty of the latter somehow ensured that of the former. I will soon return to this puzzling feature of Newton's methodological views and show a similar conjunction of mathematical and experimental certainty in the *Mathematical Lectures* of the first Lucasian Professor.

To Newton's chagrin, each of his principal critics (Hooke, Pardies, and Christiaan Huygens, all adherents of probabilism) referred to his theory of color as a hypothesis. Pardies had invoked this word in the first sentence of his first letter for Newton. Newton could not let this pass, and he concluded his reply to Pardies by insisting:

[26] *Correspondence*, 1:96–7.
[27] Quoted in full at note 41.

I do not take it amiss that the Rev. Father calls my theory an hypothesis, inasmuch as he was not acquainted with it. But my design was quite different, for it is seen to contain only some properties of light, which, now discovered, I think easy to be proved, and which if I had not considered them as true, I would choose rather to dismiss them as vain and empty speculation than to acknowledge them as my hypothesis.[28]

Newton knew that this position was heterodox and took personal responsibility for it in his covering letter to Oldenburg, perhaps because Oldenburg had deleted his last methodological declaration from the published version of the "New theory": "I herewith send you an answer to the Jesuite Pardies considerations, in the conclusion of wch you may possibly apprehend me a little too positive, but *I speak onely for my selfe.*"[29] As Pardies did not wish to debate this point or antagonize Newton, he apologized. Newton accepted the apology but not before reiterating his opposition to the current attitude toward hypotheses:

As to the Rev. Father's calling our doctrine an hypothesis, I believe it only proceeded from his using the word which first occurred to him, as a practice has arisen of calling by the name hypothesis whatever is explained in philosophy; and the reason of my making exception to the word, was to prevent the prevalence of a term, which might be prejudicial to philosophizing properly.[30]

Newton was thus quite consciously taking a position at odds with his contemporaries and was not yet prepared to abandon the quest for a more certain science. He simply did not accept the idea that "whatever is explained in philosophy" is a hypothesis and that these explanations differ only in degree of probability. To Newton, there were "rigidly" established, "true" theories – like his theory of color – and there were hypotheses, and the two must not be confused.

Newton based his claim to a more certain science on two different grounds. First, he believed that he had formulated his theory solely in terms of experimentally observed properties, or principles deduced from them, without any causal explanations (hypotheses) of those properties. Newton's critics did not accept – and perhaps did not even

[28] Newton to Oldenburg for Pardies, 13 April 1672, *Correspondence,* 1:142; Newton, *Papers and Letters,* p. 92; I have altered the translation.
[29] Newton to Oldenburg, 13 April 1672, *Correspondence,* 1:136; italics added.
[30] Newton to Oldenburg for Pardies, 10 June 1672, ibid., p. 168; Newton, *Papers and Letters,* p. 109; I have slightly altered the translation. For Newton's early optical controversies and the philosophical issues raised, see A. I. Sabra, *Theories of Light from Descartes to Newton;* and Bechler, "Newton's 1672 optical controversies."

recognize – the phenomenological nature of his theory, and he had to explain to each of them, as we saw him repeatedly do in his letter to Pardies, that he was concerned with "properties" and not "hypotheses."[31] Although this is one of the most characteristic features of his optical theories, it has also largely gone unnoticed by modern scholars.[32] Forty years later, he insisted on the same distinction in the General Scholium. By avoiding causes, he avoided hypotheses and compromising the certainty of his theory. His theory of fits would likewise be formulated in terms of properties.

One characteristic of such theories is that without an underlying physical explanation or model they suffered a loss of intelligibility. Huygens made this charge and held that until Newton had provided a mechanical explanation of color all he had taught us was "only this accident (which assuredly is very considerable) of their different refrangibility."[33] When Hooke accused Newton of supporting an emission theory of light, Newton conceded the point but insisted that that hypothesis was no part of the theory itself. He then explained to Hooke how he dealt with the problem: "I knew that the *Properties* w[ch] I declared of light were in some measure capable of being explicated not onely by that, but by many other Mechanicall *Hypotheses*. And therefore *I chose to decline them all, & speake of light in generall termes, considering it abstractedly . . . without determining what that thing is.*"[34] Newton himself recognized the subsequent loss of intelligibility caused by formulating his theories abstractedly. Three and a half years

[31] In the final draft of his reply to Hooke in June 1672, Newton clearly summarized this situation in a marginal note (omitted in the letter he sent): "That my main designe hath been mistaken w[ch] was in generall terms to declare some properties of light w[th]out respect to any Hypothesis" (Add. 3970, f. 445[v]).

[32] Cohen, in his interpretation of the three stages of the "Newtonian style" (mathematical method) of the *Principia*, sees the search for physical causes as a last, optional stage; *The Newtonian Revolution*. McMullin goes even further and argues that, not only in his optical theories, but even in his *Principia*, Newton is unconcerned with causes. McMullin emphasizes the noncausal nature of Newton's science (the mirror image of my stress on its phenomenological nature) in his "Conceptions of science in the scientific revolution," pp. 67–74, and "Newton and scientific realism," forthcoming.

[33] Huygens via Oldenburg to Newton, 18 January 1672/3; *Correspondence*, 1:256.

[34] Newton to Oldenburg for Hooke, 11 June 1672; ibid. p. 174; italics added; see also his remarks in section 5, p. 177. In a draft of this letter Newton initially wrote "theoremes" rather than "Properties," and after "in generall termes" he continued "after the mode of Mathematicians" (Add. 3970. f. 433[v]), which is yet another indication of his tendency at this time virtually to identify his mathematical and experimental approaches to nature. Newton also wrote Pardies (in the last sentence quoted at note 18) that his objection must be considered "abstractly" without regard for hypotheses.

later he justified sending his "An Hypothesis explaining ye properties of Light" to the Royal Society with the observation that some natural philosophers "when I could not make them take my meaning when I spake of ye nature of light & colours *abstractedly* have readily apprehended it when I illustrated my discours by an Hypothesis."[35] He was not concerned with the probability of his hypothesis, he wrote in the covering letter to Oldenburg, but only that it "render ye papers I send you, and others sent formerly, more *intelligible*."[36] The solution that he adopted in 1675 to the loss of intelligibility – namely, clearly labeling his hypotheses as such and setting them apart from his rigorously established discoveries – was what he continued to use throughout his career.[37] If Newton in his quest for certainty paid the price of a loss of intelligibility by adopting a phenomenological approach, he did succeed in attaining a level of formal clarity, coherence, and rigor that hitherto had been scarcely achieved in experimental science.[38]

The second, and more compelling, ground for Newton to believe that he could attain greater certainty in science was mathematics. In his *Optical Lectures* he declared his commitment to mathematical physics as a counter to the hypotheses and conjectures that he believed

[35] Add. 3970, f. 538^v^; *Correspondence*, 1:363.
[36] Newton to Oldenburg, 7 December 1675, *Correspondence*, 1:361; italics added.
[37] In his anonymous review of the *Commercium epistolicum* (1713) Newton gave his clearest account of the hypothetical nature of his most extended conjectural piece, the queries, and why he set them apart from the rest of the *Opticks*. The *Commercium* was a report (also anonymously written by Newton) issued by the Royal Society to adjudicate the priority dispute between Newton and Leibniz over the discovery of the calculus. Newton explained that in the *Principia* and *Opticks* "Mr. Newton" adopted the "experimental philosophy," in which "Hypotheses have no place, unless as Conjectures or Questions proposed to be examined by Experiments. For this Reason Mr. *Newton* in his Optiques distinguished those things which were made certain by Experiments from those things which remained uncertain, and which he therefore proposed in the End of his Optiques in the Form of Queries" ("An account of the book entitled *Commercium epistolicum*," p. 222).
[38] This is not to deny that there remain real problems with Newton's theory, such as unjustified idealization and generalizations and unproved principles. Rather, my aim is to contrast his theory with those of his contemporaries, for example, Boyle's *New Experiments Physico-Mechanical, Touching the Spring of the Air and its Effects* (1660), which Newton read in his *anni mirabiles*. Boyle's use of the newly invented air pump and introduction of the concept of the spring of the air (elasticity) to explain a broad range of pneumatic phenomena was one of the significant achievements of the Royal Society virtuosi. Yet the *New Experiments* can scarcely be judged to present a theory; rather, it consisted of a series of experiments and explanations of them, with no systematic formulation. On Boyle's conception of experimental science, see Steven Shapin and Simon Schaffer, *Leviathan and the Air-Pump*. In my consideration of the theory of fits in Chapter 4, I shall return to Newton's technique of casting his theories in an "abstracted" form and the subsequent loss of intelligibility.

were then rampant in natural philosophy. In particular, Newton vowed that he would create a new mathematical science of color:

The generation of colors includes so much geometry, and the knowledge of colors is supported by so much evidence, . . . I can thus attempt to extend the bounds of mathematics somewhat, just as astronomy, geography, navigation, optics, and mechanics are considered mathematical sciences even if they deal with physical things: the heavens, earth, ships, light, and local motion. Thus although colors may belong to physics, the science [*scientia*] of them must nevertheless be considered mathematical, insofar as they are treated by mathematical reasoning [*ratione mathematicâ*]. Indeed, since an exact science of them seems to be one of the most difficult things that a philosopher is in need of, I hope to show – as it were, by my example – how valuable mathematics is in natural philosophy.[39]

While granting that color belongs to physics, Newton is also claiming that it belongs at least as much to mathematics. Indeed, he asserts that true knowledge or *scientia* of color belongs to mathematics. He then calls for abolishing the boundary between physics and mathematics and mathematizing all of natural philosophy:

I therefore urge geometers to investigate nature more rigorously, and those devoted to natural science to learn geometry first. Hence the former shall not entirely spend their time in speculations of no value to human life, nor shall the latter, while working assiduously with an absurd method, perpetually fail to reach their goal. But truly with the help of philosophical geometers and geometrical philosophers, instead of the conjectures and probabilities that are being blazoned about everywhere, we shall finally achieve a science of nature supported by the highest evidence.[40]

Newton's own summary of his powerful appeal for the reform of natural philosophy and the creation of a mathematical science of color asserts that his theory is "to be treated not hypothetically and probably, but by experiments or demonstratively." Yet in the entire paragraph he speaks only of mathematics without mentioning experiment other than for two vague references to "evidence." And, just as in the "New theory," this affirmation to treat color mathematically is placed at the beginning of the experimental portion of his *Lectures* and not at the mathematical part. Newton's intimate association of his mathematical theory with his experimental theory of color was largely responsible for the strong claims as to its certainty.

[39] *Optical Papers*, 1:86–7.
[40] Ibid., pp. 86–9.

In fact, he conceived of them as one demonstrative science, with the experimental discoveries serving as the first principles of that science.

The successful investigation of the colors of thin films followed hard upon the theory of white light and color and was based on it. If Newton required further evidence of the power of the mathematical approach to nature, this research provided it. In the earliest version of the "Observations" that he had intended to send to the Royal Society in the spring of 1672, he made the strongest of his claims for the certainty of his theory. After presenting a physical explanation of the appearance of "Newton's rings" when viewed through a prism, he asserted: "For confirmation of all this I need alledg no more then that it is Mathematically demonstrable from my former Principles. But yet I shall add that they which please to take the paines, may by the Testimony of their senses be assured that these explications are not Hypotheticall but *infallibly true & genuine*."[41] Once again we see the intimate connection of the certainty of experiment and the existence of a mathematical theory. Newton clearly believed that a mathematical, deductive approach would lead to greater certainty and that experiment could provide the requisite certain foundations for a science. These beliefs reflect a long mathematical tradition.

The mathematization of nature is one of the defining features of the Scientific Revolution. Yet the transition from the Aristotelian intermediate sciences (*scientiae mediae*) – as the applied mathematical fields of optics, astronomy, mechanics, and harmonics were called – to mathematical physics has scarcely been studied. Thus I can only sketch some of the relevant highlights in very broad strokes.

Aristotle strictly distinguished mathematics (arithmetic and geometry) from *physike* (physics or natural philosophy), which was the science of the essences or qualities of natural bodies. He recognized the importance of mathematics for physics but treated the applied mathematical sciences as intermediate between mathematics and physics, drawing their principles from the former (being "subalternated" or subordinate to it in Scholastic terminology) and their subject

[41] Add. 3970, f. 525^r, but I cite the 1675 version, f. 510^v (Birch, 3:293), as the two are identical here; italics added. The various versions of the "Observations" are described in §2.2. In 1672, Newton originally concluded "are true & genuine & more then Hypotheticall." When he revised the "Observations" for the *Opticks*, Bk. II, Pt. II, p. $_2$44, he toned down this passage to make the unexceptionable claim that "now as all these things follow from the Properties of Light by a mathematical way of reasoning, so the truth of them may be manifested by Experiments." By this time, however, he had succeeded in creating his mathematical theory of fits.

matter from the latter.[42] The intermediate sciences did not have the same independent status as mathematics and physics. The division of competencies of the various sciences was rigidly apportioned. Mathematics treated the pure quantity (number and extension) of "intelligible matter," which was devoid of all sensible qualities; the intermediate sciences treated the quantitative aspects (like weight and rectilinear light rays) of "sensible matter"; and physics treated the qualities of matter and explained such causes as those of motion and light.

The attempt to preserve these strict boundaries can be seen in Archimedes' physical works, *On Floating Bodies* and *The Equilibrium of Planes*, which were composed about a century after Aristotle. Despite their evident relation to the real physical world, hydrostatics and statics are handled in a rigorous demonstrative mode as if they were pure mathematics, with physical causes and concepts – let alone experiments – being barely mentioned. To be sure, the disciplinary boundaries were already transgressed to some extent by Ptolemy in the second century A.D., but on the whole they were observed well into the Renaissance. This classification of the sciences required, for example, that mathematical astronomy study only the quantitative aspects of the heavenly bodies, such as their motions and periods, and not their physical aspects, such as their matter and the cause of their motions. Mathematicians and natural philosophers remained largely distinct professional groups into the seventeenth century.[43]

Mathematical demonstrations held a privileged place in Aristotelian epistemology as exemplars of absolute certainty, and the conception of science was modeled on mathematics. Physics, like all the sciences, consequently was expected to proceed from true first principles and deduce necessary truths. The principles were derived by induction from repeated experience (not experiment) of the ordinary behavior of nature and were universal and evident truths, such as heavy things

[42] For Aristotle's views on the relation of physics to mathematics, the following passages are among the most important, *Physics*, II.2, 193b22–194a12; *Posterior Analytics*, I.7, 75b14–17; *Metaphysics*, XI.3, 1061a28–b7; XIII.3, 1078a5–17. These and many other relevant passages are conveniently gathered together and translated in Thomas Heath, *Mathematics in Aristotle*. For a succinct account from the perspective of the early modern era of Aristotle's conceptions of a demonstrative science and of the mathematical sciences, see McMullin, "The conception of science in Galileo's work."

[43] The separation between the mathematical and physical differed among the various sciences. It was most rigid in mechanics and astronomy (which came to dominate the Scientific Revolution) and weakest in optics. For the disciplinary division of optics in antiquity and the Middle Ages, see Lindberg, *Theories of Vision from al-Kindī to Kepler*, Ch. 1, esp. pp. 1, 11–17.

fall downward and light rays are straight lines. In physics, it was always difficult to ensure that the principles were necessarily true and that demonstrative knowledge had been yielded. Nonetheless, the demonstrative ideal of mathematics was taken as the hallmark of certain knowledge well into the seventeenth century.

The sixteenth century saw a vigorous revival of mathematical studies stimulated by the recovery, publication, and translation of ancient Greek texts, such as Ptolemy's *Almagest*, Aristotle's *Mechanical Problems* (actually written by an early follower), and above all Archimedes' works.[44] Through the course of the century the level of mathematical sophistication increased, as did the study of "mixed mathematics" (as the intermediate sciences came to be called) – that is, of optics, astronomy, and mechanics, including statics and hydrostatics. The century even saw the effective creation (actually a restoration) of a new mathematical science of mechanics, which grew from the study of machines to encompass projectile motion and then free-fall.

For a variety of reasons, beginning in midcentury, the certainty of mathematical demonstrations became a much debated epistemological issue. One of the principal arguments used to justify the absolute certainty of pure, but not mixed, mathematics with respect to physics was Aristotle's distinction (with its Platonic overtones) between the intelligible matter of mathematics and the sensible matter of physics.[45] The Aristotelian division of scientific disciplines remained essentially unchanged, though the status of the mathematical sciences was raised. The Jesuits, under Christopher Clavius's leadership, were instrumental in this process because of the central pedagogical role they attributed to mathematics in the curriculum of their far-flung educational system.[46] Descartes and Marin Mersenne were products of the Jesuit schools; and in his optical work, Newton encountered the writings of a number of Jesuits: Christoph Scheiner, Pardies, Francesco Maria Grimaldi, and Honoré Fabri.

Many forces were at work to raise the status – epistemological as well as professional – of the mathematical sciences in the sixteenth and early seventeenth centuries. Premier must be the increasing knowledge of mathematics itself and its widespread application to physical problems, including such burgeoning fields as fortification, navigation, geography, and linear perspective. The rise of skepticism

[44] For a good overview, see Paul Lawrence Rose, *The Italian Renaissance of Mathematics;* and also W. R. Laird, "The scope of Renaissance mechanics."
[45] See Nicholas Jardine, "Epistemology of the sciences," pp. 693–7.
[46] See Peter Dear, "Jesuit mathematical science and the reconstitution of experience."

in the sixteenth century caused many moderate skeptics to seek some limited grounds of certainty against the more radical skeptical claim that no knowledge is possible. Mathematics was a natural place to seek such certainty, and along with immediate sensation and the syllogistic form, it emerged as a source of certain knowledge. Mathematical scientists outside the Scholastic mainstream were increasingly demanding a realistic description of nature. The boundary between mathematics and physics was breaking down. Mathematicians, especially mathematical astronomers like Johannes Kepler, held that their mathematical-physical descriptions of the natural world were true accounts of nature and not simply mathematical models subordinate to higher philosophical explanations.

This demand for realism, together with the confrontation with Archimedes' works and the re-creation of the science of mechanics, brought to the fore the problem of the relation, or, rather, the gap, between mathematics and physics and the need to bridge it. Some, like the mathematician Niccolò Tartaglia, were stymied by the apparent boundary between the ideal realm of mathematics and the imperfect material world. In 1546, in a discussion of simple machines Tartaglia argued that mathematicians do not accept "demonstrations made on the strength and authority of the senses in matter, but only those made by demonstrations and arguments abstracted from all matter. Consequently, the mathematical disciplines are considered by the wise not only to be more certain than the physical, but even to have the highest degree of certainty."[47] Nevertheless, Tartaglia was compelled to acknowledge that truths known "by abstraction from all material, should reasonably be verifiable in matter also by the sense of sight; otherwise mathematics would be wholly vain and useless."[48] He was unable, however, to integrate the two successfully.

Others, like Galileo, pursued the ideal of harnessing the certainty of mathematics to physical description in order to create independent mathematical-physical sciences. Galileo was committed to the mathematical interpretation of nature. In the *Assayer* (1623), he insisted that natural philosophy or the book of nature "cannot be understood unless one first learns to comprehend the language and interpret the characters in which it is written. It is written in the language of mathematics, and its characters are triangles, circles, and other geometrical

[47] Stillman Drake and I. E. Drabkin, trans., *Mechanics in Sixteenth-Century Italy*, pp. 106–7; a translation of Tartaglia's *Quesiti, et inventioni diverse*.
[48] Ibid., p. 108.

figures."[49] By his use of what we have come to call "Galilean idealization," he was able to erect one bridge between mathematical description and physical reality. In his *Dialogue concerning the Two Chief World Systems,* which Newton read in translation, he explained that the "Geometricall Philosopher [*filosofo geometra*]" can get perfect agreement between his mathematical theory and concrete nature by including "material hindrances" in his calculations. If errors remain, they "lie not in the abstractness or concreteness, not in geometry or physics, but in a calculator who does not know how to make a true accounting."[50] Later in the *Dialogue,* while discussing William Gilbert's *De Magnete,* Galileo implied that with the use of mathematics an experimental science could provide certain, demonstrative arguments. He thought that if Gilbert had possessed greater mathematical knowledge, he would not have so rashly set forth what he judged to be "true causes" (*verae causae*) for his experimental results. "His reasons," Galileo observed, "are not rigorous, and lack that force which must unquestionably be present in those adduced as *necessary and eternal scientific conclusions.* I do not doubt that in the course of time this new science will be improved with still further observations, and even more by *true and conclusive demonstrations.*"[51]

Galileo thus considered it possible for a mathematical-experimental science to be a demonstrative science. In *Two New Sciences,* which Newton did not read, Galileo constructed his new mathematical science of motion as a demonstrative science of nature and it carries an air of mathematical certainty. However, the intractable problem of producing experimental evidence capable of furnishing the requisite certain first principles eluded him.[52]

Newton was scarcely aware of the earlier efforts in Italy to understand the nature of the mathematical sciences, but, as we shall see, he was introduced to the issues and a modern resolution by Barrow. Let us therefore shift our attention to England in midcentury.

The English scientific tradition was largely experimental and observational (and long continued to be), drawing much of its vitality from the physiological school of William Harvey. Since the middle of

[49] Drake and C. D. O'Malley, trans., *The Controversy on the Comets of 1618,* pp. 183–4.
[50] Galileo Galilei, *Dialogue concerning the Two Chief World Systems,* pp. 207–8. Newton read the translation in Thomas Salusbury, *Mathematical Collections and Translations,* vol. 1 (1661), who used the phrase "Geometricall Philosopher," p. 185.
[51] Galileo, *Dialogue,* p. 406; italics added; Salusbury, *Collections,* p. 370, who translates the concluding phrase as "true and necessary Demonstrations."
[52] See Winifred Lovell Wisan, "Galileo's scientific method"; and McMullin, "Conception of science in Galileo's work."

the preceding century, mathematics had been devoted mainly to applications (instruments, navigation, and the like), but a nascent higher mathematical school was now forming around John Wallis, Wren, and Barrow (and we may add James Gregory in Scotland). Newton would successfully unite both approaches, but he drew much of his conception of science from the mathematical tradition.

The status of mixed mathematics, when discussed at all by the empirically minded members of the Royal Society, was generally held to be inferior to the absolute certainty of pure mathematics and to possess but a moral certainty like other sciences. Boyle, for instance, in discussing Descartes' demonstrations about comets, granted that "the inferences, as such, may have a demonstrable certainty"; yet because they were based on past astronomical observations, "the presumed physico-mathematical demonstration can produce in a wary mind but a moral certainty, and not the greatest neither of that kind, that it is possible to be attained."[53]

The small community of mathematical scientists held a contrary view. In his inaugural lecture as professor of astronomy at Gresham College in 1657 Wren held that

> Mathematical Demonstrations being built upon the impregnable Foundations of Geometry and Arithmetick, are the only Truths, that can sink into the Mind of Man, void of all Uncertainty; and all other Discourses participate more or less of Truth, according as their Subjects are more or less capable of Mathematical Demonstration. Therefore, this rather than Logick is the great *Organ Organωn* of all infallible Science.[54]

Thus the other sciences ("discourses") and especially mixed mathematics could approach the certainty of mathematics. Like Newton, Wren hoped that natural philosophy by "a geometrical Way of reasoning from ocular Experiment" would "prove a real Science of Nature, not an Hypothesis of what Nature might be."[55] Wren's views were unknown to Newton, so let us return to Cambridge and Newton's mentor and predecessor as Lucasian Professor of mathematics, Isaac Barrow.

Barrow devoted his *Mathematical Lectures*, delivered between 1664

[53] Boyle, *Excellency of Theology, Works,* 4:42. See also the preface to the *Hydrostatical Paradoxes* (1666), *Works,* 2:738–44. On Boyle's opposition to the use of mathematics in natural science, see Shapin, "Robert Boyle and mathematics."

[54] Christopher Wren, *Parentalia: Or, Memoirs of the Family of the Wrens* (1750), pp. 200–1.

[55] Ibid., p. 204. See also J. A. Bennett, *The Mathematical Science of Christopher Wren,* esp. pp. 118–20.

and 1666, to the philosophy of mathematics and treated both the status of the mathematical sciences and the certainty of empirical evidence. As befit the former Regius Professor of Greek, the erudite Barrow confidently explicated the classical mathematical literature, and also presented the interpretations of such Renaissance scholars as Clavius, Giuseppe Biancani (Blancanus), Pierre Herigone (all Jesuits), Francesco Maurolico, and Jacopo Zabarella. As an accomplished mathematical scientist, Barrow also called upon the work of the leading modern practitioners of the art, such as Giovanni Alfonso Borelli, Bonaventura Cavalieri, Descartes, Kepler, and Simon Stevin. Late in his life Newton recalled attending Barrow's lectures, and we can therefore safely assume that the young mathematician was regularly in the Lucasian Professor's audience.[56] Newton's philosophy of science was by no means identical with Barrow's, but there is enough common ground that a study of Barrow's position will illuminate the mathematical tradition to which they both belonged and the milieu in which Newton came to intellectual maturity.[57]

Barrow's solution to the status of mixed mathematics and its certainty was a bold one that marked off new territory for mathematics and justified the mathematization of nature that had been underway for more than a century. The distinction between pure and mixed mathematics and the classification of the mathematical sciences, which had been an academic subdiscipline for centuries, were essentially dismantled by Barrow. After reviewing various classifications that had been proposed over two millennia, he rejected the distinction between intelligible and sensible matter that served as a modern foundation for the distinction between pure and mixed mathematics: "It is a very weak and slippery Foundation to depend upon, that the Mathematics are conversant about Things intelligible and Things sensible, because in reality every one of its Objects are at the same time both intelligible and sensible . . . intelligible as the Mind apprehends and contemplates their universal Ideas, and sensible as they agree with several particular subjects occurring to the Sense."[58] The object

[56] Newton, *The Mathematical Papers*, 1:11, n. 26.

[57] Robert Kargon, "Newton, Barrow and the hypothetical physics," first suggested that Barrow influenced Newton's program for a certain, mathematical physics, though on rather different grounds than I argue.

[58] Barrow, *The Usefulness of Mathematical Learning Explained and Demonstrated: Being Mathematical Lectures Read in the Publick Schools at the University of Cambridge* (1734), Lect. II, p. 19; a translation of the posthumously published *Lectiones mathematicae XXIII; in quibus principia matheseôs generalia exponuntur: Habitae Cantabrigiae A.D. 1664, 1665, 1666* (1683), which is reprinted in Barrow, *The Mathematical Works*, 1:38. Because the translation is rather rough, I have altered it as appropriate.

of mathematics, according to Barrow, is "magnitude or continued quantity" in nature, that is, extension. The mixed mathematical sciences all deal with quantity and can be treated by strict geometrical demonstrations and are "the same in Number with the Branches of Physics." There is "no branch of natural Science" that cannot become mathematical:

For Magnitude is the common Affection of all physical Things, it is interwoven in the Nature of Bodies, blended – as it were, a substratum with all corporeal Accidents, and well nigh bears the principal Part in the Production of every natural Effect. All bodies obtain their own Figures, and execute their own local Motions; by which means, if not all, yet the most and chiefest Effects (almost whatsoever admits of a philosophical Explication) are performed, for the determining and comparing of which the Theorems of Geometry do often conduce.[59]

By pursuing to its logical conclusion the assumption of the mechanical philosophy that all natural phenomena (or at least "the chiefest") are caused by matter and its motions in space, Barrow was able to justify the mathematization of all nature. Mathematics becomes "equal to and coextensive with physics itself," and "there is no part of this [Physics]... which is not in some way dependent on [*subalternatur*] Geometry."[60] Extension, which is the object of mathematics, becomes "as it were" the new substratum of natural bodies. Mixed mathematics has become mathematical physics: Nature has an intrinsically mathematical structure and frequently requires a mathematical-physical explanation. He has even inverted the status of mixed mathematics and physics, which is now "subalternated" to mathematics. Barrow's claims were essentially a codification of the achievements of the new mathematical sciences in the preceding century.

In making his strong claims for the certainty of his "New theory," Newton did not say anything at all about the sort of mathematical theory he had created. From his *Optical Lectures* it is apparent that it would be a traditional mixed mathematical science like geometrical optics. Barrow described the nature of such sciences:

[59] Barrow, *Mathematical Lectures*, pp. 20–1; *Works*, 1:39–40. See also *Mathematical Lectures*, pp. 25–6; *Works*, 1:44. Compare Barrow's claim on the relation of mechanics to physics with Descartes' even stronger one, quoted at note 77, especially the italicized passage.

[60] Barrow, *Mathematical Lectures*, pp. 26, 22; *Works*, 1:44, 41. On the development of mathematical physics, mechanics, and the mechanical philosophy, see Alan Gabbey, "Newton's *Mathematical Principles of Natural Philosophy:* A treatise on 'mechanics'?"

In reality those which are called Mixed or Concrete Mathematical Sciences, are rather so many Examples only of Geometry, than so many distinct Sciences separate from it: for when once they are disrobed of particular Circumstances, and their own fundamental and principal Hypotheses come to be admitted (whether sustained by a probable Reason, or assumed *gratis*) they become purely Geometrical. . . . And how much the more simple and evidently possible the Hypotheses are taken, so much the nearer do these Arts approach to Geometry: From whence, for example, that Part of Mechanics treating of the Center of Gravity, and that Branch of Optics vulgarly called Perspective, are not unfitly numbered among the Parts of Geometry, because they scarce require any thing which is not granted and proved in that Science, nor use any other Principles or Reasonings than what are strictly Geometrical.[61]

It should be noted first that by "hypotheses" Barrow here means the postulates, axioms, or principles of a science (e.g., the law of rectilinear propagation in optics) and not the conjectural physical explanations that Newton so often rejected. Moreover, Barrow conceives of the mixed mathematical sciences as deductive sciences modeled on Euclidean geometry. Thus the hypothetico-deductive method does not derive from this tradition. Finally, and most importantly, according to Barrow a science like geometrical optics (of which perspective was a part) could approach the certainty of geometry. Thus, for Barrow, the only difference between pure and mixed mathematics was in the certainty of the first principles. But how can the principles of a science, especially an empirical one, be established with certainty?

Barrow answers this question in a lecture defending the certainty of mathematical demonstration against attacks by skeptics. The classical Aristotelian conception of demonstration requires universally true principles, which can be established only by induction, and this requires "perpetual observation and perfect enumeration of particulars."[62] According to the skeptics, however, universal truths cannot be established by induction because (1) it is impossible to make an infinite number of observations and (2) the testimony of the senses is unreliable. Barrow did not believe that the certain principles of mathematics derived from observation, but he felt compelled to reply to their argument for "it assails not only the certitude of mathematics but of all the sciences."[63]

Against the skeptics he argues that only fools and madmen would deny that in healthy bodies the senses often report with certainty, as

[61] Barrow, *Mathematical Lectures*, p. 27; *Works*, 1:44–5.
[62] Barrow, *Works*, Lect. V, 1:77.
[63] Barrow, *Mathematical Lectures*, p. 68; *Works*, 1:78.

when they relate "the sun is now shining." Against Aristotle's requirement of induction to establish certain principles, he objects that if it were true, no "rigid certitude [*rigidae certitudini*]" could be attained and "scarcely any human reasoning would arise above a probable conjecture [*conjecturae probabilis*]." Appealing to his natural theology, he then insists that experimental evidence may provide true principles that can serve as the foundation of a demonstrative science

where any Proposition is found agreeable to constant Experience . . . especially where it seems not to be concerned with the Accidents of Things, but pertains to their principal Properties and inner Constitution, it will at least be most safe and prudent to yield a ready Assent to it. For . . . we are guilty of the greatest Imprudence, if we shew the least Distrust, and do not yield our stedfast Assent and obstinately adhere, when we still find our Expectations most accurately answered, after a thousand Researches; and especially when we have the constant Agreement of Nature to confirm our Assent, and the immutable Wisdom of the first Cause forming all Things according to simple Ideas, and directing them to certain Ends: Which Consideration alone is almost sufficient to make us look upon any Proposition confirmed with frequent Experiments [*experimentis*], as universally true, and not suspect that Nature is inconstant and the great Author of the Universe unlike himself. Nay sometimes, from the Constancy of Nature, we may prudently infer an universal Proposition even by one Experiment [*experimento*] alone.[64]

Although still using the language of probabilism, Barrow clearly considers experimental evidence to have a higher status than the moral certainty or probable conjecture assigned to it by the Royal Society. In fact, later in his *Lectures* he rejects Aristotle's requirement for induction by enumeration even more unequivocally

since only one Experiment will suffice (provided it be sufficiently clear and indubitable) to establish a true Hypothesis, to form a true Definition; and consequently to constitute true Principles. I own the Perfection of Sense is in some Measure required to establish the Truth of Hypotheses, but the Universality or Frequency of Observation is not so.[65]

Barrow is grappling with the profound epistemological problem involved in the transition from the Aristotelian concept of universal, natural experience to the modern concept of experiment, which is singular and artificial (or unnatural). Once again he boldly asserts what the mathematical tradition required, namely, an indubitable experimental foundation, if it was to establish a certain, demonstrative

[64] Barrow, *Mathematical Lectures*, pp. 73–4; *Works*, 1:82.
[65] Barrow, *Mathematical Lectures*, Lect. VII, p. 116; *Works*, 1:116–17.

science of nature. In contrast to his empirically oriented contemporaries like Boyle, who stressed particular experiments or "matter of fact" and avoided theory and system building, the mathematical tradition stressed the generality of experiment and sought to place it within a fully articulated theory.

Barrow's exposition of a new philosophy of mathematics adequate to the mechanical philosophy helps us to understand Newton's strong claims for the certainty of his new science of color and its association with mathematical theory. We have encountered not only so much of Barrow's language ("rigid certitude," "indubitable experiment," and "true principles" as opposed to "probable conjectures") in Newton's early writings but also the same conjunction of the certainty of mathematical theory and experimental principles. Newton also adhered to the same voluntarist natural theology as Barrow. Because Newton was attempting to construct a whole experimental science and not, like Barrow, simply attempting to justify a handful of first principles, his excessive claims to certitude became all the more evident. There is no need to pursue the many other themes set out by Barrow that we have already encountered in Newton's writings.

Although Barrow's and Newton's views on methodology coincide on many points, and it is most likely that the senior mathematician helped to form the young Newton's outlook, the common source for their quest for greater certainty was undoubtedly their commitment to the mathematical sciences. Yet, if Barrow was so confident in his philosophical-mathematical lectures, in the actual scientific practice of his *Optical Lectures*, he, for the most part, made no extraordinary claims as to the certainty of his theory of image location. For the six principles or hypotheses of his geometrical optics he claimed only that they were "consistent with experiment and not repugnant to reason."[66] At the conclusion of the *Lectures*, however, he remarked that he would not go beyond mathematical optics to treat its other parts, "which are physical and consequently rather often necessarily offer plausible conjectures instead of certain principles."[67] Thus Barrow judged the principles of his mathematico-physical theory to be more certain than those of a physical theory, just as Newton would claim two years later.

Newton's claims as to the certainty and mathematical formulation of his "New theory" were immediately challenged by Hooke in his

[66] Barrow, *Lectiones XVIII, Cantabrigiae in Scholis publicis habitae; in quibus opticorum phaenomenωn genuinae rationes investigantur, ac exponuntur* (1669), in *Works*, 2:7–8.
[67] Ibid., Lect. XVIII, §XIII, 2:152.

referee's report to the Royal Society. Hooke did not recognize, or at least accept, the phenomenological nature of Newton's theory, or indeed that it was a theory. Rather, he considered it to be just a hypothesis to account for the phenomena, which his own hypothesis on light could explain equally as well. Because Newton had only "laid down" his theory without demonstration and did not even touch upon his mathematical theory or mention his *Optical Lectures*, Hooke's reaction was perfectly understandable. Even today it is not widely appreciated that before the *Principia* Newton had attempted to create a new mathematical science – a theory of color – and devoted about half of his *Optical Lectures* to that effort. Although the theory was unsuccessful and ultimately suppressed by Newton, it is historically significant as part of his program to mathematize physics.[68]

Throughout his referee's report, Hooke returned to Newton's claim to certainty and denied that the theory had been demonstrated by "any undeniable argument" or with "absolute necessity," or that it was "soe certain as mathematicall Demonstrations."[69] In his reply to Hooke, Newton rejected the attribution to him of "a greater certainty in these things then I ever promised, viz: The certainty of *Mathematicall Demonstrations.*" He then clarified his view:

> I said indeed that the *Science of Colours was Mathematicall & as certain as any other part of Optiques;* but who knows not that Optiques & many other Mathematicall Sciences depend as well on Physicall Principles as on Mathematicall Demonstrations: And the absolute certainty of a Science cannot exceed the certainty of its Principles. Now the evidence by wch I asserted the Propositions of colours is in the next words expressed to be from *Experiments* & so but *Physicall:* Whence the Propositions themselves can be esteemed no more then *Physicall Principles* of a Science. And if those Principles be such that on them a Mathematician may determin all the Phaenomena of colours that can be caused by refractions . . . I suppose the *Science of Colours* will be granted *Mathematicall* & as certain as any part of *Optiques.*[70]

In distinguishing pure from mixed mathematics, and granting that the latter can be only as certain as its physical principles, Newton was adopting the position of the mathematical tradition as expounded by Barrow. He still insisted that as a mathematical science his theory of color was as certain as geometrical optics and more certain than a

[68] For a fuller description of the mathematical theory, its limitations, and Newton's subsequent abandonment of the *Optical Lectures*, see my "Experiment and mathematics in Newton's theory of color."

[69] Hooke to Oldenburg, [15 February 1671/2], *Correspondence*, 1:110, 113.

[70] Newton to Oldenburg for Hooke, 11 June 1672, ibid., 1:187–8; Oldenburg deleted this passage from the version published in *Philosophical Transactions*.

qualitative or purely physical account. However, for the first time he admitted the contingency of his experimental principles and abandoned his strong claim for the certainty of experiment. No longer would he assert, as he did in the "New theory," that his theory was a "most rigid consequence" deduced from "experiments *concluding directly & w^th^out any suspicion of doubt*." In fact, a month later, when he closely paraphrased the methodological claims of the "New theory," he modified them: "I told you," he wrote Oldenburg in July, "that the Theory wch I propounded was evinced to me, *not by inferring tis thus because not otherwise*, that is not by deducing it onely from a confutation of contrary suppositions, but *by deriving it from Experiments concluding positively & directly*."[71]

Despite modifying his claims to certainty and abandoning his mathematical theory of color (Newton withdrew his *Optical Lectures* from publication in the spring of 1672 and never relented in this decision), his commitment to mathematical science did not in the least diminish. With the appearance of his *Principia mathematica* in the following decade, he kept his earlier vow to create a new mathematical science and to show by example "how valuable mathematics is in natural philosophy"; and the theory of fits in the next decade culminated his effort to create a mathematical theory of periodic colors.

Although Newton modified his claim to the certainty of his theory in 1672, he did not publicly adopt the probabilism of his contemporaries until his old age. In Query 23/31, which was added to the Latin translation of the *Opticks* in 1706, Newton set out what he considered to be the two acceptable scientific methods, analysis and synthesis. In analysis (which corresponds to induction) experiment and observation are used to argue from "effects to their causes, and from particular causes to more general ones"; in synthesis or composition (which corresponds to deduction) the causes discovered by analysis are taken as principles and new phenomena, which depend on those principles, are derived and proved by them.[72] For the second English edition of 1716, Newton greatly expanded his account of analysis:

This Analysis consists in making Experiments and Observations, and in drawing general Conclusions from them by Induction, and admitting of no Ob-

[71] Newton to Oldenburg, 6 July 1672, *Correspondence*, 1:209. In the style of the period the italics or underlining (probably marked by Oldenburg) indicate a quotation. Although this passage is widely cited, it is not recognized that Newton changed the phrase "w^th^out any suspicion of doubt" to "positively" and eliminated "most rigid consequence."

[72] Add. 3970, f. 286^r^; *Optice*, p. 347.

jections against the Conclusions, but such as are taken from Experiments, or other certain Truths. For Hypotheses are not to be regarded in experimental Philosophy. And although the arguing from Experiments and Observations by Induction be no Demonstration of general Conclusions; yet it is the best way of arguing which the Nature of Things admits of, and may be looked upon as so much the stronger, by how much the Induction is more general. And if no Exception occur from Phaenomena, the Conclusion may be pronounced generally. But if at any time afterwards any Exception shall occur from Experiments, it may then begin to be pronounced with such Exceptions as occur.[73]

This oft-quoted passage makes Newton seem like yet another adherent to the philosophical outlook of the early Royal Society. Yet we must not forget the more dogmatic claims nearly a half century before and his earlier deductivist goal, nor that he still does not admit that an induction from experiment can be totally erroneous – at worst, its generality may have to be restricted. If Newton finally accepted a more probabilistic view of scientific knowledge, he never relented in his rejection of hypotheses in scientific theories, for to admit them would be to compromise the more certain results founded on experiment and observation. This position had yet another important methodological consequence: He never publicly adopted the hypothetico-deductive method, one of the great methodological achievements of the seventeenth century.[74] To do so would have violated two of his fundamental tenets – that the principles of a science must be derived from experiment and observation and that conjectures and certainties must not be mixed. To be sure, as we shall see, he freely used the hypothetico-deductive method in his investigation of thin films and the creation of the theory of fits. However, when it came time to compose the formal theory, he cast aside the hypotheses and adopted an inductive mode, which, Newton considered, was the only legitimate way to argue from effects to cause.

Newton's quest for certainty is thus responsible for many of the characteristic features of his science: his total rejection of imaginary hypotheses, his strictures against mixing experimental hypotheses with more certain principles, his favoring of phenomenological theories, and his endeavor to create mathematical theories whenever possible. This quest for certainty is in fact only a mirror image of

[73] *Opticks* (Dover), p. 404.
[74] On Huygens' use of the hypothetico-deductive method and a comparison with Newton, see my "Huygens' *Traité de la lumière* and Newton's *Opticks*."

Newton's commitment to a mathematical way of thinking.[75] All of these features are found in his theories of color and of fits, but his theory of colored bodies does not altogether conform to the same mold.

1.3. TRANSDUCTION

Newton's theory of colored bodies is the only major portion of his optical theory that is explicitly cast in causal form. The properties and structure of the corpuscles that compose bodies are used to explain their color. To understand why he did not consider this theory, based on the properties of invisible corpuscles, to be a hypothesis, or how he could apply his inductive method to it, it is necessary briefly to examine his belief in the validity of transduction and the uniformity of nature. In this section we consider the formal justification of these methodological maxims that he presented in the Rules of Reasoning in the *Principia;* in Chapter 3 we will see how he actually applied them.

Transduction is a scientific method by which the laws and properties of observable macroscopic bodies are extended to the imperceptible microscopic parts of bodies.[76] The principal mechanical philosophers, Descartes, Hooke, and Boyle, whom Newton read so carefully, freely utilized that method. Indeed, it was intrinsic to the mechanical philosophy, which assumed that the entire physical world obeyed the same set of mechanical laws. In his *Principles of Philosophy* (1644), the seminal treatise of the new mechanical philosophy, Descartes explained the principal phenomena of heaven and earth by

[75] After concluding his magisterial edition of Newton's *Mathematical Papers,* Whiteside admonishes us that "we must never omit to take into account the over-riding mathematical cast of his mind, even in areas where it does not seem obviously to intrude" ("Newton the mathematician," p. 117). Richard S. Westfall's attribution of a search for Truth to Newton is no doubt just a different aspect of my quest for certainty, which lays more stress on philosophical and methodological issues; *Never at Rest.*

[76] On transduction see Mandelbaum, *Philosophy, Science and Sense Perception,* Ch. 2; Cohen, "Hypotheses in Newton's philosophy"; McGuire, "Atoms and the 'analogy of nature' "; McMullin, *Newton on Matter and Activity,* Ch. 1; and Kathleen Okruhlik, "The foundation of all philosophy." In an impressive display of scholarship, Mandelbaum was able to uncover the role of Rule III in justifying transduction from sources published before 1963, when the fruits of the modern era of Newton research had scarcely begun to appear. He used the term "transdiction" to describe the method of making inferences about the unobservable from knowledge of the observable (*Philosophy, Science, and Sense Perception,* p. 63), but McGuire adopted the term "transduction" ("Atoms and the 'analogy of nature,' " p. 3), which is now more widely used.

assigning particular shapes and motions to the invisible particles composing the aether and bodies. At the conclusion of the *Principles,* he posed the problem of "how we know the figures and movements of imperceptible particles." His response, an appeal to our knowledge of the mechanical operation of macroscopic bodies, was a justification of transduction:

> But I attribute determined figures, and sizes, and movements to the imperceptible particles of bodies, as if I had seen them; and yet I acknowledge that they are imperceptible. And on that account, some readers may perhaps ask how I therefore know what they are like. To which I reply: that I first generally considered, from the simplest and best known principles (the knowledge of which is imparted to our minds by nature), what the principal differences in the sizes, figures, and situations of bodies which are imperceptible solely on account of their smallness could be, and what perceptible effects would follow from their various encounters. And next, when I noticed some similar effects in perceptible things, I judged that these things had been created by similar encounters of such imperceptible bodies; especially when it seemed that no other way of explaining these things could be devised. And, to this end, things made by human skill helped me not a little: for I know of no distinction between these things and natural bodies, except that the operations of things made by skill are, for the most part, performed by apparatus large enough to be easily perceived by the senses: for this is necessary so that they can be made by men. On the other hand, however, natural effects almost always depend on some devices so minute that they escape all senses. *And there are absolutely no judgments in Mechanics which do not also pertain to Physics, of which Mechanics is a part or type* . . . so, from the perceptible effects and parts of natural bodies, I have attempted to investigate the nature of their causes and of their imperceptible parts.[77]

Descartes immediately conceded that the world could have been constructed differently. "For just as the same artisan can make two clocks which indicate the hours equally well and are exactly similar externally, but are internally composed of an entirely dissimilar combination of small wheels: so there is no doubt that the greatest Artificer of things could have made all those things which we see in many diverse ways." Nonetheless, he insisted that his physical explanations were at least morally certain because they accurately accounted for the phenomena and were useful for everyday life and the practice of medicine and physics.

Newton rejected both prongs of Descartes' justification of transduction (his innate principles of nature and his hypothetical method,

[77] Descartes, *Principles of Philosophy,* Pt. IV, §§203–4, pp. 285–6; italics added.

which simply required that the explanations be consistent with the phenomena), but he did not reject transduction itself. Newton treated transduction as a case of induction rather than hypothetico-deduction. Claims about imperceptible particles of matter would then have to be as thoroughly founded on experiment and observation as those about the macroscopic world. Newton's use of transduction must be distinguished from simple analogical reasoning, such as his suggestion that light particles falling on a transparent medium excite waves in it just as a stone falling on water does. It must also be distinguished from the postulation of macroscopic laws and properties to micromatter in order to derive observable laws, as in his demonstration in the *Principia* that Boyle's law follows from the assumption that air particles repel one another with a force varying inversely as their distance.[78] Newton judged both of these to be hypotheses because they were not deduced from experiment.

As a mechanical philosopher, Newton had been utilizing transduction for about a quarter of a century before he first attempted to justify it in the early 1690s. Book III of the first edition of the *Principia* began with nine hypotheses (yes, hypotheses, though none are causal physical explanations), the third of which concerned the transmutation of bodies and their qualities. By early 1690, when Newton was already planning a second edition, he replaced Hypothesis III with a new one on the universal qualities of bodies. Much later, some time after 1700, he renamed the first three hypotheses *Regulae Philosophandae* or "Rules of Reasoning in Philosophy" (but perhaps better rendered as "Rules of Philosophizing") and made a number of changes in the other six hypotheses.[79]

Rule III is now widely understood to represent Newton's general justification of transduction (though I do not believe it can support such a broad interpretation):

> The qualities of bodies, which admit neither intensification nor remission of degrees, and which are found to belong to all bodies within the reach of our experiments, are to be esteemed the universal qualities of all bodies whatsoever.
>
> For since the qualities of bodies are only known to us by experiments, we are to hold for universal all such as universally agree with experiments. . . . We are certainly not to relinquish the evidence of experiments for the sake of dreams and vain fictions of our own devising; nor are we to recede from

[78] Cohen, *Newtonian Revolution*, pp. 29–30, 76–8.
[79] *Principia*, p. 551.

the analogy of Nature, which is wont to be simple, and always consonant to itself.[80]

The first of Newton's grounds for attributing a quality to all bodies universally – namely, that it agree with all experiments "within our reach" – does not differ from what may be called "general induction" for macroscopic bodies as stated in the General Scholium: "In this philosophy propositions are deduced from the phenomena and rendered general by induction."[81] In the second place, to justify the extension of general induction to the primordial particles he appeals to the "analogy of nature," which is an assumption of the uniformity and simplicity of nature. The belief in the "analogy of nature," or that "nature is ever consonant to herself," served as a guiding maxim throughout Newton's career. In his earliest optical research he invoked it to explain vision, which, he held, occurred by means of vibrations in the nerves, just like hearing. The analogy of nature further led him to believe that there were color harmonies just as there were acoustical ones, and that the colors of the spectrum were in musical proportion. For example, in 1675, he proposed in his "Hypothesis" that light rays excite vibrations in the optic nerve and

affect ye sense wth various colours according to their bignes & mixture... much after ye manner, that in ye sense of hearing nature makes use of aereal vibrations of several bignesses to generate sounds of divers tones; for, *ye analogy of nature is to be observed.* And further, as ye harmony & discord of sounds proceeds from ye proportions of ye aëreal vibrations, so may ye harmony of some colours.[82]

Newton's faith in the analogy of nature followed from his conception of God's operation in the natural world. In 1698, he affirmed to John Harrington: "I am inclined to believe some general laws of the Creator prevailed with respect to the agreeable or unpleasing affections of all our *senses;* at least the supposition does not derogate from the wisdom or power of God, and seems highly *consonant to the macrocosm in general.*"[83] Throughout the second half of his career, after the *Principia,* Newton similarly invoked the analogy of nature to support his conviction that the microscopic world was governed by short-

[80] *Mathematical Principles,* pp. 398–9.
[81] Quoted in full at note 9; see also Query 23/31 at note 73.
[82] Add. 3970, f. 544^r; *Correspondence,* 1:376; italics added.
[83] Newton to Harrington, 30 May 169[8], ibid., 4:275; italics in original on "senses." The editors believe that the date 1693 in the initial publication of this letter (the autograph is not extant) is erroneous.

range forces between the particles of matter, just as the heavenly bodies were regulated by gravity. For example, in a suppressed conclusion to the first edition of the *Principia,* he proposed that "nature is exceedingly simple and consonant to herself [*sibi consona*]. Whatever reasoning holds for greater motions, should hold for lesser ones as well."[84] Yet, without sufficient evidence, these ideas always remained hypotheses. The analogy of nature alone was inadequate for establishing them as principles, which requires sufficient empirical evidence. It is the conjunction of the analogy of nature with experiment that makes the third Rule of Reasoning a viable maxim of scientific procedure.

To understand better how Rule III serves (or fails) to justify transduction, it is useful to turn to various drafts of it composed in the early 1690s.[85] In the earliest form it stated that "the laws and properties of all bodies in which experiments can be made are the laws and properties of bodies universally"; "and properties" was inserted above the line twice.[86] In a subsequent draft Newton changed this to "properties" alone, and in others he has "qualities" in the beginning of the statement of the rule and "properties" at the end.[87] Finally he restricted it to "qualities." Although Rule III started as a general justification of transduction, Newton narrowed its scope from laws and properties to the universal or primary qualities of matter – extension, hardness, impenetrability, mobility, inertia, and, above all, gravitational attraction, which alone concerned him in the *Principia.*[88]

[84] Newton, *Unpublished Scientific Papers,* pp. 321, 333; I have slightly altered the translation; see also pp. 304, 307. Newton publicly expressed this idea in Query 23/31 of the *Opticks,* quoted at Ch. 6, note 7. He also appealed to the analogy of nature to justify his belief that the sine law of refraction, which was discovered for the mean rays alone, would be valid for rays of each color; *Opticks,* Bk. I, Pt. I, Prop. 6. In Query 22/30 he likewise justified his belief in the mutual transmutation of light and matter on the ground that it "is very conformable to the course of nature."

[85] These drafts are described and variously published in Cohen, "Hypothesis in Newton's philosophy"; *Introduction to Newton's 'Principia,'* pp. 24–6, 184–7; *Principia,* pp. 550–5; and McGuire, "The origin of Newton's doctrine of essential qualities"; and "Atoms and the 'analogy of nature.' "

[86] From the autograph errata sheet at the back of Newton's copy of the first edition of the *Principia* in Trinity College Library, Cambridge, NQ.16.200; *Principia,* p. 552.

[87] Add. 4005, f. 81^r, Ax. 4; and Newton's copies of the *Principia* at Trinity, p. 402, and Cambridge University Library (Adv. b. 39.1), interleaf facing p. 402; published in McGuire, "Origin of Newton's doctrine of essential qualities," pp. 257–8; *Principia,* pp. 552–3; and Cohen, *Introduction to Newton's 'Principia,'* plates 1–3.

[88] Even in the limited case of universal qualities, McMullin has shown that Newton's use of transduction does not rely on Rule III alone with its general induction, intensification and remission, and "analogy of nature," but also invokes other arguments; *Newton on Matter and Activity,* pp. 22–6. Hardness, for example, is not

He added the requirement on intensification and remission in what is probably the second draft in order to eliminate secondary qualities such as "heat and cold, wetness and dryness, light and darkness, color and blackness, opacity and transparency," for only those things that cannot be intended and remitted "are usually considered to be the properties of *all* bodies."[89] In a later draft, composed after he had restricted the rule to universal qualities, Newton clearly expressed his intention to justify transduction for those qualities: "This seems to be the foundation of all Philosophy. For otherwise one could not derive the qualities of imperceptible bodies from the qualities of perceptible ones."[90] The second sentence was omitted from the published version. All of this evidence indicates that when Newton began to prepare a general justification of transduction, he soon recognized the conceptual difficulties involved and chose the narrower case of universal qualities, which served his immediate purpose in the *Principia*.

The principal problem in justifying transduction for properties like transparency and color is that for Newton they depend on the composition or hierarchical arrangement of the corpuscles that compose bodies. At some point of composition these properties vanish, and it is at a different point for each of them.[91] Such properties cannot be applied beyond a level where the corpuscles no longer possess the arrangement responsible for that property, or where the property vanishes. One cannot argue, for example, that the corpuscles or even large fragments of a glass bottle that may contain fluids are also capable of holding fluids. Newton never developed adequate criteria for deciding to what level a property could be legitimately carried or even to what properties transduction applied. As we shall see, he encountered just these problems (as well as others that he was unaware of) in his theory of colored bodies.

On the basis of copies of Newton's intended revisions of the *Principia* made by Fatio de Duillier, I. B. Cohen has shown that Newton had written the earliest draft with "laws and properties" by March 1690 and then by October 1692 had put Rule III (though still Hy-

deduced from the hardness of all bodies, for Newton recognizes that not all bodies are hard; rather, he infers that the primordials must be hard, for otherwise the hardness of the bodies they compose could not possibly be explained.

[89] Add. 4005, f. 81^r; McGuire, "Origin of Newton's doctrine of essential qualities," pp. 237, 257; italics added.

[90] Add. 3965, f. 266^r; Cohen, "Hypotheses in Newton's philosophy," p. 176.

[91] To anticipate the theory of colored bodies, transparency vanishes at the primordial particles, whereas color vanishes at compound, macroscopic corpuscles whose size corresponds to the central spot in Newton's rings.

pothesis III) in its published form.[92] In precisely the same period, most likely early in 1691, he was also preparing for publication Book II of the *Opticks* with his theory of colored bodies (§4.1). It is quite possible that it was his reencounter with transduction in that theory that gave him pause in offering a general justification for it in Rule III and caused him to restrict it to his immediate aim of establishing gravity as a universal quality. I am not proposing that Newton abandoned his belief in transduction, but only that after the early 1690s he saw problems in justifying and applying it. He had adopted it in the 1670s to explain the colors and transparency or opacity of bodies, and in the 1680s to establish universal gravitation, and there is no evidence that he ever seriously doubted the validity of either. These were not established by the simple process of Rule III but rather by a broad variety of means, including empirical evidence, mathematical theory, physical argument, and methodological rules. As a mechanical philosopher, transduction was part of Newton's intellectual heritage, and we shall soon see Hooke and Boyle similarly invoking it in their explanations of colored bodies.

The most powerful justification for transduction came not from formal philosophy or metaphysics but from recent progress in natural philosophy itself. The inventions of the telescope and microscope at the beginning of the seventeenth century opened entirely new and previously hidden vistas to direct examination. The mechanical philosophers could see with their own eyes that these new worlds were constructed and operated just like the ordinary visible world. Hooke's *Micrographia,* with its vivid text and striking illustrations of nature's hidden contrivances, widely propagated knowledge of this microscopic world. In his preface he declared:

> It seems not improbable, but that by these helps the subtilty of the composition of Bodies, the structure of their parts, the various texture of their matter, the instruments and manner of their inward motions, and all the other possible appearances of things, may come to be more fully discovered ... we may perhaps be inabled to discern all the secret workings of Nature, almost in the same manner as we do those that are the productions of Art, and are manag'd by Wheels, and Engines, and Springs, that were devised by humane Wit.[93]

Boyle similarly invoked the revelations of the microscope to argue that "the mechanical affections of matter are to be found, and the

[92] Cohen, *Introduction to Newton's 'Principia,'* pp. 184–7.
[93] Hooke, *Micrographia,* Preface, sig. a[2^v].

laws of motion take place, not only in the great masses, and the middle sized lumps, but in the smallest fragments of matter."[94]

Thus, as new and improved instruments allowed the senses to be extended so that they could see deeper and deeper into matter, the method of transduction was continually justified. Soon the atoms themselves might be seen.[95]

Finally, we should briefly look at the first two Rules of Reasoning, as Newton utilized them in his theory of colored bodies years before they were formulated for the *Principia*. Like the third rule, these, too, are statements about the simplicity and uniformity of nature. These rules already appeared in the first edition when they were still called hypotheses. Rule I is a statement of the principle of simplicity: "No more causes of natural things are to be admitted than are both true and sufficient to explain their phenomena." The justification was simplicity itself: "For nature is simple and affects not the pomp of superfluous causes."[96] The restriction to "true" causes prevents the introduction of hypotheses in place of demonstrated principles, even if their number is sufficiently small.

The second rule, which is a consequence of the first, is that essentially invoked in the theory of colored bodies:

> Therefore the causes of natural effects of the same kind are the same.
>
> As for example respiration in man and beast; the descent of stones in *Europe* and *America;* the light of a cooking fire and of the sun; the reflection of light in the earth and in the planets.[97]

This is an expression of the principle of uniformity, for by the simplicity of nature it asserts that phenomena of the same kind have the same cause. Newton does not state how we are to determine that effects are of the same kind, but in accordance with his general methodology such inferences would presumably be based on experiment and observation.

In the *Principia*, Rules I and II, together with observational evidence, play an essential role in Newton's demonstration that gravitational

[94] Boyle, *The Excellency and Grounds of the Mechanical Hypothesis*, appended to *The Excellency of Theology* (1674), in *Works*, 4:71. In *Touching Colours*, pp. 40–1, Boyle also expressed the belief that with more perfect microscopes the particles of bodies responsible for the modification of light that produces their color would become visible.

[95] See Christoph Meinel, " 'Das letzte Blatt im Buch der Natur.' "

[96] *Mathematical Principles*, p. 398; *Principia*, pp. 550–1. In the second edition Newton added an additional sentence of justification: "To this purpose philosophers say that Nature does nothing in vain, and more is in vain when less will serve."

[97] Ibid. In the third edition the rule was reworded: "Therefore the causes assigned to natural effects of the same kind must be, as far as possible, the same."

attraction is universal. He shows that the motions of falling bodies on the earth, the moon about the earth, the planets about the sun, and the satellites of Jupiter and Saturn about those planets all have the same properties and so are phenomena of the same kind (Bk. III, Props. 4, 5). Therefore, it is the same force, gravity, that causes each of them. Rule III is then invoked to extend gravitation from the cosmic and quotidian levels to the microscopic particles of bodies and establish its truly universal nature (Prop. 6). In his theory of colored bodies, Newton appeals to a variety of observations to show that the colors of thin films and of natural bodies are phenomena of the same kind and therefore, by the reasoning of Rule II, have the same cause. Inasmuch as he applied the same Rules II and III to establish his theory of gravity as his theory of colored bodies, we cannot attribute the ultimate failure of the latter theory to his method alone. Indeed, scientists still use a belief in the simplicity and uniformity of nature to guide their research.

We can note one significant difference between the two cases now. Whereas he made a very strong case that the motions of terrestrial and celestial bodies were phenomena of the same kind by appealing to Kepler's laws, his mathematical theory of motion, and astronomical observations, his demonstration that the colors of thin films and those of natural bodies were of the same kind was supported only by a few selected observations interpreted in rigidly mechanical terms. Our study of Newton's use of transduction in Chapter 3 will reveal some intrinsic problems of a method so widely used by mechanical philosophers. Let us now turn to Newton's optics and observe his methodology in practice.

2

NEWTON'S RINGS

Newton was introduced to the periodic colors of thin films through his reading of Hooke's *Micrographia* in late 1665 or early 1666, and the difficulty of understanding their complex appearance quickly captured his imagination.[1] Very soon afterward Newton devised his eponymous method to subject them to measurement and opened the way to a mathematical treatment. He was able to demonstrate with his measurements what Hooke had only conjectured, that the appearance of the rings was periodic. At this time, though, for want of sufficient experimental evidence, Newton attributed periodicity only to the appearance of the rings and not to light itself. Having gained rough intellectual control over the problem in 1666, Newton set it aside until about 1671, when he carried out an extensive series of experiments and precise measurements. By 1672, he was ready to make public his research on the colors of thin films and natural bodies, but fear of controversy caused him to delay until late 1675.

In this chapter we will follow the main outlines of Newton's remarkable research in the ten-year period from 1665 to 1675. My purpose is not to describe the development of his experimental investigation through all its stages. Rather it is to present his principal experimental results and the physical explanations and mathematical

[1] The most valuable historical research on Newton's rings and the subsequent theory of fits is by Westfall, "Isaac Newton's coloured circles twixt two contiguous glasses"; "Uneasily fitful reflections on fits of easy transmission"; and *Never at Rest*. Volume 4 of Jean-Baptiste Biot's *Traité de physique expérimentale et mathématique* (1816) contains a detailed and exceedingly useful account of Book II treated as a scientific contribution of vital contemporary interest. In addition to the biographies of Newton, see also Hanson, "Waves, particles, and Newton's 'fits' "; Sabra, *Theories of Light*, Ch. 13; and Michel Blay, *La conceptualisation Newtonienne des phénomènes de la couleur*, Pt. 3, Ch. 2.

descriptions that he proposed for them and especially the interaction of these elements. This research served as the foundation for Newton's theory of the colors of natural bodies (which was the avowed aim of his research on the colors of thin films) and, some twenty years later, for his theory of fits of easy reflection and transmission. This chapter thus serves us as the foundation for the next two chapters.

From his earliest investigations of thin films to those of thick plates some twenty-five years later, Newton relied on his hypothesis of light particles and a vibrating medium to suggest and interpret experiments and to deduce the mathematical and physical properties of periodic colors. Because his methodology required that these properties be presented "abstracted" of all hypotheses (§1.2), Newton chose to set forth his model in a separate work, his "Hypothesis," in 1675. The "Hypothesis" contains his only formal exposition of his model, so we shall examine it and its relation to his more certain scientific work in some detail.

2.1. NEWTON, HOOKE, AND THE COLORS OF THIN FILMS

The colors of thin films had, of course, been observed before Hooke presented his account of them in the *Micrographia* – most recently by Boyle – but no one before him had recognized that it was a fundamentally different physical phenomenon from dispersion and therefore required a different physical explanation. Newton built his investigation of the periodic colors of thin films on the solid foundation laid down by Hooke. Indeed, his initial investigation of them seems to have been carried out while he was reading and taking notes on the *Micrographia*.[2]

Newton's indebtedness to Hooke has been disputed ever since Hooke himself questioned Newton's priority in 1675. One cannot reasonably doubt that Newton's investigation started from Hooke's, or that his experiments and mathematical description substantially advanced that subject beyond, and was fundamentally different from, Hooke's. Newton's breakthrough came with his development of an experimental method – Newton's rings – to measure the thickness of the film producing colors. He first described this in the midst of the second essay "Of Colours," in which he was recording the experiments that formed the basis for his new theory of light and color.

[2] "Out of Mr Hooks Micrographia," Add. 3958, ff. 1^{r}–4^{r}; Newton, *Unpublished Scientific Papers*, pp. 400–13.

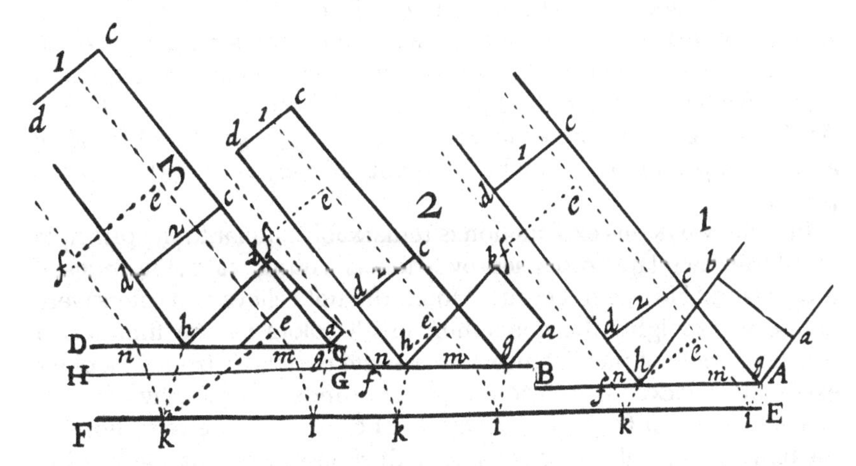

Figure 2.1. Hooke's explanation of the colors of thin films. (From *Micrographia*, Plate VI, Figure 6.)

The new theory was essential for allowing him to understand the colors of thin films.

From a careful examination of the colors produced in thin sheets of mica ("Muscovy-glass"), air, glass, soap bubbles, and other transparent, colorless substances, Hooke concluded that they were caused by "some *Lamina* or Plate of a transparent or pellucid body of a thickness very determinate and proportioned according to the greater or less refractive power of the *pellucid* body."[3] If the plates were too thin or too thick, the colors vanished.[4] His most important result was based on his observation that in films of a nonuniform thickness, such as a wedge, the succession of colors was repetitive. He deduced from this that the appearance of the colors was a periodic phenomenon.

Hooke advocated a wave theory of light and – what was not at all common in the seventeenth century – postulated the periodicity of the vibrations propagated through the aether. He explained the colors produced by thin films in terms of the varying distance between two "pulse fronts" reflected at the first and second surfaces of the film. Thus, in Figure 2.1, part 1, at the first surface of the film *AB*, the incident ray or pulse of sunlight *ab* is partially reflected into *cd* and partially transmitted. After the transmitted pulse (indicated by broken

[3] Hooke, *Micrographia*, p. 50.
[4] Ibid., p. 53.

lines) is reflected at the second surface *EF* and emerges from the first surface as *ef*, it is weakened and falls behind the stronger pulse *cd*: "this confus'd or *duplicated* pulse, whose strongest part precedes, and whose weakest follows, does produce . . . the sensation of a *Yellow*."[5] As the film gets thicker, as in part 2, the ray *ef* falls farther behind *cd* and red is produced; and when it is still thicker, as in part 3, blue is produced.

In some ways this explanation is remarkably similar to the principle of interference later proposed by Thomas Young, as Young himself recognized.[6] However, because Hooke did not believe that colors were innate to sunlight, the generation of all colors by the film was a function solely of the distance between the two pulse fronts. Hooke's explanation lacked a sufficient number of independent physical variables to account for all the observed phenomena. The refractions in the film created only one color at each thickness rather than separating a number of preexisting colors. It is difficult to understand, for example, how he could possibly explain the existence of transmitted colors – once he observed them seven years later – with only one, transmitted pulse front.[7] Hooke ended this chapter on a note of disappointment and laid down a challenge of the "greatest concern in this *Hypothesis*": to determine the thickness of the films. Despite his efforts, he had been unable to determine the particular thicknesses at which the colors appeared, because the plates are "so exceeding thin" and his microscope "so imperfect."[8]

Newton was stimulated by Hooke's account of the colors of thin films and almost immediately, or so it seems, set out to investigate them himself. The budding geometrical philosopher could not resist the challenge of determining the thickness of the film where each color appeared. It is even possible that Hooke supplied the clues to Newton's clever solution. In one experiment Hooke described the colors that appear between two small flat glass disks placed over one another and the changes they undergo when the disks are pressed together with greater or lesser force. In describing the conditions of their appearance, he explained:

> Nor is it necessary, that these colour'd *Laminae* should be of an even thickness . . . which circumstance is only requisite to make the Plate appear all of

[5] Ibid., p. 66.
[6] Young, "On the theory of light and colours" (1802), pp. 39–40.
[7] On 19 June 1672, Hooke reported to the Royal Society a number of experiments on refraction and color, including new observations on thin films; see Birch, 3:52–4, on p. 54; reprinted in *Correspondence*, 1:195–7, on p. 197.
[8] Hooke, *Micrographia*, p. 67.

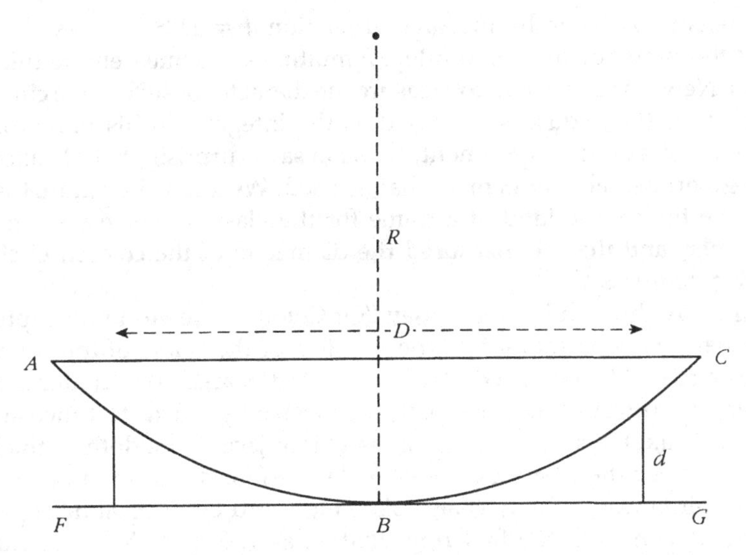

Figure 2.2. Newton's method of determining the thickness d of the air film between the spherical surface of lens ACB and the plane surface FBG in the production of Newton's rings. R is the radius of the lens and D the diameter of the rings at the thickness d.

the same colour; but they may resemble a *Lens*, that is, have their middles thicker then their edges; or else a *double concave*, that is, be thinner in the middle then at the edges; in both which cases there will be various coloured rings or lines.[9]

Newton's great insight was to realize that if, instead of forming a thin film in the shape of a lens, he were to place a real lens on a flat plate of glass, he could then readily determine the thickness of the thin film formed in the gap between them by the well-known formula for the sagitta of an arc (from Euclid's *Elements*, III, 36). If, as in Figure 2.2, a convex lens ABC is placed on a glass plate FBG and illuminated and viewed from above, a set of concentric colored circles – now aptly known as Newton's rings – produced by the thin film of air $ABCGBF$ will be seen through the upper surface of the lens. The circles will form an alternating sequence of bright and dark rings and their common center, the point of contact B, will be surrounded by a dark spot. Let the diameter of any of these colored circles be denoted by D, the thickness of the air film producing that circle by d, and the radius of

[9] Ibid., p. 53; see also p. 50.

the lens by R. Then by the sagittal relation $d = D^2/8R$. To establish that the circles do appear at integral multiples of some definite thickness, Newton simply had to measure the diameter of successive circles and see if their squares increased as the integers. In his numerous descriptions of this experiment, Newton says surprisingly little about the experimental arrangement that he used. We know little more than that he built some kind of a frame for the glasses, which were tied together, and that he measured the diameters of the colored circles with a compass.[10]

Midway through his 1666 essay "Of Colours" Newton interrupted it to set forth this solution to the problem of the colors of thin films: "The circles," he observed, "are ye broadest nearest to ye center & so beeing narrower & narrower doe (I conceive by ye exactest measure I could make) increase in number as ye interjacent aire doth in thicknesse."[11] For the first six colored circles he found that the thickness of the air between the lens and plate increased by integral multiples of the thickness at the first ring, that is, as 1, 2, 3, 4, 5, 6. Newton then calculated that "y^e^ thicknesse of y^e^ aire for one circle was $\frac{1}{64000}$ inch, or 0,000015625. [wch is y^e^ space of a pulse of y^e^ vibrating medium.]"[12] I will not analyze Newton's calculation, for, besides containing a mistake and being wide of the mark, as he later noted, it is not at all clear that Newton himself knew precisely what he had measured, maxima or minima and of what color. Problems of this sort are not unexpected at an early stage of investigation and in no way hindered Newton's unraveling the nature of the production of colors of thin films. His method for determining the thickness of the film was in principle valid, and it later allowed him to develop a mathematical theory of the appearance of periodic colors. Moreover, from his remark (cited in square brackets above), we can see that he

[10] In his various trials Newton combined glasses of different shapes to produce Newton's rings, but the fundamental geometry of Figure 2.2 and the applicability of the sagitta relation remained the same with all combinations. In "Of Colours" he used "a Sphaericall object glasse of a Prospective tyed fast to a plaine glasse" (Add. 3975, §33, p. 9; *Questions*, p. 476); in "Of ye coloured circles" a double convex lens and a plane glass; and in all versions of the "Observations," including the *Opticks*, a double-convex lens with a plano-convex lens placed on top with its plane side facing downward (Obs. 4). See also Westfall, *Never at Rest*, pp. 216–17.

[11] Add. 3975, §33, p. 9; *Questions*, p. 476.

[12] Add. 3975, §33, p. 10; *Questions*, pp. 476–7. The square brackets in this essay are Newton's way of setting off his speculative or interpretive remarks; he also uses brackets to mark deletions. Unless otherwise indicated, all square brackets in quotations are my editorial insertions.

had already begun to utilize vibrations in the aether as the physical cause of the rings.[13]

Hooke was conceptually impeded from taking the next fundamental step in Newton's investigation, examining the composition of the colored rings, because he considered the colors to be created in their passage through the thin film rather than being separated out of the incident sunlight in which they already exist. Newton, on the contrary, had just discovered that light was compounded of all colors and would quite naturally have assumed that those colors in the incident sunlight that were not reflected by the film were transmitted. By examining the transmitted rings (which Hooke did only after Newton published his theory of white light and color), Newton readily confirmed that the transmitted and reflected rings were in fact complementary.[14] And by examining the rings produced by rays of a single color, it was possible for him to understand their formation in white light when the colors are not separately visible because of their overlapping and mixing.[15] Namely, he was able to see that at the same place some rays are reflected whereas the others are transmitted, and that rays of the same color are at some places reflected and at others transmitted.

At this stage Newton had not fully elaborated these points, especially the second, which requires assigning a particular thickness or vibration length to each color. Hooke, to be sure, had conjectured that each color appears at some multiple of a definite thickness, but his conception was a fundamentally different one. In his view all the colors are not simultaneously present in the incident sunlight but each one, and only one, is successively created at these thicknesses:

> so that if there be a thin transparent body . . . in the manner of a Wedge . . . there shall be generated several consecutions of colours, whose order from the thin end towards the thick, shall be *Yellow, Red, Purple, Blue, Green: Yellow, Red, Purple, Blue, Green; Yellow, Red, Purple, Blue, Green; Yellow,* &c. and these so often repeated, as the weaker pulse does lose paces with its *Primary,* or

[13] It should be noted that the thickness of air at the first ring, which Newton here calls "ye space of a pulse," would in his mature theory from about 1670 onward be considered only half a pulse or wave; see note 40. This is another indication that he had just begun to unravel the phenomenon and develop an explanatory framework for it.

[14] Add. 3975, §37, p. 10; *Questions,* p. 477. For Hooke's observation, see note 7. Complementary colors, a term not yet used in Newton's day, are colors that compose white when added together. If a beam of white light is separated into two portions without changing or destroying any part, those two portions will be complementary.

[15] Add. 3975, §39, p. 11; *Questions,* p. 477.

first pulse, and is *coincident* with a second, third, fourth, fifth, sixth &c. pulse behind the first.[16]

In fact, this predicted sequence, in which an identical succession of colors is continually repeated, is not observed, but is only what is to be expected according to Hooke's conception of their periodicity, which does not allow for a number of different colors to appear at the same thickness. One of the reasons for suggesting that "Of Colours" originates from an early period while Newton was still studying the *Micrographia* is that Newton's record of the sequence of colors in the rings is virtually identical to Hooke's observed sequence.[17] Newton's own interpretation of the formation of Newton's rings requires that circles of different spectral colors differ in size and thus overlap and form a changing sequence of compound colors. In all of his subsequent writings on the colors of thin films, he has the actually observed changing sequence of colors with only minor variation.

We can see the undigested influence of the *Micrographia* in yet another observation. To show that air was not necessary for the appearance of colored rings between two pieces of flat glass, Hooke put water and other liquids between them. He found "much the like effect, *only* with this difference," that the greater the difference in index of refraction between the film and surrounding body the more vivid the colors appear.[18] Newton repeated this experiment by putting water in the gap between two contiguous prisms. In "Of Colours" (§41) he confirmed Hooke's report, and in the next entry he made explicit what was only implicit in Hooke's account: "The coloured circles appeared as big when there was a filme of water as when there was a film of aire betwixt ye Prismes."[19] When Newton resumed his investigation about five years later, he corrected this erroneous observation.

[16] Hooke, *Micrographia*, pp. 66–7; see also p. 65. In 1672, he still reported an identical sequence of colors in each order; Birch, 3:54; *Correspondence*, 1:196.

[17] Hooke began his chapter with a description of the colored circles observed in mica – "Blew, Purple, Scarlet, Yellow, Green; Blew, Purple, Scarlet, and so onwards, sometimes half a score times repeated" – and Newton duly entered this sequence in his notes on p. 48 of the *Micrographia;* Newton, *Unpublished Scientific Papers*, p. 403. In "Of Colours" Newton described the reflected colors after the central dark spot and a white ring as "yellow, greene; blew, purple, Red, Yellow, greene; blew, purple, Red, Yellow, Greene; blew &c." (Add. 3975, §37, p. 10; *Questions*, p. 477; punctuation altered). Aside from the ambiguity of the grouping of the colors into orders and the change of Hooke's "Scarlet" to "Red," this sequence is identical to Hooke's. It also differs radically from Newton's later observations (see note 37), where yellow always precedes red and blue precedes green.

[18] Hooke, *Micrographia*, p. 50; italics added.

[19] Add. 3975, p. 11; *Questions*, p. 477.

After Newton finally sent his papers to the Royal Society, in December 1675, Hooke immediately accused him of plagiarizing the *Micrographia*.[20] In his reply, Newton was compelled to discuss the relation of his work to Hooke's and, incidentally, to shed additional light on the development of his own research. He insisted, correctly, that their explanations differed fundamentally and had only the supposition of a vibrating aether in common, although in a draft he admitted that he borrowed this idea from Hooke.[21] Because Newton was exceedingly unlikely to grant Hooke any credit he did not deserve, I believe we must accept his acknowledgment that he borrowed a number of observations from Hooke, particularly

that of plated bodies exhibiting colours, a Phaenomenon for ye notice of wch I thank him: But he left me to find out & make such experiments about it as might inform me of ye manner of ye production of those colours, to ground an Hypothesis on; he having given no further insight into it then this y^{t} ye colour depended on some certain thicknes of ye plate: though what that thicknes was at every colour he confesses in his *Micrographia* he had attempted in vain to learn & therefore seing I was left to measure it my self I suppose he will allow me to make use of what I tooke ye pains to find out.[22]

Newton, we should note, in passing expresses his rigid requirement that hypotheses be based on experiment. In a draft he describes the two experiments, unknown to Hooke, that served as the foundation of his theory:

One of ye experim^{ts} was that each colour in ye coloured rings of those plates was reflected & refracted alternately for many successive thicknesses of ye plate. And this directly overthrows all Mr Hooks hypothetical discours about these colours & was my chief designe to explain what might be ye Mechanical cause of it. Another was that at ye same parts of ye glas divers colours alike incident were one reflected another transmitted.[23]

[20] At the meeting of the Royal Society on 16 December 1675, "After reading this discourse [Newton's "Hypothesis"], Mr. Hooke said, that the main of it was contained in his *Micrographia*, which Mr. Newton had only carried farther in some particulars" (Birch, 3:269).

[21] Newton to Oldenburg, 21 December 1675, *Correspondence*, 1:405. In his draft Newton conceded that "from Mr Hook I have taken this That aether is a vibrating Medium & whose vibrations cause ye motion of ye parts of bodies wherein heat consists & are mutually caused thereby" (Add. 3970, f. 531^{v}); Newton made a number of alterations in this passage, but I will indicate only those that I judge to be significant. He later admitted this point in a letter to Oldenburg, 10 January 1675/6, *Correspondence*, 1:408.

[22] *Correspondence*, 1:406.

[23] Add. 3970, f. 531^{v}.

Newton's acknowledgment that the *Micrographia* was his starting point, together with other evidence already alluded to, suggests that he began his experiments on the colors of thin films while he was reading the *Micrographia* and taking notes on it.[24] In his note on p. 48 he recorded in square brackets: "The more oblique position of ye eye to ye glasse makes ye coloured circles dilate."[25] This original observation goes beyond Hooke (who apparently observed the colors only with his eye almost directly over the film) and was made by Newton with a film of variable thickness such as the air between two slightly curved glass prisms (in films of constant thickness this phenomenon simply does not occur). The account of thin films in "Of Colours" is still heavily dependent on Hooke: The descriptions of the sequence of colors in Newton's rings and of their size when formed in water are Hooke's and not what Newton himself would later observe. After sketching the experimental foundation for his theory of color and for thin films in "Of Colours," Newton returned to the *Micrographia* and included more notes from it in entries 50, 54, and 56. Thus Newton was carrying out experiments on the colors of thin films while taking notes on the *Micrographia*, and after he had carried out his most important experiments, he was still taking notes on the *Micrographia*. This sequence suggests that Newton's reading of the *Micrographia* and his initial research on the periodic colors of thin films, including the discovery of Newton's rings, were roughly contemporaneous.

Newton later acknowledged that Hooke turned the colors of thin films into a "philosophical consideration."[26] The genuine animosity between the two men should not hinder us from recognizing that he

[24] Given Newton's acknowledgment that he learned of the colors of "plated bodies" from Hooke, I believe that we must reject the contention of McGuire and Tamny that Newton "discovered" the rings independently of Hooke; *Questions*, pp. 273–4. They based their judgment on Newton's description of the colors formed in the thin film of air between two superposed prisms immediately before his first observations of Newton's rings in "Of Colours," §§27–32; Add. 3975, pp. 8–9; *Questions*, pp. 474–6. At this time Newton's concern with these experiments was to demonstrate the compound nature of light by showing that total reflection occurs at different angles for different colors; see *Optical Papers*, 1:137–41. In the draft of his reply Newton was far more frank about his general dependence on Hooke: "It being my designe to describe y^t^ Hypothesis not for its own sake as intirely new but for ye sake of ye Phaenomena of light & colours in my other papers, its a convenient Hypothesis for explaining them: [*del:* I might have been excused though I had taken] I know not why I might not have ye f[r]eedom to borrow from any body what ever I should find fit for that purpose. especially where that wch I boorrow is already secured to ye Author & sufficiently known to be his" (Add. 3970, f. 531^r^).

[25] Newton, *Unpublished Scientific Papers*, p. 403.

[26] Newton to Hooke, 5 February 1675/6, *Correspondence*, 1:416.

built his research on Hooke's in a perfectly legitimate way – "standing on ye sholders of Giants," as Newton generously put it.[27] He carried out additional experiments, clarified the phenomena, and offered an alternative explanation. When Hooke had calmed down sufficiently, he graciously acknowledged that Newton had been able "to compleat, rectify and reform what were the sentiments of my younger studies."[28] Generous and gracious they might be now, but not for long.

2.2. OBSERVATIONS OF NEWTON'S RINGS

After his breakthrough in 1666, Newton did not return to the problem of thin films until 1670 or, more likely, 1671, after he had refined his theory of color for his Lucasian optical lectures. It appears that he now felt that in order to present a comprehensive account of all color that could successfully compete with Hooke's comprehensive theory in the *Micrographia,* he had to complete his investigation of the colors of thin films (and also of natural bodies). He carried out a new series of experiments that he recorded in a short paper, "Of ye coloured circles twixt two contiguous glasses," which was probably written in 1671.[29] This paper was certainly completed by the spring of 1672, when Newton wrote a discourse on the colors of thin films and natural bodies, which I call the "Observations" of 1672.[30] He had intended

[27] Ibid.

[28] Hooke to Newton, 20 January 1675/6, ibid., p. 412.

[29] The evidence indicates that this manuscript was written between the two versions of the *Optical Lectures.* The first extant version of the *Lectures* contains no reference to or any discernible influence of his investigation of the colors of thin films. Although the *Lectures* was delivered between 1670 and 1672, the evidence for when it was composed is slight. In my edition I concluded that this first version was composed in 1670 or, more likely, 1671; *Optical Papers,* 1:16–20. Newton ends the later version, which he composed in the fall and winter of 1671–2, by noting that although the colors of thin films can be treated mathematically, he feels that he has already lingered too long on optics to treat them; ibid., pp. 602–3. A more significant reference to thin films occurs in a small change that Newton made in revising the earlier version. In §84 he initially observed that green is exactly in the middle of the spectrum, but in the later version he adds that "in a certain other property (which it is not now opportune to explain)" green has a different relation to the two extreme colors; ibid., p. 527. He is alluding to his observation that the mean color or interval *I* (the "certain other property") in Newton's rings is at the border of yellow and green; see Obs. 14 of the "Observations" of 1675 (15 in the 1672 version) and note 108. These references help us date "Of ye coloured circles" to between the two versions of the *Lectures,* that is to 1670 or, more likely, 1671. This dating agrees with Westfall's. In "Newton's coloured circles," p. 184, he dates it to between 1666 and 1672, but "closer to 1672"; and in *Never at Rest,* p. 216, to no earlier than 1670.

[30] Add. 3970, ff. 519^r^–28^v^. This paper was first identified by Westfall in "Newton's reply to Hooke and the theory of colors."

to include this discourse with his reply to Hooke's critique of his theory of color. However, on 21 May 1672, he informed Henry Oldenburg that he had decided otherwise, undoubtedly because of the controversy over his "New theory," published just a few months earlier.[31] This discourse became the draft of the "Observations" that Newton sent to the Royal Society on 7 December 1675, together with his "An Hypothesis explaining ye properties of Light discoursed of in my several papers."[32] This 1675 version was later only minimally revised to form the greater part of Book II of the *Opticks*. The "Observations" of 1675 is therefore the fundamental account of Newton's investigation of the colors of thin films and natural bodies, and I will consequently base my account on it.

The three versions of the "Observations" – those of 1672, 1675, and Book II of the *Opticks* – have the identical structure for the first three parts: The first part consists of observations, many of which are quantitative; the second presents a phenomenological explanation of the observations; and the third a physical explanation of the colors and the transparency or opacity of bodies in a series of propositions. The 1672 version then proceeds to an incomplete "Hypothesis" to explain the phenomena, whereas the 1675 version ends with the third part and has the "Hypothesis" as a separate paper. In the *Opticks* the third part was expanded to include the new propositions on fits, a fourth part on the colors of thick plates was added, and the equivalent of the "Hypothesis" was now largely relegated to the concluding queries.

"Of ye coloured circles twixt two contiguous glasses" is a careful experimental study of Newton's rings, clearly intended as a working paper for his own use.[33] Throughout, Newton unabashedly works with an emission theory of light in which the light corpuscles excite

[31] *Correspondence*, 1:160.

[32] The untitled "Observations" of 1675 was published in *Papers and Letters* as "Newton's second paper on color and light"; and in the *Correspondence* it is called the "Discourse of Observations." The "Hypothesis" was read on 9 and 16 December and then the "Observations" on 20 January and 3 and 10 February 1676. Both were entered in the society's Journal Book. The Royal Society's copy of the "Observations" was published in Birch, 3:272–8, 280–95, 296–305; and it is reprinted in *Papers and Letters*. Newton sent a transcription (Add. 3970, ff. 549^r–67^v) of his autograph copy (ff. 501^r–17^r), and I will cite the latter; see *Correspondence*, 1:386–7. In revising the "Observations" for the *Opticks*, Newton made so few significant changes that the numbering in the "Observations" and Bk. II of the *Opticks* is identical from Obs. 1 in Pt. I through Prop. VIII in Pt. III. The remainder of Pt. III in the *Opticks* is new and was composed in the early 1690s (see §4.1). For the "Hypothesis" see note 59.

[33] Add. 3970, ff. 350^r–3^v; published with a valuable analysis by Westfall, "Newton's coloured circles."

vibrations in the aether, but he does not set out a systematic description of either the phenomena or his physical hypothesis. His primary aim was to examine and describe Newton's rings quantitatively through a series of mathematical propositions and supporting measurements and observations and perhaps also to confirm his belief in the corpuscular constitution of light and its interactions with the aether.

Although there are some spurious results, to which we shall return, the brief paper shows that Newton already had effective control over the phenomena. He had determined a reasonably good value for the vibration length of yellow light, which is the fundamental parameter of his explanation. Of greater theoretical significance, Newton now explicitly recognized that the vibration length varies for each color. He made "many vaine attempts to measure the circles made by coloured rays alone" in order to calculate their vibration lengths, but by a very clever technique that was based on his concept of aethereal pulses he was able to determine at least the ratio of vibration lengths for blue and red, if not their absolute values, and that the intermediate value occurred "twixt the green & yellow."[34] When there was a very small gap between the lens and plate, Newton gradually cast the spectral colors on them; and by counting the succession of light and dark rings as the colors passed between the two extremes, he was able to determine how many more violet than red pulses fit into this small space. Then, in red light, he gradually pressed the lens until it was in contact with the plate; by counting the succession of light and dark rings, he now found how many red pulses fit in this same space. He then readily determined that the ratio of "the thickness of a pulse" of blue to that of red was about 9 to 14.

Newton never succeeded in directly measuring the vibration lengths for monochromatic light, no doubt because the intensity of the rings was too low. The best he could do was to confirm the value for yellow light that he had measured in circles with white light. To determine the absolute values for the vibration lengths of all the other colors from the measured intermediate color and the ratio of the extremes, he estimated the ratios of the extent of the seven primary colors by a geometrical division in the "Observations" of 1672 and 1675 and a musical one in the *Opticks* (§2.3). These values sufficed for Newton's purposes.

In the following I shall return often to "Of ye coloured circles" in

[34] Add. 3970, f. 353^{v}; Westfall, "Newton's coloured circles," pp. 195–6.

order to show how so many of the results in the "Observations" were discovered by an imaginative utilization of aethereal vibrations, which his methodology required to be suppressed there.

In choosing to work up his recently completed "Of ye coloured circles" as a paper on the colors of thin films and colored bodies and include it as part of his reply to Hooke's critique of his "New theory," Newton was rising to the challenge laid down by Hooke: "I believe, Mr Newton will think it no difficult matter, by my hypothesis to solve all ye phaenomena, not only of ye Prisme, tinged Liquors, and solid bodies but also of ye Colors of ye Plated Bodies; wch seem to have the greatest difficulty."[35] In the draft of his reply to Hooke in the spring of 1672, when he intended to include his "Observations," Newton explained:

> It remains that I now say something of the colours of plated bodies. These I once intended wholly to passe by in silence because the most of their Phaenomena are of a different nature from the things wch I have hitherto delivered of light, & seemed not necessary to their establishment. But Mr Hook intimating that they have the greatest difficulty, it might perhaps be esteemed an objection against my Theory if I should now omit them.[36]

When Newton regained his equanimity in 1675 and decided to make his "Observations" public and sent it to the Royal Society, he substantially revised the 1672 version by including new observations and explanations, refined data, and a more thorough explanation of the causes of reflection and the colors of natural bodies. Nevertheless, for my purpose, the differences between the two versions are not great, and, when relevant, I will indicate them.

I will continue to focus on Newton's rings, as it was through this phenomenon that he gained quantitative mastery and an understanding of the colors of thin films and was then able to explain other instances, such as those of soap bubbles. Newton begins (Obs. 4) with his own, rather than Hooke's, description of the colors of the rings, in which the colors necessarily change in successive orders beyond the central transparent spot – violet, blue, white, yellow, red; violet, blue, green, yellow, red; purple, blue, green, yellow, red; green, red – and after three or four more orders of dilute colors they

[35] From Newton's copy of "Considerations of Mr Hook upon Mr Newtons Theory of Light and Colors," 15 February 1671/2, which was sent to him by Oldenburg, Add. 3970, f. 432^{r}; *Correspondence*, 1:114.

[36] Add. 3970, f. 442^{v}.

vanish into whiteness.[37] The central spot surrounding the point of contact of the glasses is "absolutely transparent," as if the glasses formed one piece. When viewed by reflected light, it is a black or dark spot, for "little or no sensible" light is reflected from it; and when looked through or viewed by transmitted light, it is white.[38]

Observation 5 is most important, for by it Newton demonstrates the periodicity of the appearance of the colored rings:

To determin the intervall of the glasses or thicknesse of the interjacent Air by w[ch] each colour was produced, I measured the diameters of y[e] first six rings at the most lucid part of their Orbits, & squaring them I found their squares to be in Arithmeticall progression of y[e] odd numbers 1, 3, 5, 7, 9, 11. And since one of the glasses was plane & the other sphericall, their intervalls at those rings must be in the same progression. I measured also the diameters of the dark or faint rings between the more lucid colours & found their squares to be in Arithmeticall progression of the eaven numbers 2, 4, 6, 8, 10, 12.[39]

If we recall the sagittal formula, then Newton has established that

$$d = \frac{D^2}{8R} = \frac{mI}{2}, \tag{2.1}$$

where I is an interval such that for m odd the ring is a bright one and for m even a dark one. The interval I, as we will soon see in the "Hypothesis," is the length of an aethereal vibration and, later, in the *Opticks*, that of a fit.[40] In neither version of the "Observations" nor in the first

[37] Compare this sequence of colors with Hooke's perfectly periodic one at note 16. Newton describes the complementary transmitted colors in Obs. 9 (13 in the 1672 version): yellowish red; black; violet, blue, white, yellow, red; violet, blue, green, yellow, red, etc. The transmitted colors have a large admixture of white light, making them "very faint & dilute" and difficult to observe.

[38] Obs. 1, Add. 3970, f. 501[r]; Birch, 3:272. The copy sent to the Royal Society omits "little."

[39] Add. 3970, f. 502[r]; Birch, 3:274–5. Newton, it should be noted, never expresses this result in the form of an equation.

[40] The thickness of air d_1 at which the first bright ring is produced is one half the physical vibration, or pulse length, that I call the interval I. All other rings, bright and dark, appear at integral multiples of this thickness. Newton chose to express this fundamental parameter of his theory in terms of a vibration length or interval, that is, $d_1 = I/2$, rather than the phenomenologically more elementary d_1 itself, on physical grounds. This choice is of no consequence for his theory, but it does indicate how his physical interpretation permeated his account of the phenomena. In contrast, Carlo Benvenuti, the most astute eighteenth-century commentator on the theory of fits, defined an interval to be one half of Newton's; *Dissertatio physica de lumine* (1761), §45, p. 26; the first edition of 1754 was entitled *De lumine, dissertatio physica.* Biot also introduced a new quantity, "the length of a fit" (*la longueur d'un accès*) i, which he likewise defined as one half of Newton's interval I; *Traité*, 4:90, 108.

two parts of the *Opticks* does Newton introduce this physical interpretation, though it is apparent from "Of Colours" and "Of ye coloured circles" that he actually arrived at these results by working with the vibrations. He treats the interval solely as an experimentally determined property of the film – "the intervall of the glasses or thicknesse of the interjacent Air by wch each colour was produced" – and not of light.[41] In the following observation he presents his measurement of the diameter of the middle of a bright circle produced in white light (which corresponds to yellow light) and calculates that $I = 1/80{,}047$ inch, "or to use a round number, the eighty thousandth part of an inch."[42] Newton made his measurements in white light where the rings are compound and one cannot observe, let alone measure, "each colour," yet he was willing to generalize the fundamental property of periodicity to all colors. When he revised the "Observations" in 1675, he attempted to rectify this intellectual leap.

Observation 7 appears to be yet another straightforward instance of Newton's experimental diligence in measuring the interval between the glasses when a ring of a given order is viewed at an increasing angle of incidence and dilates. It is, however, particularly fascinating, because it caused so much experimental and conceptual difficulty for him, and it has more fundamental implications for his theory of aethereal vibrations and fits than one might anticipate. The first hint that

[41] Add. 3970, f. 502^{r}; Birch, 3:274.

[42] Add. 3970, f. 502^{v}; Birch, 3:275. This value goes back to "Of y^{e} coloured circles," where (as he summarized his measurements in an addition to "Of Colours") he found the "thicknesse of a vibration" for yellow to be "$\frac{1}{81000}$ or $\frac{1}{80000}$ part of an inch"; Add. 3975, p. 22; *Questions*, p. 488. Because Newton made all of his measurements with white light, it was difficult for him to decide precisely what color this represents; yellow, yellowish green, and citrine yellow are all to be found in the 1672 and 1675 versions of the "Observations."

Newton's law is the same as that derived according to the wave theory except for a factor of 2, because his interval I is one half of the wavelength λ in the wave theory of light. This difference arises from a phase change that arises in the wave theory. In Newton's theory the first bright ring occurs at the first condensation (or fit of easy reflection), which is half of a vibration. As we saw in note 40, this occurs at perpendicular incidence when the thickness of the air film $d_1 = I/2$. Now, according to the wave theory and the principle of interference, the first ring occurs when the path difference Δ between rays reflected from the first and second surface is a whole wavelength λ. However, because the central spot, where the path difference is zero, is dark rather than bright, a phase change of $\lambda/2$ is assumed to occur for reflection from an optically denser medium. Because the ray reflected from the second surface traverses the thin film twice, the first bright ring appears when $\Delta = 2d_1 + \lambda/2 = \lambda$, or when $d_1 = \lambda/4$. Therefore $I = \lambda/2$. Because of this feature and the fact that it is not at all a wave theory, I have preferred to use the term "vibration length" for the wavelength of Newton's aethereal vibrations.

something is amiss is Newton's omission of any law or principle describing his observations, contrary to his usual practice. That the diameter of the rings increases with increasing obliquity of the eye is perhaps the first of Newton's own discoveries on thin films, for he entered it, in square brackets, in his notes on Hooke's *Micrographia*.[43] His first attempt to describe the variation mathematically was in "Of Colours," where he recorded that "I *observed* (though not very exactly)" how they change.[44] What followed is a proportion – later deleted with a large X through it – expressed in terms of the motion (momentum) and velocity of the incident light corpuscles. This passage is still not fully understood, but the proportion certainly does not agree with the phenomenon.

Newton made one more attempt in "Of ye coloured circles" to describe the variation of the circles in terms of the motion of the light corpuscles before he recognized that the phenomenon was simply not amenable to a corpuscular description and demanded to be described solely in terms of the vibrations set up by the corpuscles. "Of ye coloured circles" opens with six propositions to be confirmed in the subsequent observations. The properties of the circles are mathematically described and interpreted in terms of the "motion," "force," and "percussion" of the corpuscles, though no derivations are presented. The following two are typical:

Prop 2. That they [i.e., the coloured circles] swell by ye obliquity of the eye: soe y^t^ the [di]ameter of ye same circle is as y^e^ [co]secants of ye rays obliquity [in] ye interjected filme of aire, or reciprocally as ye sines of its obliqui[ty]; that is, reciprocally as y^t^ part of the motion of ye ray in ye said filme of aire wch is perpendicular to it, or reciprocally as y^t^ force it strikes ye refracting surface wth all.

Prop 3. And hence ye spaces wch ye rays passe through twixt [y]^e^ circles in one position to the said spaces in another position are as ye squares of ye said [co]secants or reciprocally as ye squares of ye sines, motion, or percussion.[45]

The obliquity q (Figure 2.3) is the angle that an incident ray makes with the refracting surface and is the complement of the angle of incidence i, which in this case is equal to the angle of refraction r in

[43] See note 25.

[44] Add. 3975, §34, p. 10; *Questions*, p. 476; italics added. Westfall considers this observation in "Newton's coloured circles," pp. 187–9, and *Never at Rest*, pp. 219–20.

[45] Add. 3970, f. 350^r^; Westfall, "Newton's coloured circles," p. 191. The edges of this badly damaged manuscript are crumbling and the ink is fading. The square brackets indicate restorations, except for "[co]secant," where Newton's slip is corrected.

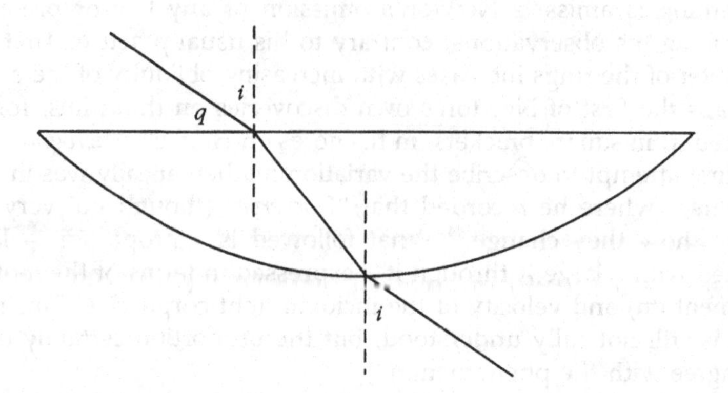

Figure 2.3. The angle of incidence i on the upper surface of the lens in the production of Newton's rings is equal to the angle of emergence i from the lower surface, that is, to the angle of refraction r into the film of air below the lens. The obliquity q is the complement of the angle of incidence i. The curvature of the lens is assumed to be negligible.

the film of air between the lenses – that is, $q = \pi/2 - i = \pi/2 - r$.[46] Prop. 2, therefore, asserts that $D \propto \csc q = 1/\sin q = 1/\cos r$. Thus $d \propto D^2 \propto 1/\sin^2 q = 1/\cos^2 r$; and the differences of d, and therefore of D^2, in successive circles at the same obliquity (which is a constant) will likewise be inversely proportional to the squares of the same trigonometric functions, which is Prop. 3.[47]

[46] Throughout Newton makes the valid approximation that for lenses of large curvature near the point of contact the curvature can be ignored, so that the angle of incidence i at the upper surface of the lens can be considered equal to the angle of refraction r into the thin film. In this respect the lens is equivalent to a flat plate in which the angle of emergence from the plate always equals the angle of incidence.

[47] Prop. 3 is, at best, ambiguous. While the thickness d does vary inversely as $\cos^2 r$, Newton's phrase "y^e spaces wch y^e rays passe through" usually means (Figure 2.4) the rays' actual oblique path s through the film, where $s = d/\cos r$; $s = d$ only at perpendicular incidence. Since $d \alpha D^2 \alpha 1/\cos^2 r$, this entails $s \alpha 1/\cos^3 r$. That this is what Newton actually meant is confirmed by his summary of "Of y^e coloured circles" at the conclusion of "Of Colours," where he corrects this proposition: "And the thicknesse belonging to each vibration is as the squares of those [co]secants of celeritys, And y^e lengths of y^e rays belonging to each vibration as their *cubes*" (Add. 3975, p. 22; *Questions*, p. 489; italics added).

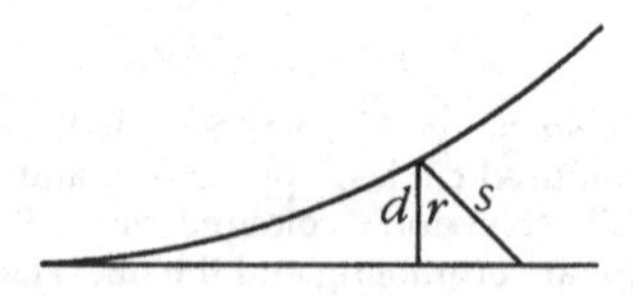

Figure 2.4. A ray is refracted into a thin film of thickness d at angle r and traverses the path length s. The curvature of the lens is assumed to be negligible.

Both of these propositions are erroneous. The correct law is $d \propto 1/\cos r$, as Newton would soon unwittingly prove. He then proceeded to a series of measurements of Newton's rings, which he presented in tabular form, in order to confirm his various laws.[48] He measured the diameters of the first six circles at five obliquities. In the first column he entered sin q (or, equivalently, $1/\cos r$) with the first two entries being 1,000 and 667 (corresponding to q equal 90° and about 42° or r equal to 0° and 48°). It is important to note that Newton did not enter $\sin^2 q$. In the fourth column he formed the differences of D^2 for the first six circles and found their average values to be 647 and 969 for r equal to 0° and 48°. He then observed: "y[t] y[e] figures in y[e] first column are reciprocally proportionall to the differences in the fourth very nearely: as 1000.664::969.647 ferè [i.e., approximately]. &c. wch proves y[e] second rule, & y[e] 3[d]."[49] Despite his declaration that he had confirmed his two propositions, Newton, on the contrary, had just succeeded in proving (within half a percent) the correct law, $d \propto 1/\cos r$, for he had tabulated $\sin q = \cos r$ and $D^2 \propto d$. His fascination for an emission theory led him to postulate a hypothetical property of light, a practice that he later cautioned Pardies should not be adopted. After correcting this simple error, Newton committed himself to a still deeper hypothetical property, a vibration length that varied with direction.

By the time Newton composed the "Observations" in 1672, he recognized the glaring contradiction between his theoretically derived and experimentally established laws. Although he abandoned the former, why did he not adopt the latter? In Observation 7 he now undertook a more thorough examination by measuring the diameters at thirteen obliquities. The new data supported the inverse-cosine law until about 60°; at greater angles deviations from that law grew continually, until at grazing incidence the discrepancy was impossibly large. The problem is an experimental one, for at large angles of incidence it is exceedingly difficult to measure the diameters accurately. Newton had recognized this problem but thought that he had overcome it by using the film between two prisms for measurements at large angles. Because he did not have the "correct" law deduced from the wave theory to guide him, he did not attempt to refine his measurements further. Instead, he searched for a law that would

[48] Add. 3970, Expt. 1, f. 350[v]; Westfall, "Newton's coloured circles," p. 192. Westfall, p. 188, recognized that Newton proved the correct law.

[49] Add. 3970, f. 350[v]; Westfall, "Newton's coloured circles," p. 192. Instead of 667 in his proportion, Newton mistakenly (and inconsequentially) used 664, which was the value for the sine that he had in his table before he changed it to 667.

satisfy the data, and in the *Opticks* he proposed his ungainly law of 106-mean-proportionals (§4.4). The theoretical significance of Newton's discovery that the diameters of the rings vary with obliquity is that it broke the "analogy of nature" and required the interval of the aethereal vibrations, that is, the wavelength, to vary with direction. No such phenomenon is observed in sound, for pitch does not vary with direction.

In Observation 10 (9 in the 1672 version), Newton summarized his careful measurements from "Of ye coloured circles" that showed that when the film is of water the diameters of the circles, and thus the thickness of the film, for a given order decrease in the ratio of the index of refraction. He has thus corrected his earlier erroneous observation, in which he implicitly followed Hooke, that the size of the rings remains unchanged with water.[50] Newton suggests of this discovery that "perhaps it may be a generall Rule," and on the basis of his measurement with this one substance he later assumes it to be true.[51] Our earlier eq. 2.1 therefore becomes

$$d = \frac{D^2}{8R} = \frac{mI}{2n}, \tag{2.2}$$

where n is the index of refraction of the film. Newton was probably led to accept this as a general rule (it became Prop. 17 in the *Opticks*), because he was able to deduce it from his model of light particles and aethereal vibrations. In "Of y^e coloured circles" he had stated this law in Prop. 4 in terms both of the index of refraction (the "subtilty" of the medium) and "ye motions of ye rays in that medium."[52] Like all the other propositions in "Of ye coloured circles," no derivation was presented, but it is readily motivated. If the particles move faster in water in proportion to the increase of index of refraction (as the emission theory required), then they would more quickly reach the lower surface of the film and encounter the first aethereal condensation. The vibration length would then be shorter in the inverse proportion.

Newton begins a sequence of observations with monochromatic light in Observation 12 (13 in the 1672 version). The simple qualitative observation of the rings when illuminated with each of the spectral colors in Observation 15 (16 in the 1672 version) was crucial for es-

[50] See note 19.
[51] Add. 3970, f. 503; Birch, 3:277.
[52] Add. 3970, f. 350^r; Westfall, "Newton's coloured circles," p. 191.

tablishing their cause. The reflected rings – sometimes more than twenty – were all of the same color as the incident light and the transmitted rings corresponded to the dark bands in the reflected rings, that is, they were "of the same colour wth those wch were reflected, & of the bignesse of their intermediate spaces." Thus he was able to conclude:

> And from hence the origine of these Rings is manifest, namely that ye aereall intervall of the glasses, according to its various thicknesse, is disposed in some places to reflect, & in others to transmit the light of any colour, & in the same place to reflect one colour where it transmits another.[53]

Newton was careful not to attribute periodicity to light, for that would introduce a hypothesis. His experiments and observations had shown only that light exhibits periodic behavior in thin films. In fact, as we shall see in the "Hypothesis," he did not at this time believe that periodicity was a universal property or even a property of light itself. He held that periodicity was a restricted phenomenon produced when light encountered the surface of a thin film, or any refracting surface. Despite his care in avoiding the hypothesis of the periodicity of light and in attempting to describe only observed properties, he had introduced his own hypothesis about the cause of the colored rings by attributing to the "aereall interval" a disposition to transmit or reflect light. Thus periodicity was a property of the air, or rather, as he explained in the accompanying "Hypothesis," of the aether in its pores. That Newton should have unconsciously slipped in his hypothesis of aethereal vibrations is not surprising. He had conceived of the phenomenon of colored rings in this way since his breakthrough in 1666 and had successfully utilized the concept throughout his investigations. The aethereal vibrations were so fruitful that it had already become difficult, if not impossible, for Newton to conceive of the colors of thin films without them. It was only in the early 1690s with his research on the colors of thick plates that he felt he had sufficient evidence to justify attributing periodicity (fits of easy reflection and transmission) to light.

In Observation 13 Newton once again draws on the earlier "Of ye coloured circles" and reports that the ratio of the thicknesses of the films producing circles of the same order for the extreme red and violet was 14 to 9. Although he had derived this result by cleverly using pulses, they have been carefully suppressed here. Thus the

[53] Add. 3970, f. 504^{r}; Birch, 3:278.

interval *I* is not a constant but varies with color. In "Of ye coloured circles" Newton conceded that he was unable to measure these values for monochromatic colors, and in the "Observations" he initially relates only some qualitative judgments of the relative proportions of the circles (Obs. 14; 15 in the 1672 version). Yet Newton knew that his law of arithmetic progression, which he established with white light, actually applied only to monochromatic colors and should be experimentally demonstrated. This was the most fundamental difference between Hooke's interpretation and his own and a necessary consequence of his theory of the compound nature of white light. In revising the "Observations" in 1675, Newton evidently judged that his inability to demonstrate quantitatively the law of arithmetic progression for each color apart left a crucial gap in his work. He thus added a new observation (number 16) confirming that that law applied to each of the prismatic colors. His experimental support consists solely of the report that his measurement of the diameter of the sixth yellow circle agreed with that made in white light. I strongly suspect that Newton was able only to confirm his previous value and not to make an independent measurement, because the single measured value that he reports is identical to the measurement he made in white light.[54] Despite his best efforts, technical limitations (insufficiently intense rings of monochromatic colors) were responsible for his failure to confirm his law of arithmetic progression for each color. Nonetheless, he declared it to be a demonstrated property because it was a necessary consequence of his theory of white light and color. It certainly was not contradicted by any evidence, and it was supported by his one measurement and various qualitative observations.

Newton's willingness to consider a law to be established with just one or two measurements or observations is not unusual for him, and has been noted by others. Sometimes he was hampered by technical limitations, but it also arose because he had such confidence in his deductions, especially when they had quantitative or mathematical support, that he simply did not consider it worth the bother to confirm the obvious. For example, he could have further tested his proposed "general rule" that the interval varies inversely with index of refrac-

[54] In revising Obs. 6 for the *Opticks*, Newton remeasured the diameter of the sixth ring and found 55/100 inch instead of the earlier 58/100 inch. When he reported his "confirmation" in Obs. 16 with "citrine yellow" light, he did not change the 58/100 inch. He later noticed the discrepancy and in the manuscript added "or a little less," which he still considered "agreable to the 6^{t} observation"; Add. 3970, f. 145^{r}; *Opticks*, p. 219.

tion by measurements with, say, spirit of wine, linseed and olive oil, or turpentine (all of which have a large index of refraction) inserted between the lenses, and thereby satisfied modern methodological canons. But modern canons did not exist, and that simply was not Newton's style.[55] This law, as we saw, readily followed from his physical model of vibrations. In both of the present examples his confidence in his physical deductions and insight led him to correct laws, as it often did, but it could also lead him astray, as in the notorious case of the dispersion law that implied the impossibility of constructing an achromatic lens in the *Opticks* (Bk. I, Pt. II, Prop. III, Expt. 8).[56]

In any case, these matters of Newtonian style should not hinder us from recognizing the magnitude of his accomplishment in the first part of the "Observations." Not only had he uncovered the exact laws of the appearance of the relatively complex phenomena of Newton's rings, but his extensive series of experiments and measurements had no precedent in the physical sciences and set standards for precision that were not to be matched for a century. Only in astronomy were similar standards achieved. To appreciate Newton's achievement we should compare the first part of the "Observations" with the corresponding portion of Hooke's *Micrographia,* which was itself a fine piece of work that greatly assisted Newton. When we do this we can appreciate one of the reasons for Newton's outrage when Hooke accused him of plagiarism. Lest we attribute Hooke's failure to erect a quantitative treatment to his inferior mathematical abilities, we need only note that at virtually the same time that Newton was carrying out his successful investigation, Huygens likewise tackled it after reading the *Micrographia* and failed.[57] Newton's reputation as a quantitative experimentalist rests largely on this work.

Besides observations of Newton's rings and laws deduced from them, Newton describes related observations of the colors seen in soap bubbles and in the air film between two superposed prisms. Only one of these observations concerns us now. Although Newton did not at this time develop a mathematical analysis of the colors of films of uniform thickness, like mica or glass sheets, he did observe

[55] Newton's minimal requirements for establishing a law may reflect his belief in the constancy of God's action as expressed, for example, by Barrow; see Ch. 1, note 64. Nonetheless, even by contemporary standards, Newton's requirements for confirmation were lax.

[56] See Bechler, " 'A less agreeable matter' "; and my "Newton's 'achromatic' dispersion law."

[57] Westfall, *Never at Rest,* pp. 217–18.

(Obs. 19) that the colors change when the film is viewed at different obliquities; this does not occur in Newton's rings.[58] This property would be important in his analysis of the colors of thick plates. Before I turn to the second part of the "Observations," with Newton's phenomenological explanation of the colors of thin films, I shall first treat his physical explanation in the accompanying "Hypothesis."

2.3. THE HYPOTHESIS OF AETHEREAL VIBRATIONS

Newton explained that he wrote "An Hypothesis explaining ye properties of Light discoursed of in my several papers"

> because I have observed ye heads of some great virtuosos to run much upon hypotheses as if my discourses wanted an hypothesis to explain them by, & found that some when I could not make them take my meaning when I spake of ye nature of light & colours *abstractedly* have readily apprehended it when I illustrated my discours by an Hypothesis.[59]

Nonetheless, he insisted on his standard distinction "that no man may confound this wth my other discourses, or measure ye certainty of one by th'other." The composition of the conjectural "Hypothesis" is entirely consistent with Newton's methodology, which allowed hypotheses to be used to explain properties, provided they were based on empirical evidence and not mingled with more certain principles, such as those in the accompanying "Observations." Newton no doubt sent his "Hypothesis" to the Royal Society because it would enable his other work to be more readily accepted, but when he told Oldenburg that these were hastily recollected ideas he was misleading.[60] The underlying model of vibrations in the aether was adopted as early as 1666 in "Of Colours," and he continued to utilize it decades after he had abandoned the aether itself.

[58] In a draft of his reply to Hooke on 21 December 1675, Newton considered this experiment – " y^{t} ye colours of a thin plate change by ye obliquation of ye eye" – to be one of his three more important ones that "directly overthrow all Mr Hooks hypothetical discours" (Add. 3970, f. 531^{v}); see Eq. 2.3 in §2.3. Hooke had independently made this observation by June 1672, when he reported it to the Royal Society; *Correspondence*, 1:197.

[59] Add. 3970, f. 538^{v}; *Correspondence*, 1:363–4; italics added. Newton sent his autograph copy of the "Hypothesis" (Add. 3970, ff. 538^{r}–47^{r}) to the Royal Society. There is also an incomplete transcription in his papers (Add. 3970, ff. 573^{r}–81^{r}). Birch, 3:247–60, 262–9, published the copy in the society's Journal Book; and it is reprinted in Newton, *Papers and Letters*. *Correspondence*, 1:362–86, publishes the version in the Journal Book "supplemented and corrected" by Newton's autograph copy.

[60] Newton to Oldenburg, 7 December 1675, *Correspondence*, 1:361.

Newton's allusion to a pulse in "Of Colours" in his initial measurement of Newton's rings marks the first appearance in his works of aethereal vibrations to explain the properties of light.[61] In a short essay on the optic nerves, which was appended to "Of Colours" and also written in 1666, he utilized the aethereal vibrations to explain sensation, which was always one of the primary functions to which he applied the aether and other subtle media. He argued here that vision occurs by vibrations in the aether that are propagated from the retina to the brain, or sensorium:

> Light seldom striks upon ye parts of grosse bodys (as may bee seen in its passing through them), its reflection & refraction is made by ye diversity of aethers, & therefore it effect on the Retina can only bee to make this vibrate wch motion then must bee either carried in ye optick nervs to ye sensorium or produce other motions that are carried thither.[62]

When he wrote "Of ye coloured circles" some five years later, it is clear from the sophisticated way in which he used aethereal pulses to determine the ratio of the red and violet vibration lengths that he had already sufficiently articulated that concept to explain the colors of thin films. It was, however, only in the incomplete conclusion of the 1672 "Observations," entitled "An Hypothesis hinted at for explicating all the afforesaid properties of light," that Newton began systematically to set forth his model of the aether to explain the properties of light. When he returned to the "Hypothesis" in 1675, he greatly expanded it to give an explanation of most known optical phenomena. He was evidently then in one of his periodic speculative phases, for he did not long remain content with its restricted scope. As the successive drafts show, after the optical part was complete he continued to add sections on the general role of the aether in nature, including large sections on electricity and muscular contraction.

The aether in Newton's natural philosophy played two, divergent roles that merged into one another. On the one hand, in a Cartesian manner he invoked the mechanical properties of an invisible, subtle fluid that acted only by contact to explain such physical phenomena as cohesion, electrical attraction, and the refraction and reflection of light. On the other hand, Newton also imagined the aether to be a source of activity in nature for such vital processes as fermentation,

[61] Quoted at note 12.

[62] Add. 3975, p. 19; *Questions*, p. 487. McGuire and Tamny date this piece on the optic nerve "almost certainly" to 1666; *Questions*, p. 223.

putrefaction, and the generation, growth, and decay of living matter ("vegetation") that could not, he felt, be explained by mere mechanical causes. Under the influence of the Cambridge Platonists, especially Henry More and Ralph Cudworth, Newton believed that matter is passive. To attribute active powers or self-movement to matter would ineluctably lead to atheism, for what need would there then be for God? Alchemy was one source that Newton hoped would lead to revelations about the origin of active powers in nature.[63] After the *Principia* and the introduction of force and a new ontology, this distinction between the mechanical and nonmechanical is no longer a meaningful one, for the mechanical philosophy's requirement of contact action was abandoned and force could serve the role of an active principle. This profound reorientation of natural philosophy nonetheless had little effect on Newton's optical theories, for they were all developed one to two decades before the *Principia*, except for the theory of fits. Even that theory did not utilize the concept of force, for it was essentially a reworking of the old theory of aethereal vibrations. In fact, a major aspect of eighteenth-century optics was the clash between Newton's seventeenth-century mechanical theories and his new, eighteenth-century physics of force.

A few years before he wrote the "Hypothesis," Newton composed an untitled alchemical paper known by its opening line, "On nature's obvious laws and processes in vegetation," or simply as "On vegetation."[64] Here he proposed that the earth is a living being that imbibed a vital aethereal spirit, which he described as "nature's universal agent," "her secret fire," and the "material soul of all matter."[65] Scholars recognize that this vegetative aether is reflected in the "Hypothesis," where Newton suggests that "perhaps ye whole frame of nature may be nothing but various contextures of some certain aethereal spirits or vapours condens'd . . . ," and that "Nature is a perpetual circulatory worker."[66] They differ, though, as to whether Newton

[63] Newton's assignment of a passive role to matter in favor of various active principles is widely discussed; see, for example, McGuire, "Force, active principles, and Newton's invisible realm"; and McMullin, *Newton on Matter and Activity*. On the controversial issue of the role of alchemy in Newton's natural philosophy, see, for a negative assessment, McGuire, "Neoplatonism and active principles"; and for a positive assessment, Betty Jo Teeter Dobbs, "Newton's alchemy and his theory of matter."

[64] Dobbs has published a facsimile of the manuscript (Smithsonian Institution Libraries MS 1031 B) as an appendix to her *Alchemical Death and Resurrection*. She there dates the manuscript to about 1672 (p. 36), and in "Newton's alchemy" (p. 517) to about 1674.

[65] Dobbs, "Newton's alchemy," p. 517, which presents a useful summary of the manuscript.

[66] Add. 3970, 538^v, 539^v; *Correspondence*, 1:364, 366.

intended the aether of the "Hypothesis" to be a strictly mechanical one. The nature of the aether in the "Hypothesis," as well as other active principles that Newton utilized throughout his career, is, I believe, genuinely ambiguous and perhaps irresolvable. The dividing line between the passive and active, and even their nature, remained ill defined for Newton. "On vegetation" can, however, illuminate for us Newton's conception of mechanical chemistry, which is a foundation of his theory of colored bodies.

Newton introduced a distinction in "On vegetation" between vegetative and "purely mechanicall" actions in nature. "Vulgar," or "common," chemistry involves the grosser particles of matter and "are but *mechanicall* coalitions or seperations of particles . . . & y^t w^{th}-out any vegetation." These processes take place "by ye sleighty transpositions of ye grosser corpuscles, for *upon their disposition only sensible qualitys depend.*" Besides these "*sensible* changes wrought in ye textures of ye grosser matter," there are vegetative processes: "a more subtile *secret* & noble way of working in all vegetation which makes its products distinct from all others [,] & ye immeadiate seate of thes operations is not ye whole bulk of matter, but rather an exceeding subtile & inmaginably small portion of matter diffused through the masse. . . ." Moreover, the vegetative spirit acts through the grosser particles (not "upon" them) and is "very apt" to make them "put on various external appearances" like such ordinary substances as bone, flesh, and wood.[67]

Newton's characterization of mechanical and vegetative processes as "sensible" and "secret," respectively, is an important distinction. It helps to explain why, aside from various social pressures, he did not appeal to alchemical processes in such scientific writings as his theory of colored bodies, but rather common, mechanical chemistry. In the first place, his theory of colored bodies is built on the physical disposition of the same grosser corpuscles as common chemistry. Second, it is exceedingly difficult, if not impossible, to establish that inner vegetative processes are at work, as they operate through the grosser corpuscles and mimic their sensible appearance. Newton demanded that his natural philosophy be founded on sensible appearances, or experiment and observation, and alchemical processes thus provided a shaky foundation. Consequently, he confined his alchemical views to his more speculative papers. Finally, transduction is inapplicable for knowledge of the secret, inner works of matter, for

[67] SIL MS 1031 B, ff. 5^v–6^r; italics added.

by definition they operate in a fundamentally different way from the larger mechanical world.

The absorption and emission of light, which take place in the inner parts of matter, were for Newton not purely mechanical processes. He had little knowledge of these processes other than that they do occur and that they are correlated with the presence of heat. Whether Newton was drawing on the alchemical tradition or his physical intuition, or both, he believed that light was incorporated within matter and contributed to its activity. In "On vegetation" he conjectured that the body of light is the vegetative spirit:

> Note that tis more probable ye aether is but a vehicle to some more active spirit . . . & in y^t aether ye spirit is intangled. This spirit perhaps is the body of light because both have a prodigious active principle, both are perpetuall workers. 2 because all things may bee made to emit light by heat. 3 ye same cause (heat) banishes also the vitall principle. 4 Tis suitable wth infinite wisdom not to multiply causes w^{th}out necessity. 5 Noe heat is so pleasant & bright as the sun's. 6 Light & heat have a mutuall dependance on each other, & noe generation w^{th}out heat. heat is a necessary condition to light & vegetation. [heate exites light & light exites heat. . . .][68]

A few years later, in the "Hypothesis," Newton did not invoke his idea that light is the principle of activity in nature. He allowed the possibility that the "material principle of light" may be carried in the aether, but he did not attribute any particular activity to that principle.[69] Newton, in fact, said very little about light in the preliminary speculations on the aether that were all added after the original optical portion of the "Hypothesis" was composed. In the optical section, light is treated as particles of matter that obey mechanical laws – if not always the ordinary laws – just as the aether is treated as a mechanical fluid. Nonetheless, Newton never abandoned his belief that light plays an active role in the economy of nature (§3.3).

Newton set forth his physical model of light and the optical aether in the "Hypothesis" in five hypotheses. The essence of his model is the idea that light consists of particles emitted by luminous sources, and when these particles encounter the aethereal surface of a body, they stir up "vibrations in it, as stones thrown into water do in its surface."[70] These vibrations interfere with the motion of the light

[68] Ibid., f. 4^r; Dobbs, "Newton's alchemy," p. 521; the brackets are Newton's and his "spt" is expanded to "spirit."
[69] Add. 3970, f. 539^v; *Correspondence*, 1:366.
[70] Add. 3970, f. 543^r; *Correspondence*, 1:374.

particles by reflecting them at a condensation and transmitting them at a rarefaction. The aether, which is diffused through all space, is "much of ye same constitution wth air but far rarer subtiler & more strongly elastic" and "a vibrating Medium like air; only ye vibrations far more swift & minute" (Hypotheses 1 and 2).[71]

In order to treat light in mathematical and mechanical terms, Newton was compelled to assume its finite speed of propagation even before Ole Rømer demonstrated it in 1676.[72] The aether behaves almost like a resistanceless medium, for it resists the motion of light particles only initially, at their emission from a luminous source, and at the boundaries of diverse bodies, where its density changes. When light particles are emitted they are accelerated "by a principle of motion . . . till ye resistence of ye aethereal Medium equal ye force of that principle." Henceforth the aether offers as little resistance as a vacuum. This is contrary to the principles of Galilean mechanics, and Newton knew it: "God who gave Animals self motion beyond o^{r} understanding is wthout doubt able to implant other principles of motion in bodies wch we may understand as little. Some would readily grant this may be a spiritual one, yet a mechanical one might be shown did not I think it better to pass it by."[73] It is ambiguities such as this that suggest that Newton's model may not be entirely mechanical. Newton emphasizes that he considers the particles to be light and not the vibrations, "I suppose light is neither ye aether nor its vibrating motion" – which is simply an effect of light (Hypothesis 4).[74] If light were the vibrations or waves themselves (as Hooke and Huygens held), he could not imagine how such phenomena as rectilinear propagation, opacity, and the colors of thin films could be explained.

For millennia reflection was a nonproblematic phenomenon explained by analogy to the mechanical rebound of balls from planes. Newton had encountered this approach in his readings of Descartes, Boyle, and Hooke and had adopted it in his first "Of Colours."[75] However, by the second "Of Colours," he was already moving toward his mature view in attributing reflection to "ye diversity of Aether," although still conceding that "ye parts of ye Glasse must necessarily

[71] Add. 3970, ff. 538^{v}, 539^{v}; *Correspondence*, 1:364, 366.

[72] Newton estimated that light may "be an hour or two, if not more, in moving from ye Sun to us" (Add. 3970, f. 545^{r}; *Correspondence*, 1:378).

[73] Add. 3970, f. 541^{r}; *Correspondence*, 1:370. This problem, of course, vanished with the abandonment of the dense aether after the *Principia*.

[74] Add. 3970, f. 539^{v}; *Correspondence*, 1:370.

[75] Add. 3996, f. 123^{r}; *Questions*, p. 434.

reflect some rays."[76] By the early 1670s, Newton totally rejected a collision model of reflection because of its incompatibility with numerous phenomena. In the "Hypothesis" he appeals specifically to the regularity of reflection from polished surfaces. The collision model, he argues, requires the altogether implausible condition that every corpuscle in the surface of a mirror be exactly plane and face the same direction.[77] To explain the regularity and other properties of reflection, he assumed that light is reflected from an aethereal surface that "eavenly overspreads" the corpuscles of the body.[78] With Hypothesis 5 – "light & aether mutually act upon one another" – he justified invoking the aether to explain all optical phenomena except absorption.

Like Descartes, Newton assumes that the aether is rarer in denser bodies because it cannot so readily penetrate their narrower pores (Hypothesis 3).[79] It does not terminate in a sharp mathematical surface at the interface of two different media but gradually varies in density through a small physical region. Because "ye densest aether acts most strongly" on light and presses it toward the rarer aether, this density gradient is the cause of refraction and total reflection.[80] To explain the more general case of reflection (from dense to rare aether and at any angle of inclination), Newton proposes an additional mechanism, a stiff aethereal surface: "You are further to consider how fluids neare their superficies are less pliant & yeilding then in their more inward parts & if formd into thin plates or shels they become much more stiff & tenacious then otherwise."[81] The stiffness, and consequently reflective power, is assumed to increase with the difference of density of the media, as is found by observation. He elaborates this idea still further and supposes that the constant bombardment of light particles on this rigid surface excites vibrations in it that are propagated throughout the aether, "but yet continue strongest where they began, & alternately contract & dilate ye aether in that physical superficies."

There is experimental evidence for the existence of these vibrations: "For its plain by ye heat wch light produces in bodies, that it is able

[76] Add. 3975, §55, p. 14; *Questions*, p. 481.

[77] In the accompanying "Observations," Prop. 8, Newton gave many other cogent reasons for rejecting reflection from the parts of bodies; see §3.3.

[78] Add. 3970, f. 543^v; *Correspondence*, 1:375.

[79] Add. 3970, f. 539^v; *Correspondence*, 1:366–7.

[80] Add. 3970, f. 542^r; *Correspondence*, 1:371. See Bechler, "Newton's search for a mechanistic model of colour dispersion," esp. pp. 11–14, for Newton's explanation of refraction and reflection by means of the aether's density gradient.

[81] Add. 3970, f. 543^r; *Correspondence*, 1:373.

to put their parts in motion, & much more to heat, & put in motion the more tender aether. . . ."[82] By appealing to a number of analogies that show the ability of vibratory motions to move large massy bodies, he concludes that aetheral vibrations are the "best meanes by wch such a subtile agent as light can shake ye gros particles of solid bodies to heat them."[83] Partial reflection, namely, that at any point of a transparent surface some rays are reflected while others transmitted, can now be explained by the vibrations of the aethereal surface: "If a ray of light impinge upon it while it is much comprest, I suppose it is then too dense & stiff to let ye ray pass through, & so reflects it; but ye rays y^t impinge on it at other times when it is either expanded . . . or not too much comprest & condens'd, go through & are refracted."[84] The hitherto simple phenomenon of reflection was becoming rather complex, for yet another cause of reflection, that from the corpuscles to produce color, must be added.

To apply this hypothesis of aethereal vibrations to the periodic colors of thin films, Newton had to assume that the vibrations stirred up by the incident light particles move faster than the particles themselves: "ye waves excited by its passage through ye first superficies, overtaking it one after another till it arrive at ye second superficies, will cause it to be there reflected or refracted accordingly as ye condensed or expanded part of ye wave overtakes it there."[85] If the vibrations moved slower than the particles, they could not affect the particles at the second surface; and if they moved at the same speed as the particles, they would always be in the same state (rarefaction) at the second surface. It is also necessary to assume that the particles are affected by that wave which they themselves stir up at the first surface and not, as in partial reflection, by some preceding wave, for otherwise the state of the vibration at the second surface relative to the light particles would be independent of the distance between the surfaces. The vibrations of the aether occur only when light encounters a reflecting or refracting surface, and so do not accompany light from its first emission as is the case in a wave theory of light. This feature of his model is consistent with and follows from his observations, which established periodicity only in light's interactions with bodies.

[82] Add. 3970, f. 543^r; *Correspondence,* 1:374.
[83] Add. 3970, f. 543^v; *Correspondence,* 1:374.
[84] Ibid.
[85] Add. 3970, f. 545^r; *Correspondence,* 1:378. This passage originally continued before deletion: "to compres or relax that physical superficies & thereby augment or diminish its reflecting power."

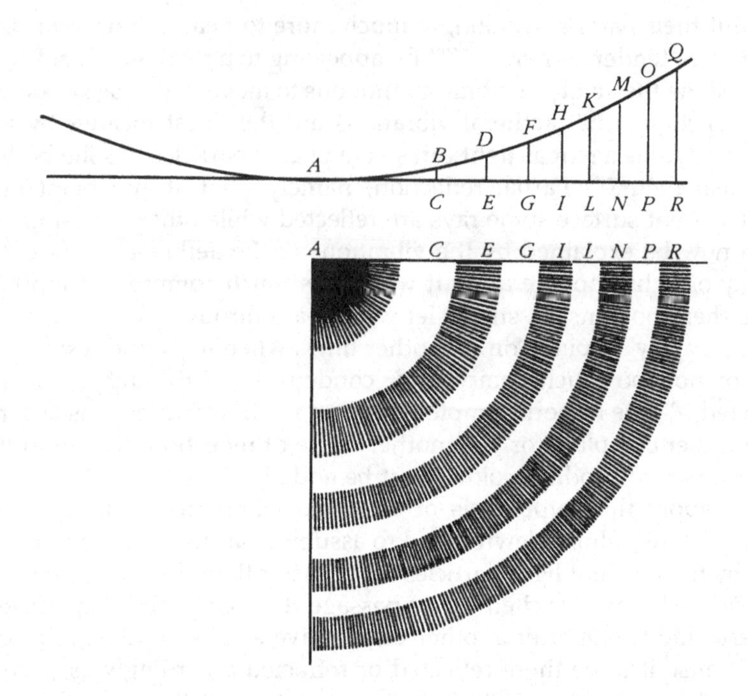

Figure 2.5. Newton's explanation of the periodicity of the circles formed with light of a single color in Newton's rings. (After his drawing in the "Hypothesis," Add. 3970, f. 545r.)

The periodicity of Newton's rings is now readily explained. In Figure 2.5 let *ARQ* be a thin film of air formed between the lens *ABQ* and the plate *ACR*, and let the rings be viewed from above the lens. At the center *A* the particle will be transmitted because the aether in the two glasses is continuous "as if but one uniform medium," so that a central dark spot will be seen in reflected light. At the thickness *BC* the particle will encounter the condensed part of the first overtaking "wave" at *C* and be reflected, so that a bright ring will be seen; at double that thickness *DE*, it will encounter the rarefied part of that wave at *E* and be transmitted, and thus a dark ring will be seen; at triple the thickness *BC*, or *FG*, it will encounter "ye condensed part of ye second wave" and be reflected, so that a second bright ring will be seen; and so on in arithmetic progression, in agreement with observation.[86] The interval *I* of a whole vibration or wave – one complete

[86] Add. 3970, f. 545r; *Correspondence*, 1:378. According to Newton's hypothesis that the

condensation and rarefaction – will equal length *DE*. This interval (which I call the "vibration length") Newton called the "bigness" or "thickness" of a vibration and later, in the *Opticks*, the "interval of the fits." His physical explanation of the formation of Newton's rings differs fundamentally from the principle of interference, even in the primitive form advocated by Hooke. Newton places the cause of the reflection and transmission at the second surface, where a particle and a wave interact, rather than at or above the first, where two waves interfere: "the second superficies being made able or unable to reflect accordingly as it is condenst or expanded by y^e^ waves."[87]

As Newton's diagram of the reflected rings shows, he does not assume that reflection occurs only at the greatest compression but also at thicknesses on either side of it, extending as far as a quarter interval ($I/4$) on either side.[88] The reflection is assumed to be greatest at the thickness corresponding to the greatest compression and to decrease gradually as the wave is less condensed (and so resists the particles' motion less); the possibility of reflection ends altogether at thicknesses corresponding to rarefactions. Newton grants:

> Yet all ye rays w^th^out exception ought not to be thus reflected or transmitted, for sometimes a ray may be overtaken at ye second superficies by ye vibrations raised by another collaterall or immediately succeeding ray; wch vibration, being as strong or stronger then its own may cause it to be reflected or transmitted when its own vibration alone would do ye contrary. And hence some little light will be reflected from ye black rings wch makes them rather

vibrations move faster than the light particles, these cannot be the first and second waves. Any number of waves, 2, 13, or 1,013, for example, could have overtaken the particles before they reach the second surface. These, however, will be the first and second waves that encounter the particles at the second surface and "interfere" with them. This is a subtle but important point, because the unknown speed of the aethereal vibrations produces an arbitrariness in their mathematization. This problem was eliminated in the theory of fits, where the vibrations move at the speed of light, just as in a wave theory. Because Newton has left few traces of how he performed his calculations with these faster, overtaking waves, I can only specify the conditions that they must satisfy: At the thickness d_1 of the first bright ring, which is also the first condensation, there must be an odd number of half intervals; and all succeeding rings – bright and dark – will occur at integral multiples (2, 3, 4, 5, . . .) of $I/2 = d_1$. For example, if the condensation of the thirteenth wave causes the first bright ring, succeeding bright rings will be caused by the condensations of the fourteenth wave ($= 3d_1$), the fifteenth ($= 5d_1$), and so on. By subtracting the physically intractable thirteen waves from each, these become the first, second, and third waves, as Newton calls them.

[87] Add. 3970, f. 545^r^; *Correspondence*, 1:378.

[88] The specification of one-quarter interval follows directly from the theory and is explicitly indicated in the nomograph in the second part of the "Observations" (described in §2.4).

black then totally dark; & some transmitted at ye lucid rings, wch makes ye black rings, appearing on ye other side of ye glasses, not so black as they would otherwise be.[89]

Thus, from observation, Newton had to conclude that the aethereal vibrations could not be totally effective in causing reflection and transmission. In the *Opticks* he refrained from explicitly invoking any model to justify the theory of fits, but he carried over the idea that they are not necessarily effective. He expresses this by describing the fits as states that *dispose* the rays to be *easily* reflected and transmitted, and not as states in which the rays are reflected and transmitted (§4.4). The problem confronting Newton was that his theory (and all other optical theories until the beginning of the nineteenth century) could treat intensities only qualitatively, and, in any case, intensities could not be measured yet.

To extend this model to white light, Newton only had to introduce the idea that the rays or particles of different color vary in "magnitude, strength or vigour" and so excite vibrations of different size.[90] The red vibrations are assumed to be larger than the violet ones and thus will form larger circles than they do, as is found by observation. Could they be seen alone, the rings in white light would be red on the outside, violet on the inside, and intermediate colors in the intermediate parts. Except at the edges all colors would be compound, because each color is reflected in a ring with some breadth. In reality, when all the rings are present, they quickly overlap one another, become increasingly compound, and after eight or nine orders vanish into white. Newton presents a particularly lucid explanation of this process.[91]

In Figure 2.6 let the solid lines *CB*, *GD*, *LF*, *PM*, *RN*, and *SX* represent quadrants of the first six circles made by the extreme red alone, and the dashed lines $\eta\beta$, $\gamma\delta$, $\lambda\phi$, $\pi\mu$, $\rho\nu$, and $\sigma\xi$ represent the quadrants made by the extreme violet. In sunlight, then, the first bright ring would be $\eta\beta BC$, the second $\gamma\delta DG$, the third $\lambda\phi FL$, the fourth $\pi\mu MP$, the fifth $\rho\nu NR$, and the sixth $\sigma\xi XS$. The slender dark ring $D\phi\lambda G$ between the second and third orders already vanishes, because the figure does not include the breadth of the rings of the simple colors as in Figure 2.5. In the third ring $\lambda\phi FL$ the violet of the fourth $\mu\pi$ already intrudes substantially; and the fifth ring $\rho\nu NR$ is com-

[89] Add. 3970, f. 545^r,v^; *Correspondence*, 1:378–9.
[90] Add. 3970, f. 544^r^; *Correspondence*, 1:376.
[91] Add. 3970, ff. 545^v^–6^v^; *Correspondence*, 1:380–2.

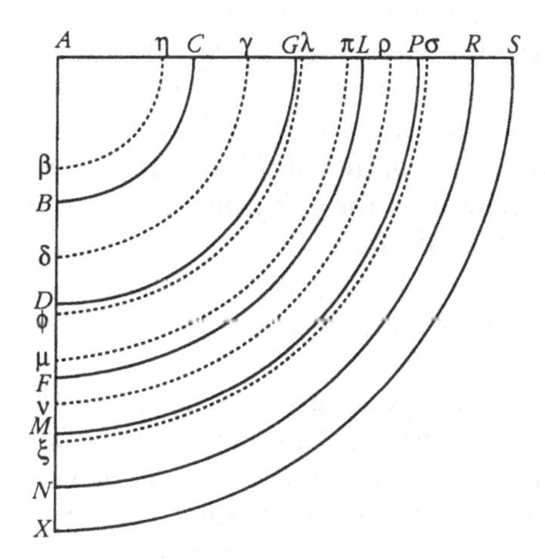

Figure 2.6. Newton's explanation of the formation of the overlapping circles of different color seen in Newton's rings with white light. (After his drawing in the "Hypothesis," Add. 3970, £ 546′.)

pletely overlapped by the violet end of the sixth $\xi\sigma$ and the red end of the fourth *MP*. Thus, by the eighth or ninth ring, the colors will be so intermingled that the rings vanish and form an even white. Newton tells us that this can also be verified by computation. In the "Observations" it was found that the squares of the diameters of the bright circles, both the outer ones *AC*, *AG*, *AL*, . . . and the inner ones $A\eta$, $A\gamma$, $A\lambda$, . . . , increase as 1, 3, 5, . . . ; and also that the ratio of the square of the diameter of the inner, violet circles to that of the outer, red ones is 9 to 14. Thus we can determine the ratio of the diameters of the red of the third order *AG* to the violet of the second $A\lambda$, for $AG^2/A\lambda^2 = (3/5)(14/9) = 14/15$, or $AG/A\lambda = .966$, just as is represented in the figure.

Newton's explanation of the formation of Newton's rings in white light is exceedingly clear, but there is no equivalent in the *Opticks*. I suspect that it was omitted inadvertently, simply because it was in the "Hypothesis" rather than the "Observations," which served as the text for the *Opticks*. Yet there is nothing hypothetical about this explanation. The vibrations play only a superficial, introductory role, and the explanation could have been easily transformed into a phenomenological one utilizing only earlier observations. Understanding the formation of the rings in white light is not easy. Until the end of the eighteenth century most scientists failed to recognize that the colors in Newton's rings are compound, showing that few were able to comprehend the composition of the rings on their own.

Only one more point in the "Hypothesis" need concern us now, namely, the increase in the rings' diameters as they are viewed more obliquely. As we already saw, Newton encountered great difficulty in discovering the law governing this variation. In films of uniform thickness, on the contrary, the same ring, or rather line, changes color as the obliquity is varied, as Newton reported in Observation 19. These phenomena are readily explained by the principle of interference. A bright ring or line occurs in a thin film when

$$2nd\cos r = (m + \frac{1}{2})\lambda, \qquad (2.3)$$

where m is an integer and λ is the true wavelength. In a thin film of constant thickness d, when the angle of incidence (and hence the angle of refraction r) varies, a ring of the same order m and color λ cannot be seen. In Newton's rings, on the contrary, the thickness varies continuously. Therefore, as the angle of incidence increases, the path difference $2nd\cos r$ can remain constant and so a ring of a given order can always be seen with the same color.[92] Newton, however, considers not the path difference of two rays but the actual path traversed by a single ray, and this increases with increasing obliquity. In Newton's theory a ring of the mth order is produced when m vibrations overtake the light particle. It was therefore natural for him (if somewhat odd for us) to assume that as the obliquity and path increased, the vibration length increased too: The same number of vibration lengths had to be fit into an increasing path. In the "Hypothesis" he considered another possible cause to operate simultaneously with increasing vibration length (and decreasing frequency), namely, that the velocity of propagation of the vibrations decreases:

> And of this ye reason may be, *partly* that an oblique ray is longer in passing through ye first superficies & so there is more time between ye waving forward & backward of that superficies & consequently a larger wave generated, & *partly*, that ye wave in creeping along between ye two superficies may be impeded & retarded by ye rigidnes of those superficies bounding it at either end, & so not overtake ye ray so soon as a wave that moves perpendicularly cross.[93]

In the *Opticks* Newton abandoned the idea of any variation of velocity with direction and in Prop. XV invoked only a variation of vibration length, now the interval of the fits. Newton's treatment of

[92] Recall from note 46 that for the thin film in Newton's rings we can set $\cos r = \cos i$ to a very good approximation.
[93] Add. 3970, f. 546^v^; *Correspondence*, 1:382.

this variation is probably the strangest feature of his explanation of the colors of thin films because it so strikingly violates the analogy to the propagation of sound. Newton, it must be concluded, was not as committed to the analogy of nature as is sometimes supposed.

Perhaps one of the reasons that Newton was not troubled by the violation of the analogy of nature – besides the law evidently being demanded by observation – is that though his model of optical phenomena is mechanical, it was not derived from the principles of mechanics. Even after the *Principia*, neither Newton nor anyone else could, for example, deduce the wavelength stirred up by the impact of the light particles on the aethereal surface; nor could he even begin to analyze how the compressions and rarefactions of the waves could alter the light particles' motion.[94] Although his hypothesis was essentially qualitative, Newton skillfully exploited the one mathematical property, periodicity, that he could treat to discover so many properties of the colors of thin films. We saw how he was able to apply his hypothesis of "pulses," "vibrations," or "waves" to his astutely designed observations and experiments to deduce such properties as the law of arithmetic progression for the succession of the rings, their dependence on the index of refraction, and the ratio of the intervals of the red and violet rays. Having deduced the properties, he eliminated all traces of the hypothesis and related only the discovered properties in the "Observations." This was a consequence of his methodology: his refusal to mix principles and hypotheses and the consequent requirement that properties be "abstracted" from hypotheses (§1.2). In this case, the experiments and observations are related to the properties in a relatively direct way, and so little difficulty in comprehension resulted, especially as the "Observations" was supplemented by the explanatory "Hypothesis." When he would follow the same method of abstraction some twenty years later with the phenomena of thick plates and the theory of fits, he would not be so successful. The experiments and the properties are not directly and evidently linked.

We should pause here to examine a surprising consequence of Newton's aethereal model of optical phenomena when combined with the theory of colored bodies set forth in the third part of the accompanying "Observations." This theory demonstrated that the colors of bodies

[94] Bechler makes and discusses this point with examples in "Newton's search for a mechanistic model," pp. 21–3.

are produced by their corpuscles in exactly the same way that a thin film of the same size and density produces color. Newton's model is based almost entirely on the aether, with little role assigned to the solid parts that compose bodies. The only function assigned to the parts, besides defining the pores occupied by the aether, is to stop any light particles that collide with them. If we consider a plate of glass so thin (approximately 1/160,000 inch) that it exhibits a yellow of the first order, according to Newton this plate consists primarily of aether with some interspersed solid parts. The vibrations of the aether cause mostly yellow light to be reflected while allowing all other colors to be transmitted. If we now assume, as Newton does, that yellow bodies of the first order are composed of corpuscles of the same size as the thickness of this plate – indeed, they can be considered to be fragments of the plate – then those corpuscles must likewise primarily consist of aether and some parts. Thus the corpuscles' composing bodies already have a structure and are themselves composed of parts.

Newton's optical model, then, entails two of his most characteristic doctrines, namely, that bodies consist mostly of pores with little matter and that they possess a compound structure. To explain how apparently solid matter could consist mostly of pores, Newton proposed his hypothesis of the hierarchical structure of matter in the Latin translation of the *Opticks*, though he had held this idea for many decades. After observing that magnetism, gravity, and light can pass freely through bodies, he admitted: "How Bodies can have a sufficient quantity of Pores for producing these Effects is very difficult to conceive, but perhaps not altogether impossible. For *the Colours of Bodies* arise from the Magnitudes of the Particles which reflect them."[95] If now we imagine a body to consist of an equal quantity of parts of that magnitude and pores, and then imagine those parts to be composed of an equal quantity of much smaller pores and parts, and then imagine this process to proceed until solid particles (the primordials) are reached, bodies would consist mostly of pores. A body, for ex-

[95] *Opticks* (Dover), Bk. II, Pt. III, Prop, VIII, p. 268; italics added. This new concluding paragraph to Prop. VIII was added in the "Errata, corrigenda, & addenda" of the *Optice*, sig. b^v. It was included in the text of the subsequent English editions. I illustrate a corpuscle with three levels of composition in Figure 3.2. S. I. Vavilov, "Newton and the atomic theory," first recognized the importance of Newton's hierarchical matter theory and argued that it was a consequence of his concept of force. See Joshua C. Gregory, "The Newtonian hierarchic system of particles"; and Arnold Thackray, " 'Matter in a nut-shell' "; and *Atoms and Powers*. On the origin of the phrase "matter in a nut-shell," see Ch. 5, note 30.

ample, with four such compositions would have fifteen times more pores than solid parts, and with ten compositions above one thousand times more pores than parts.

In his "Hypothesis," in 1675, Newton did not explicitly call attention to the great porosity of bodies, nor did he mention his compositional theory of matter. However, in the incomplete "Hypothesis" of 1672, he began to discuss the composition of corpuscles just as the manuscript broke off uncompleted, at the end of the page. He was about to describe how light particles that collide with any of the "parts" of a body will "stick fast" to them. He attributed this to the pressure of the aether, which

> seemes to be the principall cause of the cohesion of the parts of all naturall bodies, composing them for the most part first into those very small clusters wch I have hitherto called corpuscles, & then aggregating those clusters into greater clusters. And much after the same manner it is that when a corpuscle of light[96]

In this account Newton assigns at least two compositions (parts into corpuscles and corpuscles into clusters) with aether filling the interstices. His analogical theory of colored bodies requires that the corpuscles be composed of parts and not be solid, for otherwise the analogy completely collapses. It does not, however, require (or exclude) additional compositions, that is, a full hierarchical theory of matter with many levels of structure. Newton, in fact, postulated a compound structure for matter as early as "Of Colours" in 1666. In some brief notes on reflection, transparency, and opacity, he refers to substances such as metals and marble "whose pores betwixt their parts admit a grosser Aether into y^{m} y^{n} y^{c} pores in their parts."[97] Insofar as I am aware, these passages are the earliest in which Newton adopts a compositional theory of matter. In the same essay and its appendixes his tendency to minimize the role of solid particles in optical phenomena was already underway; he had concluded from the nature of transparency and reflection that light "seldom" strikes the parts of bodies and that reflection and refraction are caused by the aether.[98]

There can, I believe, be little doubt that Newton first invoked his compositional theory of matter to resolve various optical problems and still less doubt that that idea was in his mind intimately related

[96] Add. 3970, f. 528^{v}; "for the most part" is an insertion.
[97] Add. 3975, §56, p. 15; *Questions*, p. 481.
[98] See notes 62 and 76.

to optical matters. Figala has suggested that Newton drew the hierarchical matter theory from alchemy, and this may well be a parallel or independent source.[99] However, there is strong evidence that he drew the less complex compositional theory that he was using in his early optical theory from Boyle's *Origine of Formes and Qualities, (according to the Corpuscular Philosophy)* (1666). The notebook (Add. 3975) that contains "Of Colours" consists mostly of Newton's notes on chemical subjects. Among these notes are numerous references to the *The Origine of Formes*, and some of them appear to match the handwriting of "Of Colours" – that is, they would also be from 1666.[100] In the section "Of Generation, Corruption, and Alteration," Boyle first assumes that "there are in the world great store" of undivided, solid particles that are too small to be sensible, "and these may . . . be called *minima* or *prima naturalia*." Then he assumes that these particles are further compounded:

> That there are also multitudes of corpuscles which are made up of the coalition of several of the former *minima naturalia;* and whose bulk is so small and their adhesion so close and strict, that each of these little primitive concretions or clusters (if I may so call them) of particles is singly below the discernment of sense. . . . And these are, as it were, the seeds or immediate principles of many sorts of natural bodies, as earth, water, salt, &c.[101]

In the 1672 version of the "Hypothesis," Newton appears to have adopted Boyle's terms "corpuscle" and "clusters" for the compound particles. He did not, however, follow Boyle in the close packing of particles with "their adhesion so close and strict," for he had already concluded from the nature of reflection and transmission that bodies must be porous and filled with aether. Boyle, moreover, held that there can be higher order compositions, for he notes "that even grosser and more compounded corpuscles may have such a permanent texture" as the "primitive" composition.[102] He then gives an example of the various "disguises" that quicksilver may assume, such as a red powder or "fugitive smoke," and yet remain recoverable mercury, an example that Newton duly summarized to begin the section in his

[99] See Karin Figala, "Newton as alchemist"; and " 'Die exakte Alchemie von Isaac Newton.' " Dobbs attributes a broader alchemical influence on Newton's analogical theory of colored bodies; see Ch. 3, note 48.

[100] For instance, the entry on p. 61 cited at note 103. See also Westfall, *Never at Rest*, p. 282, n. 6.

[101] Boyle, *Works*, 3:29–30.

[102] Ibid., p. 30.

notebook entitled "Of forms & Transmutations wrought in them."[103] Newton, in fact, had a year or so earlier already encountered the idea of the compound structure of the coloring corpuscles of bodies in Boyle's *Touching Colours*, where it was more sketchily set out.[104] Like so many of the concepts that Newton drew from his readings, the hierarchical theory of matter became his own when set within the context of his natural philosophy. We will return to its role in Newton's theory of colored bodies (§3.3), but let us now resume our account of the "Observations."

2.4. DESCRIBING NEWTON'S RINGS: THE NOMOGRAPH

Newton's aim in the second part of his "Observations" was to present a phenomenological account of the observations in the preceding part. In contrast to the causal, physical explanations in the "Hypothesis," those in this part are all given by means of principles and laws deduced from the observations themselves or other previously established principles. Some of the latter – especially the claims of the "New theory" on the unequal refrangibility of rays of different color and the compound nature of white light – were not in 1675 generally accepted, but that does not alter the nature of his approach. I will continue to be concerned solely with Newton's rings and especially the nomograph that he utilizes to explain the phenomenon. To my knowledge there is no modern description of Newton's nomograph, whose odd appearance and oracular directions for construction bely its intrinsic simplicity. Yet the nomograph is essential for understanding his theory of colored bodies as well as its later history, for it led directly to the discovery of absorption spectroscopy at the turn of the eighteenth century.

The nomograph is Newton's way of graphically representing with empirically derived values the law $d = ml/2n$ (Eq. 2.2) that he had derived in the first part. He never expresses the law as an equation.

[103] Add. 3975, p. 61. Newton's reference to "pag 72 of Formes, Mr Boyle," as his other page references, shows that he read the first Oxford edition of 1666 of *The Origine of Formes* and not the second Oxford edition of 1667, which is paginated differently. This is consistent with a date of 1666 for these notes. If Newton ever owned a copy of this book, it was not in his library at his death; see John Harrison, *The Library of Isaac Newton*. Gregory, "Newtonian hierarchic system," first suggested that Boyle was the source for Newton's hierarchical theory (although, rather oddly, he does not indicate that his quotations are from *The Origine of Formes*), but without the manuscript evidence just cited it remained a very perceptive conjecture.

[104] See Ch. 3, note 8.

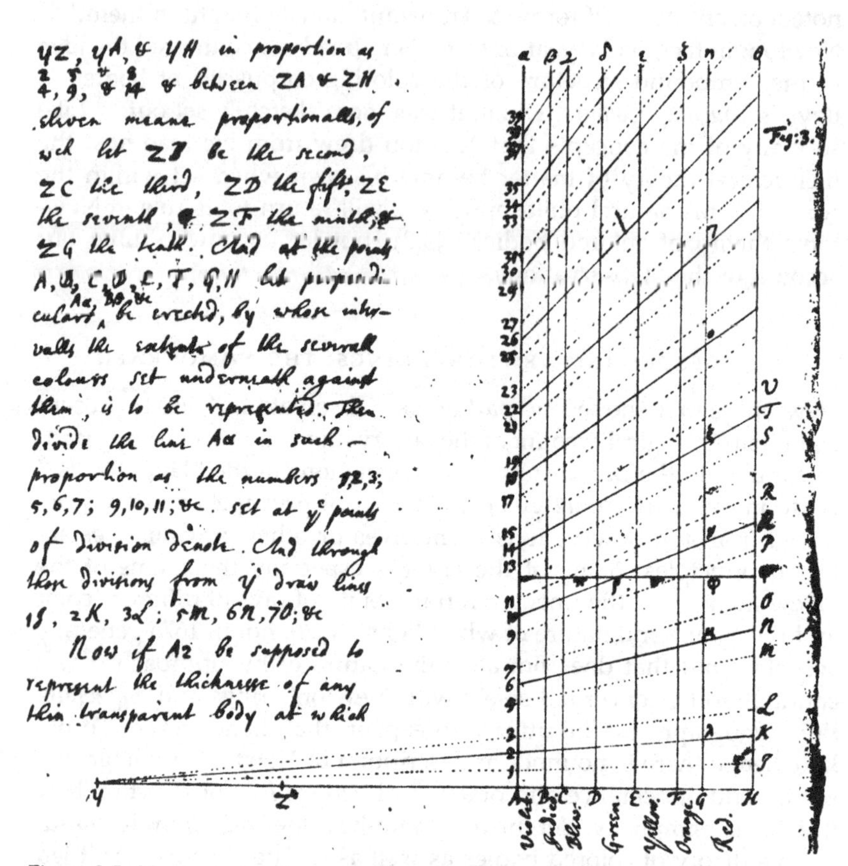

Figure 2.7. Newton's drawing of his nomograph in the "Observations" of 1675, Add. 3970, f. 507r. (By permission of the Syndics of Cambridge University Library.)

Let us recall that *d* represents the thickness of the film; *m* the integers 1, 2, 3, . . . ; *I* the interval or vibration length for a particular color; and *n* the index of refraction, which for the time being we will set equal to 1 inasmuch as we are dealing with thin films of air. There are differences in the numerical values and the way in which he apportions the colors in the three versions, but these do not alter the fundamental structure of the nomograph.

Newton begins the construction of the nomograph (Figure 2.7) by telling us that in the horizontal axis *YH* we are to take "the lengths *YZ*, *YA*, & *YH* in proportion as 4, 9, & 14 & between *ZA* & *ZH* eleven meane proportionalls, of wch let *ZB* be the second, *ZC* the third, *ZD* the fift, *ZE* the seventh, *ZF* the ninth, & *ZG* the tenth."[105] The division of line *AH* by the mean proportionals represents the distribution of the spectral colors by the ratio of their vibration lengths *I*, with the segments *AB*, *BC* . . . *GH* corresponding to the seven principal colors from violet through red. In particular, the division of *AH* (which is half of *ZH*) by eleven mean proportionals is that of equal temperament, with $2^{1/12}$ being the equal-tempered half note. In the revision of his *Optical Lectures*, completed some months before the 1672 version of the "Observations," Newton had considered equal temperament as a possible representation of the division of the colors of the spectrum cast by a prism. However, he rejected it in favor of a just diatonic scale, even though he conceded that the difference between the two was not observationally distinguishable.[106]

It is not apparent why Newton here chose the equal-tempered scale, unless it was to indicate clearly that the two phenomena have different physical origins: The spectrum is formed because of the unequal refrangibility of the light particles, whereas Newton's rings arise from the different intervals of the vibrations.[107] In any case, he abandoned the equal-tempered scale in the *Opticks*. The observational grounds for the division of the colors by their vibration lengths are far poorer than those of the spectrum. The division is based entirely on qualitative evidence reported in Observation 14; for instance, that the midpoint of the colors was at the boundary of the yellow and green and that the extent of the red was almost double that of the violet. This, as Newton notes, is "contrary to what happens in the colours made by the refraction of a Prism," that is, the order of the colors is reversed in the two divisions.[108] The other division in the horizontal line, that $YA:YH = 9:14$, represents Newton's determination of the ratio of the violet to the red pulse lengths, that is, that $I_v:I_r = 9:14$, where the subscripts v and r indicate violet and red.

In the manuscript of the *Opticks*, Newton initially had the same

[105] Add. 3970, f. 507^r; Birch, 3:286.

[106] *Optical Papers*, 1:545.

[107] In his *Optical Lectures* Newton refered to the equal-tempered scale as a "geometrical ratio" to distinguish it from a "musical ratio," which he then preferred because it allowed for an analogy between color and musical harmonies. However, a few years later in the "Hypothesis" he considered the vibrations excited by light particles in the retina to follow this geometrical ratio while still being harmonious; Add. 3970, f. 544^{r,v}; *Correspondence*, 1:376–7.

[108] Add. 3970, f. 504^r; Birch, 3:278; this corresponds to Obs. 15 in the 1672 version, which was substantially expanded in 1675.

division of the horizontal axis, but he replaced it with a new instruction that we are to take "from the point *Y*, the lengths *YA*, *YB*, *YC*, *YD*, *YE*, *YF*, *YG*, *YH*, in proportion to one another, as the Cube-roots of the Squares of the numbers $\frac{1}{2}, \frac{9}{16}, \frac{3}{5}, \frac{2}{3}, \frac{3}{4}, \frac{5}{6}, \frac{8}{9}$, 1, whereby the lengths of a musical Chord to sound all the Notes in an Eighth are represented."[109] The principal import of this new distribution of the vibration lengths of the colors, which Newton first used in his investigation of the colors of thick plates (Bk. II, Pt. IV, Obs. VIII), was to introduce the just diatonic scale and thereby bring it in line with the musical division of the prismatic spectrum and all the others invoked in the *Opticks*. It should be noted, though, that the 2/3 power of the string lengths has neither musical nor physical significance. Nor does this change of scale have much empirical significance: The ratio of I_v to I_r now becomes 9 to $14\frac{7}{24}$ rather than 9 to 14, and the relative position of the borders of the colors is shifted by about 1.5%.

Let us now turn to the vertical lines erected at *A*, *B* . . . *H*, the borders of each color. We are to divide the first line *A*α "in such proportion as the numbers 1, 2, 3; 5, 6, 7; 9, 10, 11, &c. . . . And through those divisions from *Y* draw lines 1*I*, 2*K*, 3*L*; 5*M*, 6*N*, 7*O*; &c."[110] The vertical lines, Newton tells us, represent the thickness *d* at which any order of a color is reflected. The units of the axis *A*α are quarter vibration intervals $I_v/4$. Similar triangles then ensure that the verticals for all other colors will represent the thickness $d = mI$, and the oblique lines simply represent different orders *m*. Newton's use of quarter intervals of the vibrations, rather than half intervals, as in Part I, necessitates rewriting our earlier equation as $d = mI/4$ (recalling that we have set $n = 1$ for air); reflection will now occur at $m = 2, 6, 10. \ldots$ Thus the first three orders of violet will be reflected at thicknesses of *A*2, *A*6, and *A*10, and all colors of the first three orders will be reflected when their thicknesses lie on the oblique lines 2*K*, 6*N*, and 10*Q*. Newton chose quarter vibration intervals because he wanted to illustrate not just the maxima but the entire range in which a color could be reflected. Thus the colors will be alternately reflected in the spaces 1*IL*3, 5*MO*7 . . . , and transmitted in *AHI*1, 3*LM*5, 7*OP*9. . . .

Now that we understand how Newton constructed his nomograph, we can learn how to use it. To know what colors of white light are

[109] *Opticks*, p. ₂31; Add. 3970, f. 155ʳ.
[110] Add. 3970, f. 507ʳ; Birch, 3:286.

reflected at any thickness, all you need do is place a ruler parallel to the horizontal line *AH* at a point on the vertical line that represents that thickness. The color reflected will be compounded of those colors through which the ruler passes. To appreciate this better, let us consider an example that ultimately became a sort of *experimentum crucis* for Newton's theory of colored bodies:

> Thus if the constitution of the Green in the third series of colours be desired; apply the Ruler as you see at πρσ,φ. And by its passing through some of the Blew at π, & yellow at σ, as well as through the green at ρ, you may conclude that the green exhibited at that thicknesse of the Body is principally constituted of originall green, but not without a mixture of some Blew & Yellow.[111]

If, therefore, the light of that green ring were viewed through a prism or passed through it, you would see a spectrum consisting of three contiguous colors. Now if you slide the ruler down a bit so that it passes through the middle of the violet of the third order, you will see that it also passes through the upper part of the red of the second order. This violet will incline to a "reddish purple," because it is compounded of violet and red.[112] If its light is passed through a prism, its spectrum will consist of two separate bands of red and violet. Thus the nomograph directly yields the general appearance or composition of the spectrum of any color of a thin film.

I need not describe the other clever ways in which Newton used his nomograph, for once its construction is understood these are straightforward. It is easy to convert the relative values of the nomograph to absolute ones, as is required for determining the thickness of the air film or corpuscle that reflects a particular color. In Observation 6 the thickness Gλ exhibiting orange-red of the first order was found to be 1/160,000 inch.[113] From this one measured value, and the similar triangles built into the structure of the nomograph, any other thickness or vibration interval may be determined from a single proportion.

By restricting himself thus far to a thin film of air, Newton did not have to incorporate the index of refraction into his nomograph. To

[111] Add. 3970, f. 507v; Birch, 3:287.

[112] Add. 3970, f. 508r; Birch, 3:288.

[113] Newton has in fact made a slip here, for in Obs. 6 he determined the thickness for yellow, and not orange-red. In the *Opticks*, he corrected his slip and shifted that thickness to the border of yellow and orange, to correspond with his measurement of citrine yellow in Obs. 6. This mistake in part reflects the difficulty (described in note 42) of determining what color has in fact been measured. Moreover, in the *Opticks* he had redetermined that thickness to be 1/198,000 inch.

apply it to a film of water, the thicknesses (and consequently the vibration intervals) are reduced by 3/4, because he found $n = 4/3$ for water; for glass they are reduced by 20/31, because $n = 31/20$; "and the like of other Mediums."[114] Newton then used this rule and his observations of the colors of the rings to construct a table (Figure 2.8) "wherein the thicknesse of Air, water, & glasse at wch each colour is most intense & specific, is expressed in parts of an inch divided into ten hundred thousand equall parts [i.e., 10^{-6} inch]."[115] The table plays a central role in the theory of colored bodies, where it is used to determine the size of the corpuscles of bodies from their color. It must be stressed that this table gives the compound color seen at a particular thickness in white light, and not the thickness at which a monochromatic color is "most intense & specific."[116] Newton tells us that this table should be compared with the nomograph so that "you will there see the constitution of each colour as to its ingredients or the originall colours of wch it is compounded, & thence be enabled to judg of its intensenesse or imperfection."[117]

An important consequence now comes to pass. We just saw that the nomograph also represents the general appearance of the spectrum of each color. According to Newton's theory of colored bodies, if the light reflected from or transmitted through a colored body were passed through a prism, then its spectrum should appear exactly like that of a thin film of the same color. Now what is truly remarkable is that when techniques – most notably absorption spectroscopy – were developed in the late eighteenth century for studying the spectra of colored bodies, they looked just like those predicted by Newton's theory: They consisted of groups of colors that were frequently separated by dark bands of missing colors. It is hard to believe that this

[114] Add. 3970, f. 508^v^; Birch, 3:288. Newton implicitly assumes that I/n is linear in all substances, which is consistent with his thinking at about this time in his new "linear" or "achromatic" dispersion law that follows from his musical division of the spectrum; see Shapiro, "Newton's 'achromatic' dispersion law," pp. 109–13.

[115] Add. 3970, f. 508^v^; Birch, 3:288. In the 1672 version of the "Observations," this table was included in Prop. 5 of the third part (Prop. 7 in the 1675 version).

[116] For example, it was just determined that the thickness for monochromatic light exhibiting the color at the border of orange and red is 1/160,000 inch, or $6\frac{1}{4} \times 10^{-6}$ parts, whereas according to the table of colors that compound color is produced at a thickness of about $9\frac{1}{2}$ parts. Even when the color is corrected to yellow or yellow-orange (see note 113), the table indicates 8 or $8\frac{1}{2}$ parts.

[117] Add. 3970, f. 508^v^; Birch, 3:289.

1675/6.] ROYAL SOCIETY OF LONDON. 289

		The thickneſs of		
		Air	Water	Glaſs
The colours of the firſt order	Black	2	$1\frac{1}{2}$	$1\frac{1}{4}$ or leſs.
	Blue	$2\frac{2}{5}$	2	$1\frac{3}{4}$
	White	$5\frac{1}{3}$	4	$3\frac{1}{2}$
	Yellow	8	6	$5\frac{1}{4}$
	Orange	9	$6\frac{3}{4}$	$5\frac{4}{5}$
	Red	10	$7\frac{1}{2}$	$6\frac{1}{2}$
Of the ſecond order	Violet	12	9	$7\frac{3}{4}$
	Indico	$13\frac{1}{4}$	$9\frac{11}{12}$	$8\frac{1}{2}$
	Blue	$14\frac{3}{4}$	11	$9\frac{1}{2}$
	Green	16	12	$10\frac{1}{3}$
	Yellow	$17\frac{1}{2}$	$13\frac{1}{8}$	$11\frac{1}{3}$
	Orange	$19\frac{1}{4}$	$14\frac{1}{2}$	$12\frac{2}{5}$
	Bright red	20	15	13
	Scarlet	$21\frac{1}{4}$	16	$13\frac{2}{3}$
Of the third order	Purple	23	17	$14\frac{4}{5}$
	Indico	24	18	$15\frac{1}{2}$
	Blue	$25\frac{1}{5}$	19	$16\frac{1}{4}$
	Green	$27\frac{1}{5}$	$20\frac{2}{3}$	$17\frac{1}{2}$
	Yellow	$29\frac{1}{3}$	22	19
	Red	31	$23\frac{1}{4}$	20
	Bluiſh red	$33\frac{1}{3}$	25	$21\frac{2}{3}$
Fourth order	Bluiſh	36	27	$23\frac{1}{4}$
	Green	$37\frac{2}{3}$	$28\frac{1}{4}$	$24\frac{1}{7}$
	Yellowiſh green	$39\frac{1}{3}$	$29\frac{1}{2}$	$25\frac{1}{3}$
	Red	44	33	$28\frac{1}{3}$
Fifth order	Greeniſh blue	$50\frac{2}{3}$	38	$32\frac{2}{3}$
	Red	$57\frac{1}{3}$	43	37
Sixth order	Greeniſh blue	64	48	$41\frac{1}{3}$
	Red	$70\frac{2}{3}$	53	$45\frac{2}{3}$
Seventh order	Greeniſh blue	$77\frac{1}{3}$	58	50
	Red or White	84	63	$54\frac{1}{2}$

Figure 2.8. Newton's table of colors, showing the order of colors in Newton's rings and the thickness of a thin film of air, water, or glass in millionths of an inch producing that color. (Reproduced from Birch [3:289], who published the copy of the "Observations" of 1675 in the Royal Society Journal Book.)

was sheer coincidence. Though Newton did not describe the spectra of colored bodies or compare them with those of thin films, he did observe the colors of each through a prism. In the twenty-fourth of his "Observations" he describes viewing Newton's rings through a prism; and in the *Optical Lectures* he explains that to test his theory of color successfully, the experiments must be carried out using bodies with simple colors: "You will identify these if, with the use of a prism, you choose those bodies that appear more distinct and less variegated at their edges that are bordered by blackness."[118]

Newton's primitive method of observing the spectra of colored bodies explains why he omitted a description of them. Interpreting the composition of the spectrum by viewing reflected boundary colors is extremely difficult, and he knew that. It was sufficient for his purpose that he could tell how compound or variegated that color was and deduce some of its components. Thus Newton knew that any successful theory of colored bodies had to produce compound colors. This feature undoubtedly contributed to his confidence in it. When techniques were developed to observe true absorption spectra, they did show Newton's prescience, but they also provided an opportunity to test his theory by comparing the spectra in detail. It is at this point that Newton's theory breaks down, for it turns out that the correspondence between the two sorts of spectra is only in their general appearance. This exciting story is treated in Part II of this book.

After Newton had explained his observations, he turned in the concluding sections of the second part to show how his explanation of the colors of thin films extended and confirmed his theory of color. This led him back to his earlier theme on certainty and the mathematical nature of his theory. We already encountered his claim here that his "explications are not Hypotheticall but infallibly true & genuine."[119] He ended this part by insisting that he had shown how "the Science of colours becomes a Speculation more proper for Mathematicans then naturalists, & deserves rather to be esteemed Mathematicall then Physicall, as I told you in my former letter [i.e., the 'New theory']."[120] His research on the colors of thin films reinforced his conviction that he had created a new, mathematical science of

[118] *Optical Papers*, 1:510–11.
[119] Ch. 1, note 41.
[120] Add. 3970, f. 511^v; Birch, 3:300. Newton deleted from "& deserves" to the end in the copy sent to the Royal Society. In the *Opticks* he revised the remainder to the unobjectionable "the Science of Colours becomes a Speculation as truly mathematical as any other part of Optiques" (*Opticks*, Bk. II, Pt. II, p. $_2$48).

color. It was, he held, mathematical and based on properties, not hypotheses, though we have seen how much he relied on his hypothesis of aethereal vibrations to derive those properties, which were presented "abstractedly." His own copy of the "Observations" concluded with the declaration that hypotheses were "no part of my designe. I undertook only to discover ye properties of light so far as I could derive them from experiments; & therefore content my self wth having shewn those properties."[121] When he did decide to compose his "Hypothesis," he deleted this conclusion. This decision did not entail any weakening of his methodological strictures, for we must recall his insistence that his "Hypothesis" should not be confused with his experimentally established "Observations," as well as how the two papers of 1675 marked the emergence of his technique of composing distinct works for his conjectures and certain principles. Let us now turn to the remaining, third part of the "Observations" and the theory of colored bodies.

[121] Add. 3970, f. 517^{r}.

3

THE COLORS OF NATURAL BODIES

Newton repeatedly declared that one of the principal aims of his investigation of the colors of thin films was to form a foundation for an explanation of the colors of bodies. In 1672 he announced that he had intended to include his "Observations" with his reply to Hooke's critique of his theory of white light and color because they "may conduce to further discoveries for completing the Theory [of color], especially as to the constitution of the parts of naturall bodies on which their colours or transparency depend."[1] He repeated this claim in 1675 and again in the *Opticks.*[2] Still, historians have failed to take Newton's declaration seriously.

The intimate connection that Newton draws here between his theory of white light and color and that of colored bodies is essential for appreciating the development of the latter. From his theory of color he deduced (and subsequently demonstrated) that bodies acquire their color by reflecting some colors in the light falling on them in greater quantity than others. I call this his "phenomenological theory" of colored bodies, for it does not offer any physical explanation for the origin of this property of bodies. In a further development of his phenomenological theory, Newton argued that all bodies are one color by reflection and the complement by transmission. When he made this overgeneralization, Newton may very well have already had in mind his analogical theory of colored bodies. If bodies acquire their

[1] Add. 3970, f. 442^{v}. By specifying that bodies be natural, Newton did not intend to exclude the artificial from his theory, for he treats such manufactured substances as paper, linen, and glass. Rather, his intention was to exclude bodies and colors that are the products of the imagination; compare his statements at the end of Prop. VII in the *Opticks*, Bk. I, Pt. II, and at the end of Bk. II, Pt. II.

[2] Add. 3970, f. 501^{r}; Birch, 3:272; and *Opticks*, Bk. II, Pt. I, p. $_{2}$1.

color from the transparent corpuscles composing them in exactly the same way as thin films produce colors – which I call his "analogical theory" – then the reflected and transmitted colors must be complementary. Thus the two components of his theory of colored bodies go hand in hand.

Newton's demonstration of his analogical theory by means of transduction provides a particularly fascinating insight into seventeenth-century natural philosophy, and it will be a principal focus of my account of his theory. Transduction is a method of making inferences about the unobservable, microscopic components of bodies from knowledge of the observed properties of macroscopic bodies, and, for Newton, it was a form of induction (§1.3).

Bodies, alas, can produce colors in a great variety of ways: refractive dispersion (as in diamonds), interference (soap bubbles), diffraction (feathers), scattering (the sky), selective absorption (grass), selective reflection (copper), and fluorescence (lignum nephriticum), to mention only some of the processes. When interest turned to the colors of bodies in England in the 1660s, only the first two were understood as distinct physical processes and sometimes correctly applied to explain coloration; and diffraction had become known in England only after Hooke and Newton had formulated their theories. The most common way in which bodies acquire color is by selective absorption. In the process of selective absorption, incident white light actually penetrates a body, where some colors are absorbed, and rays of the remaining colors, which together determine the body's color, are then diffusely reflected outward from the body's flaws, impurities, and sides. In metals alone are colors selectively reflected at the surface, where the others are absorbed. The various physical processes involved in the coloration of bodies are very complicated. It was only at the end of the eighteenth century that these began to be recognized, sorted out, and understood, an accomplishment largely of the nineteenth century. Much of the progress in the nineteenth century in fact emerged from testing, refuting, and replacing Newton's theory. Hooke and especially Newton were, in retrospect, attempting to encompass too many different phenomena within a single explanation.

3.1. BOYLE'S CONSIDERATIONS AND HOOKE'S HYPOTHESIS

Newton's interest in the colors of bodies was stimulated and perhaps first aroused by his encounter with Boyle's *Touching Colours* in late 1664 or early 1665. Boyle's goal was to show that the colors of bodies could best be explained by the tenets of the mechanical philosophy, namely, that they arise from some modification of the incident light caused by the corpuscles of bodies and that they are not inherent qualities of bodies. He was chiefly concerned with the corpuscular properties of bodies, not with optical theory and the particular mechanisms by which these modifications occurred and caused the appearance of a particular color. Boyle's goal of establishing a connection between the color and the corpuscular structure of bodies became the subject of Newton's own theory of colored bodies.

About one year after reading *Touching Colours*, Newton encountered Hooke's theory – or, rather, as he called it, his "hypothesis" – of colored bodies in the *Micrographia*. This was the first detailed mechanical explanation of colored bodies. Mechanical philosophers had hitherto devoted their attention to explaining colors produced by refraction – "the celebrated Phaenomena of Colours," as Newton would call them in 1672 – and had only casually treated the ways in which bodies modified incident white light to generate their colors. Hooke's optical theory served as a challenge to Newton, for Hooke had presented a comprehensive account of all the ways then known in which color was produced: by refraction, by thin films, and by bodies.[3] Any subsequent theory would have to be equally comprehensive, and, in Newton's eye, it would have to be an emission theory, not Hooke's misguided wave theory. Newton not only drew a challenge from Hooke's exposition, but many of its features would reappear in his own theory.

Boyle attributed the colors of bodies "chiefly" to reflection from their superficial parts, for in this way he could explain the colors of opaque bodies as well as those of slicks and films, which are manifestly restricted to the surface. He described in great detail the innumerable ways in which corpuscles of various sizes and shapes could form surfaces with different structures that would variously reflect and modify the incident light to produce all the variety of colors. Although Boyle readily conceded that color could be caused otherwise, for example, by refraction in transparent bodies, he insisted that whatever the cause it must be interpreted in terms of the corpuscles that compose the bodies. As befit a chemist, he devoted much of *Touching Colours* to describing colors brought about by chemical operations, and it was this aspect of the book that most intrigued Newton, as is evident from his notes "Of Colours."

[3] See Newton's statement in Ch. 2, note 36.

According to Boyle, the most common way of producing such color changes is by liquids: "and these for the most part abounding with very Minute, Active, and Variously Figur'd Saline Corpuscles . . . may well enough very Nimbly alter the Texture of the Body they are imploy'd to Work upon, and so may change the form of Asperity, and thereby make them Remit to the Eye the Light that falls on them, after another manner than they did before, and by that means Vary the Colour."[4] Boyle then enumerated the various mechanisms by which liquids may bring about such color changes. Because Newton incorporated Boyle's mechanical mode of chemical explanation in his theory of colored bodies, let us pause over a few examples. "A Liquor may alter the Colour of a Body by making a Comminution of its Parts," Boyle explained, "by Disjoyning and Dissipating those Clusters of Particles . . . which stuck more Loosely together . . . [or] by Dividing the Grosser and more Solid Particles into Minute ones, which will be always Lesser, and for the most part otherwise Shap'd than the Entire Corpuscle so Divided." Liquids also act by the contrary way, "by procuring the Coalition of several Particles that before lay too Scatter'd and Dispers'd to exhibit the Colour that afterwards appears."[5]

Newton's extensive notes on *Touching Colours* are concerned solely with Boyle's experiments and observations, especially the chemical ones. A number of his notes record Boyle's discovery of the color indicator test for acids and alkalies. Boyle found that when acids are added to blue vegetable solutions ("as syrrup of violets impregnated wth ye tincture of ye flowers, juice of blew bottles, or coneweede, juice of ripe privet berrys," to use Newton's summary) they turn red; and when an alkali (sulfurous salts) is added, they turn green.[6] "Which," Newton jotted down, "may bee usefull to ye finding whither bodys abound more wth acid or sulphureous Salts."[7] These observations would reappear in Newton's theory of colored bodies, and his mechanical interpretation would still be disputed 150 years later.

Boyle himself posed some questions about his explanation of the colors of bodies that indicated the direction of future research. Hitherto he had assumed that the corpuscles are opaque, but he now suggests that it should be seriously investigated

[4] Boyle, *Touching Colours*, Pt. I. Ch. 3, p. 54.
[5] Ibid., pp. 57–9.
[6] Add. 3996, §27, f. 133^v^; *Questions*, p. 456. Boyle, *Touching Colours*, Pt. III, Expt. 20, pp. 245–8. See Marie Boas, "Acid and alkali in seventeenth century chemistry."
[7] Add. 3996, §47, f. 134^v^; *Questions*, p. 460. See also §48.

whether or no Particles of Matter, each of them singly Insensible, and therefore Small enough to be capable of being such Minute Particles, as the *Atomists* both of old and of late have (not absurdly) called *Corpuscula Coloris,* may not yet consist each of them of divers yet Minuter Particles, betwixt which we may conceive little Commissures where they Adhere to one another, and, however, may not be Porous enough to be, at least in some degree, Pervious to the unimaginably subtile Corpuscles that make up the Beams of Light, and consequently to be in such a degree Diaphanous.[8]

This is perhaps the first time that Newton encountered the compositional theory of matter, for Boyle has assumed that the coloring corpuscles are themselves composed of smaller particles. To support the possibility that the "coloring corpuscles" may be transparent, Boyle shows that bodies considered to be opaque, such as shells, wood, and marble, when made thin enough, become partly transparent. Yet he also found metallic substances, such as iron filings and minium (red lead), to be opaque even when viewed through a microscope. With his assumption that according as a substance can or cannot be seen through its corpuscles is similarly transparent or opaque, Boyle is arguing by transduction. Hooke and Newton would adopt the identical argument.[9] Boyle concludes his considerations in a genuine quandary and urges further investigation. Both Hooke and Newton recognized that a more complex process than a simple superficial reflection was required to account for the colors of bodies and held that the corpuscles, or at least the higher order ones, must be transparent so that light could penetrate them.

Hooke began his "hypothesis" on the colors of bodies in "Observ. X. Of metalline, and other real colours" in his *Micrographia* with an explanation of transparent colored bodies, but he held that it applies as well to opaque ones. Bodies, he assumes (Figure 3.1), consist of transparent "tinging" particles, *E, F, G, H,* of a "determinate bigness" that are regularly dispersed throughout a transparent medium with a different refractive power.[10] To explain how colors are generated, he adopted the same explanation as he used for thin films, that of a duplicated or split pulse front in which the weaker pulse either precedes the stronger one to produce blue or follows it to produce red

[8] Boyle, *Touching Colours,* Pt. I, Ch. 3, pp. 68–9.

[9] In his discussion of Boyle's use of transduction, Mandelbaum identifies two principles underlying his method. He characterizes one as "the extension of sense knowledge by analogy," which Boyle applies here, and the other as "the translation of explanatory principles from the observed to the unobserved"; *Philosophy, Science and Sense Perception,* pp. 107, 110–11.

[10] Hooke, *Micrographia,* p. 68.

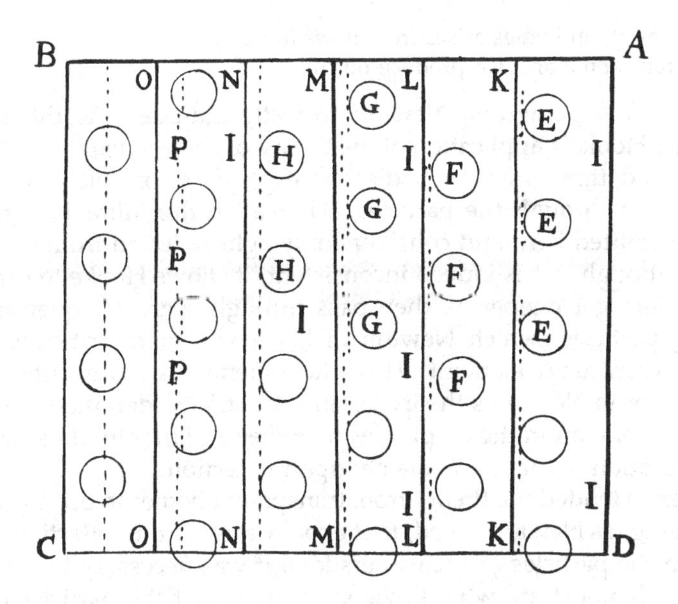

Figure 3.1. Hooke's explanation of the colors of bodies. (From *Micrographia,* Plate VI, Figure 7.)

and yellow (§2.1). Hooke thus implicitly associated the colors of bodies and those of thin films. If the tinging particles slow down the light pulses (as is illustrated), then red is generated, for as a pulse proceeds through the tinging particles of the body, a weak pulse *PP* (dashed) is gradually formed following the stronger one *OO;* but if light moves more swiftly through the tinging particles, then blues are generated.

In a draft of his response to Hooke's charge of plagiarism in December 1675, Newton characterized Hooke's explication as "difficult" and "peculiar," and it does have obvious problems.[11] Most of these arise from Hooke's assumptions that colors are not innate to white light but modifications of it and that there are only two primary colors, red/yellow and blue. While preparing his response to Hooke, Newton also made some notes on Hooke's explanation of colored bodies – a topic barely touched on in the letter he actually sent – that serve to bring out the differences, and problems, of their theories:

> He explains ye colours of tinged liquors by transmitted light [,] of plates by reflected. Of those he supposes one plate to make a full colour & many particles of these to go to a full colour. . . .

[11] Add. 3970, f. 532^{r,v}.

That opacity in bodies arises from flaws in them
That reflexion alone can produce no colours.[12]

In the first paragraph Newton correctly indicates the difference between Hooke's application of his theory of color to explain colored bodies and thin films. In bodies he explains color solely by light transmitted through the particles, whereas in thin films he ignores the transmitted light and considers only light reflected from the surface. Although this is indeed inconsistent, it allows Hooke to explain how colors get deeper as they pass through a greater quantity of tinging particles, which Newton in his next remark criticizes as a further discordant element. This phenomenon is more difficult to account for in Newton's theory because he strictly identified the process of coloration in the corpuscles of bodies and in thin films, where full coloration occurs immediately upon reflection.

Hooke extended the theory from transparent bodies to opaque ones by arguing, as Newton noted, that opacity arises from reflection from flaws in the particles or their interstices. It was necessary in the first place to demonstrate what Boyle suspected, that the particles of all bodies, even the most opaque metals, are transparent. He marshals a variety of evidence to support this contention – for example, when metals are made thin enough, they become transparent, like gold leaf; that metals are used to color transparent glass; and above all, that the microscope reveals virtually all particles to be transparent.[13]

> To summ up all therefore in a word, I have not yet found any solid colour'd body, that I have yet examin'd, perfectly *opacous*. . . .
>
> And indeed, there seem to be so few bodies in the world that are *in minimis* opacous, that I think one may make it a rational *Query,* Whether there be any body absolutely thus *opacous?* For I doubt not at all . . . that could we very much improve the *Microscope,* we might be able to see all those bodies very plainly transparent, which we now are fain onely to ghess at by circumstances.[14]

The second proposition of Newton's theory would be based directly on Hooke's demonstration that the parts of all bodies are transparent.

Just like Boyle before him, Hooke is arguing by transduction, for he supposes that the macroscopic concept of seeing through something (or transparency) is the same as the ability of micromatter to transmit light. He assumes his tinging particles to be like little crystal

[12] Add. 3970, f. 532^{v}.
[13] Hooke, *Micrographia*, pp. 72–3, 78–9.
[14] Ibid., p. 78.

balls and does not consider the possibility that they could be opaque, and yet the body that they compose could still transmit light waves. Undoubtedly the power of the microscope to make the hitherto invisible visible vividly supported this transduction, especially to the author of the *Micrographia* and its captivated readers.

To justify his claim that the opacity of bodies arises from flaws in their transparent particles, Hooke again argues by transduction. For example, when a large transparent piece of colored glass is flawed and cracked by heating and quenching, it becomes opaque and retains the same color except that it is whiter; the same thing happens when a transparent colored liquid is whipped into a froth.[15] They become whiter because so many rays are reflected from the surfaces without passing through the tinging particles of the body. Hooke had deduced, no doubt from theory and experience, that, as Newton summarized it, "reflexion alone can produce no colours." He was so convinced of the generality of this principle that he attempted to extend it to metals despite Boyle's observations: "for it seems very probable, that those Rays that rebound from them ting'd, with a deep Yellow, or pale Red, as from Copper . . . have past through them; for I cannot conceive how by reflection alone those Rays can receive a tincture, taking any *Hypothesis* extant."[16] Hooke's zeal to create one theory for all colored bodies led him into error, for metals are the one class of substance that he studied that are colored by selective reflection. He explained that two sorts of reflection occur in colored bodies: a specular reflection from the surface and a diffuse reflection or scattering from the inner flaws or the particles' surfaces.[17] Bodies are colored solely by the second sort of reflection, after the rays have passed through the tinging particles.

Although Newton adopted various elements of this explanation, the most fundamental, testable difference between their theories was Hooke's principle that "reflexion alone can produce no colours." Newton held, on the contrary, that bodies acquire their color by a specular reflection after passing through just one particle. Even if Newton had not freely admitted that he drew upon Hooke's observations, it is evident from his writings on the colors of bodies.[18] We shall also be able to discern Boyle's influence, but Newton's theory was anything but a pale imitation of either.

[15] Ibid., pp. 72–77.
[16] Ibid., p. 73.
[17] Ibid., p. 76.
[18] See Newton's public acknowledgment at note 44.

3.2. NEWTON'S PHENOMENOLOGICAL THEORY

As an immediate consequence of his theory of color, Newton included a proposition on the coloration of bodies (the phenomenological theory): "The colors of natural bodies are derived from the sort of rays that they reflect most."[19] This represented a radical break with all preceding theories, which assumed that bodies either modify the color of the light incident upon them or impart their intrinsic color to it. For Newton, bodies can appear only of the color(s) with which they are illuminated. They cannot create, impart, or modify colors but only selectively reflect those incident upon them, some in greater quantity than others. The proposition is easily verified by illuminating bodies with lights of different color: A leek shines much more brightly in green light than in red or violet, and cinnabar (mercuric sulfide, or vermilion) shines brightest in red light and least in violet. This explains why in daylight a leek ordinarily appears green and cinnabar red. Newton's phenomenological explanation is unexceptionable, provided that we interpret "reflect" in its general meaning of to turn or cast back. In fact, as we shall see, Newton had a separate process of coloration by reflection in mind, distinct from that by transmission, and is advocating that bodies are colored by selective reflection.

Newton soon extended his phenomenological theory to include the principle that bodies are one color by reflection and the complement by transmission. This was a new principle and not a consequence of his color theory, but it prepared the way for his analogical theory. In his *Optical Lectures* he clearly motivated this principle

> because bodies become colored by reflecting some kinds of rays and letting in others, if they are in some measure transparent the conclusion seems unavoidable that those colors are transmitted the most that are reflected the least, and therefore that they are one color when they are looked through and another when they are seen by reflected light.[20]

To support this claim he introduced a variety of phenomena: a solution of lignum nephriticum (the wood from a Mexican shrub that was used for treating kidney ailments), which is blue when viewed by reflection and yellow or red by transmission; gold leaf, which conversely is yellow by reflection and blue by transmission; stained glass, which frequently varies in color according to the observer's

[19] *Optical Papers*, 1:508–9. This proposition was included in the "New theory," Prop. 13, and the *Opticks*, Bk. I, Pt. II, Prop. X.
[20] *Optical Papers*, 1:510–11.

position; and some very clear optical glass, which is blue by reflection and yellow by transmission. These "archetypal phenomena," as I call them, which Newton came upon in Boyle's *Touching Colours*, are invoked in all his writings on colored bodies.[21]

These bodies do indeed appear of one color by reflection and their complement by transmission. Newton was, however, led astray by them in his quest for a general explanation of colored bodies, for – what he could not have known – they involve different physical processes: Lignum nephriticum and probably the clear glass are colored by fluorescence, gold leaf by selective reflection, and stained glass most likely by scattering. As early as his second essay "Of Colours" in 1666 Newton had related his archetypal phenomena to the colors of thin films, where complementary colors are manifestly produced by reflection and transmission: "The flat pieces of some kinds of Glase will exhibit ye same Phaenomena wth Lignum Nephriticum. And these Phaenomena of Gold & Lignum Nephriticum are represented by ye Prisme in ye 37th experiment . . . ," which relates the sequence of reflected and transmitted colors produced by the thin film of air between two superposed prisms.[22] Conceiving of the colors of thin films and his archetypal instances of colored bodies as phenomena of the same kind was the initial step to conceiving of the former as the cause of the latter. It is quite possible, and I consider it likely, that Newton already had that idea in mind, but the analogical theory does not make an appearance until the "Observations," in the spring of 1672.

After describing the archetypal phenomena in the *Optical Lectures*, Newton concedes that it does not always turn out that bodies are one color by reflection and another by transmission

> because there is in bodies not only a power to reflect or transmit rays, but also one to stifle and terminate them within themselves. Thus, some bodies stop and retain all kinds of rays, and in that way they become completely black; others reflect some and suppress the rest, as opaque colored bodies; others suppress some and partly reflect and partly transmit the rest, as transparent colored bodies that are the same color all around; and others reflect some and transmit the rest, as is established by the examples just related.[23]

Newton has here introduced into optics the concept of selective absorption (*potestas suffocandi et in se terminandi*) as an independent

[21] Boyle describes the colors of gold leaf, lignum nephriticum, and glass in *Touching Colours*, Pt. III, Expts. 9, 10, 11, pp. 198–219. Newton's notes on these phenomena are in Add. 3996, §§39, 47, 51, ff. 134^{r,v}, 135^{r}; *Questions*, pp. 458, 460, 462.
[22] Add. 3975, §3, p. 1; *Questions*, p. 466.
[23] *Optical Papers*, 1:512–15.

process of coloration. Rays that enter a body are either absorbed or transmitted. Selective absorption is a consequence of his theory of color: because all colors are already present in sunlight and not created by bodies, some process must necessarily be posited to eliminate those colors not exhibited by a body. It is also evident from this passage that Newton has placed reflection alongside transmission as an independent process of coloration. His second and especially his fourth categories, opaque colored bodies and the archetypal phenomena, assume the existence of selective reflection and not, as in Hooke's theory, the scattering or diffuse reflection of light transmitted within the body. Indeed, the analogical theory based on Newton's rings is meaningless if coloration by reflection and transmission are not distinct.

Although Newton had almost certainly not yet read Kepler's *Ad Vitellionem paralipomena* (1609), Kepler had already "established by an easy experiment that light is not colored by reflection [*repercussu*]."[24] He reflected sunlight from various colored fluids placed in a basin and found that the reflected beam was uncolored and white, which demonstrated that "colors are not in the surface but rather in the depth" of bodies.[25] He also related, however, that metallic mirrors (which had not yet been replaced by the looking glass) did color reflected light, though weakly. In midcentury, Niccolò Zucchi and Francesco Maria Grimaldi (whose works I am also rather certain Newton did not know) confirmed that reflection does not produce colors by reflecting sunlight from colored glasses.[26] In the *Opticks* Newton proposed a more careful version of these experiments to examine the color of reflected light.

Newton's entire theory of coloration would ultimately founder on the third category that he describes, transparent colored bodies. These, as he correctly notes, reflect and transmit the same color (whatever is not absorbed) and therefore violate his scheme that different colors (complements) are reflected and transmitted. Newton's rings never exhibit the same color by reflection and transmission. We can see here the problem confronting him in his attempt to bring the coloration of all bodies into one comprehensive scheme. Newton would be engaged in a long struggle to bring theory and observation in agreement on this point, but in the end the phenomena would

[24] Kepler, *Gesammelte Werke*, 2:368.
[25] Ibid.
[26] Zucchi, *Optica philosophia* (1652), pp. 278–9, 283–4; Grimaldi, *Physico-mathesis de lumine* (1665), p. 97.

yield to theory; that is, he would violate his own methodology and create a hypothetical property. In "Of Colours" he had already recognized the conflict and attempted to reconcile the two:

> But Generally bodys wch appeare of any colour to ye eye, appeare of ye same colour in all positions; Nay Gold if it bee not soe very thin as to bee transparent appeares onely yellow & perhaps ye yellow colour of Lignum Nephriticum would vanish if ye tincture bee strong & ye liquor of a greate thicknesse. And perhaps there are many coloured bodys wch if made so thin as to bee transparent would appeare of one colour when looked upon & of another when looked through.[27]

Newton is suggesting here that the reflected and transmitted colors are in fact complementary in all bodies, with the only difference being their thicknesses and relative absorptive powers. Why he temporarily abandoned his unifying scheme for colored bodies in his *Optical Lectures* and let the phenomena reign is not apparent.

By the early 1690s, when he composed the *Opticks*, Newton returned to that scheme and the argument of "Of Colours." He asserted – "I question not" – that if the bodies exhibiting the archetypal phenomena were made so thick as to become opaque, they would exhibit only one color "like all other opake Bodies . . . though this I cannot yet affirm by experience." Similarly, "so far as my Observation reaches," all opaque bodies when made sufficiently thin become "in some measure transparent, and differ only in degrees of transparency from tinged transparent Liquors."[28] In going over the proof sheets of the *Opticks*, Newton must have realized that his argument that all bodies exhibit complementary reflected and transmitted colors was apparently contradicted by simple observation of transparent colored bodies and that he would have to justify it. He then added an experiment (which is not in the printer's copy of the manuscript of the *Opticks*) suggesting that the apparent uniform color of such bodies is an artifact:

> A transparent Body which looks of any Colour by transmitted Light, may also look of the same Colour by reflected Light, the Light of that Colour being reflected by the further surface of the Body, or by the Air beyond it. And then the reflected Colour will be diminished, and perhaps cease, by making the Body very thick, and pitching it on the back-side to diminish the reflexion of its further surface, so that the Light reflected from the tinging particles

[27] Add. 3975, §4, p. 1; *Questions*, pp. 466–7.
[28] *Opticks*, Bk. I, Pt. II, Prop. X, p. 140.

may predominate. In such cases, the Colour of the reflected Light will be apt to vary from that of the Light transmitted.[29]

Newton's suggestion that some transmitted light is scattered back out from transparent bodies is correct. But when the experiment was finally carried out in 1784, by carefully eliminating the scattering of all transmitted light, these bodies were found to be colorless – that is, *all* colored light reflected from bodies is actually transmitted light that has been scattered after passing through it. The elaborate series of experiments carried out by Edward Hussey Delaval confirmed the conclusion of Kepler, Zucchi, Grimaldi, and Hooke that there is no colored reflected light and marked the beginning of the long process of the rejection of Newton's theory of colored bodies.[30] After Newton presented this experiment in the *Opticks,* he concluded the proposition with the promise that "whence it is that tinged Bodies and Liquors reflect some sort of rays, and intromit or transmit other sorts, shall be said in the next Book. In this Proposition I content my self to have put it past dispute, that Bodies have such Properties, and thence appear coloured."[31]

Now that he has established, at least to his own satisfaction, the phenomenological basis for his analogical theory, let us follow his attempt to demonstrate that theory.

3.3. NEWTON'S ANALOGICAL THEORY

As late as 1827, John Herschel described Newton's analogical theory of the colors of natural bodies as one of "extraordinary boldness and subtilty, in which great difficulties are eluded by elegant refinements, and the appeal to our ignorance on some points is so dexterously backed by the weight of our knowledge on others, as to silence, if not refute, objections which at first sight appear conclusive against it."[32] In following Newton's formulation of his theory, we should bear in mind Herschel's apt characterization and balance our occasional skepticism by an appreciation of its audacity, imagination, and explanatory power, just as much of the scientific community did for nearly a century.

Newton's strategy in the third part of his "Observations" is – beginning with the assumption of the corpuscular nature of bodies – to

[29] Ibid., pp. 140–1; Add. 3970, f. 127r.
[30] On Delaval's experiments, see §7.1.
[31] *Opticks*, p. 141.
[32] Herschel, "Light," §1134, p. 580.

demonstrate in nine propositions the optical properties of the corpuscles and pores necessary to explain the transparency or opacity and, especially, the colors of bodies.[33] The most fundamental (and ultimately notorious) of these properties is that the corpuscles produce the colors of bodies in exactly the same way that thin films produce the colors of Newton's rings. The essential features of the theory are established by transduction; the causes of reflection, opacity, and color are all carried over directly from the macroscopic to the corpuscular level.[34] To demonstrate the central proposition, that the colors of bodies are caused by their corpuscles just as in thin films, he also invoked what would become the second Rule of Reasoning in the *Principia,* that they are phenomena of the same kind. Just as Newton founded his theory of color on experiment and observation, so he utilized them in applying transduction to establish his theory of colored bodies. That the latter theory is not as compelling as the former is, I believe, in the first place, because it is a causal theory and not a phenomenological one like the theory of color. In the second place, it is less compelling because we do not find transduction to be a universally convincing method.

One of my aims in presenting Newton's actual use of transduction is to illuminate the nature as well as the pitfalls of a method that was so widely used by the mechanical philosophers to gain an understanding of the hidden properties and structure of matter. Newton's assumption of corpuscularity and his strict mechanical interpretation of chemical processes and the interactions of light and matter play an essential role in his use of transduction and are largely responsible for the problems that he encountered in its application. Because his theory of colored bodies was a structural theory in which particular optical properties are attributed to different arrangements and levels of particles and pores, he had difficulty in projecting the properties to the proper level of organization. Moreover, because he had essen-

[33] As in the preceding chapter, I will follow the 1675 version of the "Observations," while indicating differences with the 1672 version; see Ch. 2, note 32. It should be recalled that the propositions are numbered identically in the *Opticks* and the 1675 "Observations" through Prop. 8.

[34] Readers who want an overview of the theory before following it in detail can turn first to the summary in §3.4. McGuire, in his important study of transduction and the third Rule of Reasoning, "Atoms and the 'analogy of nature,' " pp. 10–11, recognized that Newton's theory of colored bodies is an instance of transduction, but needless to say in two paragraphs he scarcely analyzed that theory or the role of transduction. Dobbs adopted McGuire's insight but also attributed a significant role to alchemical ideas in the development of the theory of colored bodies; *Foundations of Newton's Alchemy*, pp. 221–5; see also note 48.

tially reduced optical properties to mechanical ones, he was not in fact always carrying the same observed property to lower levels of matter.

For about a century after Newton proposed his theory, the method of establishing it went virtually unquestioned. If we enter into this mode of thinking, we, too, can be seduced by the persuasiveness of his argument and appeal to experimental evidence. Even when Newton's theory came under widespread criticism, the method of transduction was so deeply embedded in the methodological armament that it was not directly attacked. To my knowledge, only one person, Gibbs Walker Jordan, in 1800, presented a thoroughgoing critique of Newton's use of transduction. Although his book was published in the midst of the great controversy over Newton's theory and was known in Britain, there is no evidence that it played a major role in the debate. Because a running presentation of Jordan's critique alongside Newton's theory would tend to undermine it, I have placed a fuller account of his criticism in Appendix 1. Strictly considered, Jordan belongs in Part II, but his arguments are better grasped with Newton's theory fresh in mind.

Newton began his theory with three propositions that demonstrate the cause of reflection and opacity much as in Hooke's theory:

> Prop: 1. Those superficies reflect the greatest quantity of light wch have the greatest refracting power; that is, wch intercede Mediums that differ most in their refractive densities: And in the confines of equally dense Mediums there is no reflexion.
>
> Prop 2. The least parts of naturall bodies are in some measure transparent. And the opacity of those bodies ariseth from the multitude of reflexions caused in their internall parts.
>
> Prop: 3. Between the parts of opake or coloured bodies are many interstices replenished with Mediums of other densities: as water between the tinging corpuscles wherewith any liquor is impregnated; Air between the aqueous globules that constitute Clouds or mists; and for ye most part spaces voyd of both Air & Water but yet perhaps replenished wth some Subtiler Medium between the parts of hard bodies.[35]

Prop. 1 states that reflection arises from a difference in refractive or optical density. Newton demonstrates this by first showing that total reflection occurs soonest when the difference in refractive density

[35] Add. 3970, f. 512^r,v^; Birch, 3:297, 298. The 1672 version of the "Observations" consisted of only five propositions corresponding to 2, 3, 4, 5, and 7 of the 1675 version. Prop. 2 was virtually the same as 3; the contents of the others will be noted below.

is greatest. Then he relates a series of observations showing that the quantity of light reflected at the interface of two media increases as the difference of density increases; for example, that more light is reflected from the interface of glass and air than at water and air, and still less at water and oil.[36] Newton states and demonstrates this proposition for macroscopic surfaces but, in a fine example of transduction, freely applies it in the subsequent propositions to reflection from the corpuscles' surface.

Prop. 2 actually consists of two distinct propositions. The first part, on the transparency of the least parts, Newton tells us "will easily be granted by them that have been conversant with Microscopes" and also from the observation that all bodies when made sufficiently thin become transparent.[37] Although he is following Hooke here, Newton did recognize that "Metalline Bodies" are an exception, for they "seem to reflect all the light incident on their first superficies."[38] As I have already observed concerning Boyle's and Hooke's similar arguments by transduction, it does not necessarily follow that if one can see through a body then one can see through its parts.[39] In fact, Newton himself would ultimately reject this conclusion. I shall return to the problem of the transparency of the parts and show that Newton has been overzealous in his application of transduction and mistakenly carried it to the "least parts" rather than to the highest order, or largest corpuscles in his compositional theory of matter; in the 1672 version of the "Observations" he more accurately refers to the "small parts."[40]

In the second part of Prop. 2 (which, we should note, remains

[36] Newton had already arrived at the crux of Prop. 1 in "Of Colours," Add. 3975, §55, p. 14; *Questions*, p. 481. It was not yet a formal proposition in the 1672 version but was invoked in the proof of Prop. 2, which became Prop. 3 in 1675.

[37] Add. 3970, f. 512^v; Birch, 3:298. In the 1672 version this first part alone was Prop. 1: "That the small parts of all naturall bodies (those of metalls perhaps & some other ponderous minerall substances wch are of a mercuriall originall being excepted) are transparent" (Add. 3970, f. 525^r). The second part on opacity, though not yet granted the status of a proposition, was invoked in the demonstration of Prop. 2. In the *Opticks* Newton slightly modified Prop. 2 to apply to "almost all naturall bodies."

[38] Add. 3970, f. 512^v; Birch, 3:298. In the *Opticks*, p. 252, Newton added in continuation (once again drawing on the *Micrographia*): "unless by solution in menstruums they be reduced into very small particles, and then they become transparent." Compare Hooke's observation at note 15. Because metals do appear to be optically different from other substances, Newton had to struggle to bring them into agreement with the rest of his theory, as this addition shows. Inasmuch as it offers no new insights into his theory or method, I will not pursue the special case of metals.

[39] Jordan vigorously made this point; *New Observations Concerning the Colours of Thin Transparent Bodies* (1800), pp. 55–6; see Appendix 1.

[40] Add. 3970, f. 525^r.

unproved), Newton has implicitly introduced corpuscularity with his assumption that bodies have parts and pores. This universal assumption of the mechanical philosophy is absolutely essential for the establishment of his theory. Although Newton presents evidence that opacity is caused by the parts of bodies in the following proposition, it is all drawn from macroscopic bodies. He argues, on the one hand, that opaque substances become transparent when their pores are filled with a fluid of the same or nearly the same density as their parts, for example, when paper is dipped in water or oil, or the oculus mundi stone (opal) is soaked in water. On the other hand, transparent substances become opaque when their parts are separated, for example, when glass is pulverized or water whipped into a froth; this instance of transduction is drawn directly from Hooke.[41] Newton, we should note, has presented evidence showing only that the "parts" cause opacity, but not that light is reflected from them.[42]

In Prop. 3, Newton appears to accomplish the impressive feat of demonstrating from optical properties the corpuscular nature of bodies or, at least, of opaque bodies.[43] He proves the proposition by simple deduction from the preceding two: By Prop. 2, opacity arises from reflections from the internal parts of bodies, and by Prop. 1, reflections can occur only if there are interstices of different densities from the parts. The feat becomes somewhat less impressive when we recall that he had already introduced corpuscularity in the preceding proposition.

Before proceeding to see how Newton extends the corpuscular structure to transparent bodies, we should pause to note how much of his theory thus far is drawn from Hooke. At the conclusion of Prop. 3, Newton acknowledged his debt to Hooke's *Micrographia:* "In wch Book he hath also largely discoursed of this & the precedent Proposition, & delivered many other very excellent things concerning the colours of thin Plates & other naturall Bodies wch I have not scrupled to make use of so far as they were for my purpose."[44] Prop. 1 is

[41] See Newton's notes on the *Micrographia*, pp. 68, [72], 83 in *Unpublished Scientific Papers*, pp. 404, 405; and also "Of Colours," Add. 3975, §56, p. 15; *Questions*, p. 481. Jordan criticized this argument; *New Observations*, pp. 69–71; see Appendix 1.

[42] See Jordan, *New Observations*, pp. 57–8.

[43] In the *Opticks* Newton made two changes in the statement of Prop. 3. At the beginning he replaced "interstices" by "spaces, either empty or," and at the end he replaced "replenished wth some Subtiler Medium" with "not wholly void of all substance."

[44] Add. 3970, f. 513^{r}; Birch, 3:299. This acknowledgment was in the 1672 version (f. 525^{v}) but was removed in the *Opticks*. If it had appeared earlier and so were read at the first instead of the last of the five meetings of the Royal Society at which the

Newton's and not to be found in the *Micrographia*, although there can be little doubt that Hooke would have readily accepted it. If the remainder of Newton's theory is altogether original, it nonetheless draws its inspiration from Hooke and, to a lesser degree, Boyle.

Prop. 4 – "The parts of bodies & their interstices must not be lesse then of some definite bignesse to render them opake & coloured" – establishes the conditions for opacity and, from its converse, some of those for transparency.[45] Once again Newton's principal proof is by transduction, and he shows that when bodies are very small, reflection ceases: Opaque bodies become transparent when their parts are subtly divided ("as Metalls by being dissolved in Acid Menstruums"); at the central spot in Newton's rings, in the region where the lenses are in contact and the film is thinnest, there is no reflection; and the same thing occurs in the thinnest parts of soap bubbles.

The statement of the proposition implies that if the parts and pores are less than a certain size, a body will be transparent, but thus far Newton has established in Prop. 3 only that opaque bodies have pores. To extend corpuscularity to transparent bodies, such as water and glass, he had to make a vague appeal to nonoptical evidence: "For upon divers Considerations they seem to be as porous as other bodies, but yet their pores & parts too small to cause any opacity."[46] Such a move was reasonable, for it would be very odd indeed if suddenly for one sort of body and at a particular magnitude, nature

"Hypothesis" and "Observations" were presented, it is possible that Hooke would have been placated and not accused Newton of plagiarism; see Ch. 2, note 20.

[45] Add. 3970, f. 513^r; Birch, 3:299. This was Prop. 3 in 1672.

[46] Add. 3970, f. 513^v; Birch, 3:299. In the preceding sentence Newton states: "On these grounds I conceive it is that water, salte, glasse, stones & such like substances are transparent." This explanation of transparency appears to conflict with that at the end of Prop. 1, where transparency is attributed to the uniformity of the medium, so that there are no reflections from the internal parts: "Whence it comes to passe that uniforme Mediums have no sensible reflexion but in their externall superficies where they are adjacent to other Mediums of a different density" (Add. 3970, f. 512^v; Birch, 3:297). This statement is ambiguous, for it is not clear at what degree of fineness uniformity applies. Does it apply at the corpuscular level? In the *Opticks* Newton removed all ambiguity and answered yes. I indicate his additions to the preceding sentence with italics: "*So then the reason why* uniform *pellucid* mediums, (*such as Water, Glasse, or Crystal*) have no sensible reflexion but in their external superficies, where they are adjacent to other mediums of a different density, *is because all their contiguous parts have one and the same degree of density*" (*Opticks*, pp. $_2$51–2). This now conflicts directly with the explanation of Prop. 4, for contiguous parts have no pores. Newton seems to have carelessly and unnecessarily carried his transduction to too fine a level, for by the transduction in Prop. 4 he experimentally restricted the validity of Prop. 1 only insofar as "some definite bignesse." As Newton argued in the "Hypothesis," the two glasses at the central spot behave "*as if*" they were "one uniform medium"; see Ch. 2, note 86.

abandoned corpuscularity; and it was necessary, for if transparent bodies were not corpuscular, the entire analogical theory would break down. Moreover, no contemporary mechanical philosopher would have objected to Newton's claim, for it was widely assumed that such transparent substances as water and glass consisted of corpuscles and pores that were either empty or filled with aether. Otherwise, how could light pass through them?

A still more fundamental problem with his model confronted Newton, namely, distinguishing between colorless opaque – that is, black – and transparent bodies and also between colored opaque and transparent bodies. Prop. 4 laid down the conditions only for the appearance of colored opaque and colorless transparent bodies. Black and colorless transparent bodies literally represent the two sides of the central spot in Newton's rings, which is transparent by transmission and black by reflection. Consequently, such bodies have parts of the same size. Likewise, a transparent and an opaque body of the identical color must have corpuscles of the identical size, corresponding to a particular thickness in Newton's rings. In Prop. 7, Newton deduces that in a black body "the corpuscles must be lesse then any of those wch exhibit colours."[47] At any greater size there is too much light reflected to constitute black. The corpuscles must, therefore, be less than the thickness of the central spot, or the same size as that of a colorless transparent body.[48] He suggests that the corpuscles of black bodies "may perhaps variously *refract* it [light] to & fro within themselves so long untill it happen to be stifled & lost," but he does not describe the structural property responsible for this.[49]

[47] Add. 3970, f. 515^{r}; Birch, 3:302.

[48] Dobbs argues that Newton attributed black to the smallest particles because of his alchemical views: "There was really no justification for equating black with the smallest particles, except that unquestioned assumption from alchemy." "That assumption" she explains, was "that the black matter of putrefaction was in a relatively unformed condition, or, in mechanical terms, that it was composed of matter in particles smaller than those produced later in the alchemical process as the matter 'matured' " (*Foundations of Newton's Alchemy*, p. 225). Newton equated black with the smallest corpuscles, it should be perfectly clear, solely on the analogy to the colors of Newton's rings and the observation that the central spot is black; no other thickness is black. That this conclusion may agree with Newton's alchemical ideas does not alter the grounds on which he in fact deduced it. On the assignment of particular colors to particular corpuscle sizes, Newton's theory is totally consistent. The only "difficulty," as he grants (see note 61), is in deciding which order of a particular color should be assigned to a given body, but for black only one size is possible.

[49] Add. 3970, f. 515^{r}; Birch, 3:302; italics added.

In the "Hypothesis" Newton resolves his problem by attributing transparency and opacity to the size of the pores and color to that of the corpuscles. Because every body has pores of various sizes, the aether's density should vary from pore to pore "& so light be *refracted* in its passage out of every pore into ye next, wch would cause a great confusion & spoile ye bodies transparency." However, the constant aethereal vibrations drive the aether from one pore to another, and "this must eavenly spread ye aether into all ye parts not exceeding some certain bignes, suppose ye breadth of a vibration, & so make it of an eaven density throughout ye transparent body."[50] Thus, if the pores are less than approximately one vibration length I, the aether will be uniform and the body transparent; but beyond this size its varying density will refract and reflect the light to and fro within the body and so make it opaque.[51]

The difficulty with Newton's neat solution to differentiating transparent and opaque bodies is its hypothetical nature, for it attributes certain properties to the aether without any evidence. When he composed the *Opticks* and had abandoned the aether, this explanation depending on pore size and aether density also had to be abandoned. Apparently, he could not devise an alternative using forces, for this idea did not find its way into the queries.[52]

In the next trio of propositions Newton gets to the essence of the analogical theory, that the corpuscles of bodies produce colors just as a thin film of the same thickness and density does:

Prop: 5. The transparent parts of bodies according to their severall sizes must reflect rays of one colour & transmit those of another, on the same grounds that thin Plates or Bubbles doe reflect or transmit those rays. And this I take to be the ground of all their colours.

Prop: 6. The parts of Bodies on which their colours depend are denser then the Medium wch pervades their interstices.

[50] Add. 3970, f. 544^r^; *Correspondence,* 1:375; italics added.

[51] This solution appears to conflict with Prop. 2, which attributed opacity to "*reflexions* caused in their internall parts." Moreover, as the theory unfolds in Props. 7–9, opacity turns out to arise from *refractions* within the pores and ultimately absorption by the (least) parts. We can eliminate much of this tension, if we loosely interpret "reflection" as "turned back." True reflection *within* the parts will be reserved to explain the colors of bodies. These inconsistencies highlight the continuing problem that Newton had with his multiple mechanisms for reflection and maintaining transduction at the proper level, problems that were never fully eliminated. For Jordan's criticism of this point, see *New Observations*, pp. 58–9.

[52] Nearly forty years later (in a manuscript now in private possession) Newton once again toyed with this idea, but with the electric spirit replacing the aether.

Prop: 7. The bignesse of the component parts of Naturall Bodies may be conjectured by their colours.[53]

Newton begins by affirming that Prop. 5 is to be taken literally, for he sees "no reason why" if a thin plate of constant thickness and uniform color were shattered, the heap of fragments would not form a powder of the same color as the unbroken plate. "And the parts of all naturall bodies being like so many fragments of a Plate must on the same grounds exhibit the same colours. Now that they doe so will further appeare by the affinity of their Properties."[54] This would be a perfect example of transduction had it been possible to acquire such a thin uniform plate and actually carry out the experiment.[55]

At this point Newton invokes what would become his second Rule of Reasoning – "the causes of natural effects of the same kind are the same" – and endeavors to show that the colors of natural bodies and thin plates are of the same kind and therefore have the same cause. He did not have to show that their properties were absolutely identical, as new properties could emerge in colored bodies when the thin transparent corpuscles are compounded, as turns out to be the case for opacity. For example, he notes that a solution of lignum nephriticum and many other substances[56] – the archetypal phenomena – exhibit one color by reflection and another by transmission; painters' pigments change their color "a little" when they are finely ground and their corpuscles crushed; flowers when bruised also change their color; and as the size of water vapor increases, clouds change their appearance from transparent to variously colored by reflection and transmission. All of these phenomena exhibited by colored bodies occur in thin films. Drawing upon his earlier study of Boyle's *Touching Colours*, Newton also appealed to color changes in chemical reactions:

Nor is it much lesse to my purpose that by mixing divers liquors very odd & remarqueable productions & changes of colours may be effected, of wch no cause can be more obvious & naturall then y^{t} the saline corpuscles of one liquor do variously act upon or unite with the tinging corpuscles of another, so as to make them swell or shrink (whereby not onely their bulk but their

[53] Add. 3970, ff. 513^{v}, 514$^{r, v}$; Birch, 3:299, 300. The 1672 version contained only Props. 5 and 7, then numbered 4 and 5; Prop. 6 is new.

[54] Add. 3970, f. 513^{v}; Birch, 3:299.

[55] For an eighteenth-century attempt to carry out this experiment, see Ch. 5, note 39; see also Jordan, *New Observations*, pp. 73–4.

[56] In the *Opticks*, Bk. II, Pt. III, Prop. V, p. 256, Newton specifically mentions stained glass and gold leaf.

density also may be changed,) or to divide them into smaller corpuscles, or make many of them associate into one cluster.[57]

Just as with the examples of painters' pigments and bruised flowers, Newton has adopted a strict version of the mechanical philosophy and assumed that changes in the properties of bodies – specifically, their color – can be explained solely by changes in the arrangement, motion, or size of the corpuscles composing them. This interpretation of chemical change, which is drawn directly from Boyle's writings, represents the mechanical philosophy in the 1670s, before Newton himself introduced the concept of force. This simple mechanical chemistry – or, as Newton called it, "vulgar," or "common," chemistry – was carried over into the *Opticks* unchanged, and a century later it would be the focus of vigorous attacks by chemists, who considered it not to be chemistry at all.[58]

Although Newton had to show an "affinity" or close relation of the properties of thin films and colored bodies rather than an identity, the relation of the two is nonetheless a weak one. In our discussion of the phenomenological theory, we have already seen him struggling with the fact that most colored bodies are unlike the archetypal phenomena and are not one color by reflection and another by transmission, as the analogy to thin films requires. Nor are most colored bodies seen by a specular reflection, as occurs in thin films. However, to the objection that unlike a thin film the color of a body does not change when viewed from different angles, Newton counters with Prop. 6, that the corpuscles are denser than their pores. Once again he argues by transduction, for the color of a thin film that is denser than the surrounding medium, such as water in air, changes little when the angle of the incident light varies (Obs. 19). Moreover, the colors are also brisker and more vivid in this case (Obs. 22).

Prop. 7 represents the culmination of the theory. It partakes of the excitement of Newton's new, mathematical science of color, for by it he is able to make predictions as to the size of the corpuscles of various bodies on the basis of their color. Assuming that the corpuscles are of the same optical density as water or glass, "as by many circumstances is obvious to collect," then their size is immediately found from the color of the body by the table in the second part (Figure 2.8) that gives the thickness of a film producing that color.[59] If, for ex-

[57] Add. 3970, ff. 513^{v}–14^{r}; Birch, 3:300. See Jordan's criticism of Newton's reasoning, *New Observations*, pp. 79–80, and Appendix 1.
[58] See, for example, Bancroft's comments quoted at Ch. 5, note 55.
[59] Add. 3970, f. 514^{v}; Birch, 3:301.

ample, the corpuscles of a body of a green of the third order are of the same optical density as glass, the table tells us they will be $17\frac{1}{2} \times 10^{-6}$ inch in diameter. Despite their apparent smallness, we should recognize how relatively large they are. If we consider a thin sheet of colorless glass that produces this green color, it must contain a number of the corpuscles (and aether-filled pores) that make it glass. A fragment of this glass sheet must be considered a macroscopic body. It is a small, albeit very small, piece of glass with all the properties of glass. Green glass (or grass) will be composed of corpuscles of the same size as these fragments, each of which is composed of the corpuscles that compose colorless glass.[60]

"The greatest difficulty," Newton allows, "is here to know of what order the colour of any Body is."[61] This is determined by carefully comparing the color of the body with those observed in Newton's rings. Most of this proposition is devoted to Newton's analysis of the qualities of the various orders of each color, and it is compelling reading. Blues and violets, for example, of the second and third order are good, but the best are those of the third. Newton buttresses his argument by once again appealing to chemical evidence, the color indicator test for acids and alkalies that he had learned from Boyle's *Touching Colours*. Because it later greatly exercised chemists, I will cite it in full:

> Thus the colour of *violets* seems to be of that [third] order, becaus their Syrup by acid liquors turnes red and by urinous & alcalizate turne green. For since it is of the nature of acids to dissolve or attenuate, & of Alcalies to precipitate or incrassate, if the purple colour of ye Syrup was of the second order, an acid liquor by attenuating its tinging corpuscles would change it to a red of the first order, & an Alcaly by incrassating them would change it to a green of the second order; wch red & Green, especially the Green, seem too imperfect to be the colours produced by these changes. But if the said purple be supposed of the third order, its change to red of the second & green of the third may without any inconvenience be allowed.[62]

Newton's argument has a testable consequence. If it is assumed, as he does, that chemical reactions involve only a change in the size or density of the corpuscles, and that colors of bodies are produced

[60] For Jordan's analysis of this point, see his *New Observations*, pp. 67–71, and Appendix 1.
[61] Add. 3970, f. 514^v; Birch, 3:301.
[62] Add. 3970, ff. 514^v–15^r; Birch, 3:301. See note 6 for Newton's early notebook entries on these color changes.

by those corpuscles just as in thin films, then the sequence of color changes in a chemical reaction should follow the order of the colors in Newton's rings (Figure 2.8). This may be tested. Although we may now consider such a correspondence to be fanciful, it is truly surprising how many chemical processes later seemed to support Newton's theory.

We should pause here to grasp fully what Newton has wrought. He has reduced the property of the color of a body solely to the size and density of its corpuscles and made it completely independent of its chemical composition. As he summarized his theory for Boyle, "ye colours of all natural bodies whatever seem to depend on nothing but ye various sizes & densities of their particles."[63]

Newton's description of the green of vegetation, the most common of all colors, attracted the most attention:

> There may be good *Greens* of the fourth order but the purest are of the third. And of this order the Green of all Vegetables seemes to be, partly by reason of the intensenesse of their colours, & partly because when they wither some of them turne to a greenish Yellow, & others to a more perfect Yellow or Orange or perhaps to Red, passing first through all the afforesaid intermediate colours. Which changes seem to be effected by the exhaling of the moisture wch may leave the tinging corpuscles more dense, & something augmented by the accretion of the oyly & earthy part of that moisture. Now the Green without doubt is of the same order with those colours into wch it changeth because the changes are graduall, & those colours though usually not very pure yet for the most part are too pure & lively to be of the fourth order.[64]

In Chapter 2 (§2.4) we encountered a detailed account of the composition of this green. When it was tested, more than 150 years later, it ultimately proved to be the grounds for the decisive refutation of Newton's theory.

Newton concludes the proposition by informing us that he has entered into such detail "because it is not imposible but that Microscopes may at length be improved to the discovery of the corpuscles of Bodies on wch their colours depend."[65] If their magnification could be increased five or six hundred times, we might be able to see some of the largest corpuscles, and if "thre or four thousand times perhaps they might all be discovered but those which produce blacknesse." Newton stops at the corpuscles producing black, because they are

[63] Newton to Boyle, 28 February 1678/9; *Correspondence*, 2:291.
[64] Add. 3970, f. 514^v^; Birch, 3:301.
[65] Add. 3970, f. 515^v^; Birch, 3:303.

perfectly transparent and reflect no light. They are, in principle, invisible, he claims, whatever the magnification. The discussion ends with Newton's oft-quoted lament:

However it will add much to o^{r} satisfaction if those corpuscles can be discovered with Microscopes, wch if we shall at length attain to I fear it will be the utmost improvement of this Sense. For it seems impossible to see the more secret & noble works of Nature within those corpuscles by reason of their transparency.[66]

Once again Newton has taken a stand in direct opposition to Hooke, who in his *Micrographia* had hoped that with improvements in the microscope we might "perhaps" see the "secret workings of Nature."[67] For Newton, the "secret & noble works" within the corpuscles implicitly refer to his compositional theory of matter, which he had explicitly discussed in 1672 and would again discuss in the *Opticks* in this context of the colors of natural bodies. Alas, Newton's fears were unfounded, for he would soon have to introduce absorbent, opaque, and thus visible, primordials. He had carried transduction too far, beyond the compound higher order corpuscles that produce blackness and colors. In Prop. 2 he had demonstrated that the "least parts" are transparent, but let us see how he had to revise that claim in the final two propositions of the 1675 version of the "Observations."

Having concluded his account of "the constitution of naturall bodies on wch their colours depend," Newton subjoined two propositions on reflection:

Prop: 8. The cause of Reflexion is not the impinging of Light on the solid or impervious parts of bodies, as is commonly supposed.

Prop: 9. Tis most probable that the rays wch impinge on the solid parts of any body, are not reflected but stifled & lost in the body.[68]

[66] Add. 3970, f. 516^{r}; Birch, 3:303. In the 1672 version, Newton had written "secret & noble contrivances" before replacing "contrivances" with "workes"; Add. 3970, f. 528^{r}. To be more precise, Newton should have attributed invisibility to "their *perfect* transparency." All corpuscles, according to him, are transparent, and invisibility arises not from transparency itself but from lack of reflection. The corpuscles of all colored bodies, that is, those greater than "some definite bignesse," reflect and transmit light; but the smaller corpuscles of colorless bodies reflect no light and are perfectly transparent. Thus the corpuscles of clear and black bodies will be invisible because they reflect no light back to the eye and will be seen through as if they were not there.

[67] Quoted in full at Ch. 1, note 93.

[68] Add. 3970, f. 516^{r,v}; Birch, 3:303, 304. These propositions are new and were not in the 1672 version.

Prop. 8, which denies that light is reflected by the "solid or impervious parts," appears to conflict with Prop. 5, which states that for the production of colors light is reflected by the "transparent parts." Confusion may be avoided by bearing in mind that in Prop. 5 Newton is referring to higher order corpuscles, and in 8 to primordials. This, however, is only a temporary solution, for there is a genuine problem of keeping to the proper level of structure with his use of transduction.

Props. 8 and 9 are essential for completing the theory of colored bodies, as Newton had not yet introduced the process of absorption. The analogical theory could not provide a solution to the problem, for there appears to be no absorption in Newton's rings. All the light is either reflected or transmitted. In his earlier treatment of opacity in Prop. 2, he explained that opaque bodies prevent the transmission of light by reflecting it to and fro from their internal parts. If some process of absorption within them were not assumed, the light would ultimately escape and make the body appear luminous, and "there would be no Principle of the obscurity or blacknesse wch some bodies have in all positions of the eye."[69]

In the 1672 version of the "Observations," just as the manuscript broke off, Newton was about to offer an hypothesis for absorption by means of his compositional theory of matter and the aether.[70] In the 1675 "Hypothesis," however, he explains that a ray is absorbed when it hits the "solid parts" because they do not have "sufficient elasticity or other disposition to return nimbly enough the smart shock of ye ray back upon it."[71] As we have seen (§2.3), Newton did not consider the absorption of light to be simply a mechanical process, but one that contributed to the activity of nature in its secret, inner parts. He always restricted this idea to his speculative writings. In Query 22/30, added to the first Latin edition of the *Opticks* (1706), he finally publicly conjectured that absorbed light enters into the composition of bodies: "Are not gross bodies & light convertible into one another: & may not bodies receive their activity from the particles of light wch enter their composition?"[72] In about 1698, Newton had ex-

[69] Add. 3970, f. 517^{r}; Birch, 3:305; cited in full at note 75.
[70] See Ch. 2, note 96.
[71] Add. 3970, f. 546^{v}; *Correspondence*, 1:383.
[72] Add. 3970, f. 259^{r}; *Optice*, p. 319. In an earlier draft, then Query 21, Newton had written "most active powers" rather than "activity"; Add. 3970, f. 292^{r}. In the second English edition he modified his claim to "much of their Activity"; *Opticks* (Dover), p. 374. Echoing the "Hypothesis," Newton also asserted in this query that "the changing of bodies into light & light into bodies is very conformable to the course of nature, which seems delighted with transmutations."

plained, according to David Gregory's record, that the rays of various colors that are not reflected from a plant's leaves and flowers are "busy within" it:

> It seems that the rays of light of different Colours have the greatest share in Natural effects. that for example while the green rays alone are reflected from most plants the rest of the rays are bussy within, and while the blue (or otherwise coloured) rays are alone reflected from the flower the rest are busy within.[73]

Newton undoubtedly would have felt that his speculations were vindicated by the discovery of photosynthesis a century later, and he would have been truly excited by the spate of photochemical reactions then being discovered.

Newton notes that Prop. 9 is in accordance with ("consentaneous to") Prop. 8, and in revising them for the *Opticks* he eliminated the former and incorporated it within number 8. The conjectural nature of Prop. 9 is awkward in a work that in the next paragraph declares that it is derived from experiment and disavows hypotheses, for it assumes an emission theory of light; the conclusion need not at all follow in a wave theory. Subsuming Prop. 9 within Prop. 8 made its nature less obvious, but it did not alter the facts that it is not derived from experiment and that it is essential to his theory of colored bodies.

In the *Opticks* Newton deleted the very brief paragraph stating that these propositions were not part of the theory of colored bodies, but we should note that in demonstrating them he no longer pursued the method of transduction. He establishes Prop. 8 by appealing to a cogent sequence of phenomena (to which still more were added in the *Opticks*) that show that light cannot be reflected from the parts of bodies. For instance, when light passes from glass to air at an angle of incidence greater than 40° or 41°, it is totally reflected, and it is unimaginable that at a slightly smaller angle it should suddenly encounter enough pores to be mostly transmitted, when just before it encountered only parts to reflect it. And, a fortiori, why should it always find enough pores when it goes the other way, from air to glass? To take just one more argument from Newton's rings: If reflection were from the parts, it would be "impossible" for thin films at one place to reflect rays of one color and transmit those of another. It is simply inconceivable that the blue rays, for example, "should have the fortune to dash upon the parts" and the red the pores,

[73] Christ Church Library (Oxford), MS 346, p. 90. Judging by its position among other dated memoranda, this was written between August 1697 and April 1699.

whereas at a thicker or thinner portion of the film the contrary should happen.[74]

Prop. 9 completes the theory by postulating a way for the required property of absorption to occur. It pursues the argument of Prop. 8:

If all the rays should be reflected which impinge on the internall parts of clear water or crystall, those substances should rather have a cloudy then so very clear transparency. And further there would be no Principle of the obscurity or blacknesse wch some bodies have in all positions of the eye. For to produce this effect it is necessary that many rays be retained & lost in the body, & it seems not probable that any rays can be stopt & retained in it wch do not impinge on its parts.[75]

Thus the "solid," "internall" parts must be opaque and not transmit light. Something is awry: Earlier (in Prop. 2), Newton had demonstrated that the "least" parts are transparent, and that opacity arises from reflection from the "internall" parts.

We are dealing with a real problem, whether the least parts or primordials are transparent or opaque, but let us first deal with a terminological issue. Newton does not use the term "part" consistently. Generally, by "part" he means the compound particles or clusters that are composed of primordials and, in turn, compose bodies; often he also calls these "corpuscles." Thus, in Prop. 5, he calls the parts that cause color "transparent parts." In Prop. 2, he adds the adjective "least" for the uncompounded, solid primordials out of which all other bodies are composed, and in Props. 8 and 9, he adds "solid or impervious." With various levels of composition and properties to follow, one can easily get confused. And this seems to have happened to Newton, for we have encountered a genuine contradiction. The least parts cannot be both transparent and impervious.[76]

The theory of colored bodies began by demonstrating by transduction that bodies consist of transparent parts and pores, and the analogy to thin films led to the conclusion that all parts less than a

[74] Add. 3970, f. 516^{v}; Birch, 3:304.

[75] Add. 3970, ff. 516^{v}–17^{r}; Birch, 3:305. In the manuscript Newton later added after the penultimate sentence: "For if all or almost all the light incident on those bodies did after many reflexions & refractions come out again it would make those bodies as lucid as other bodies are."

[76] That the emergence of absorption was a genuinely confusing point even to a serious student of Newton's theory of colored bodies becomes apparent from Thomas Melvill's query, "Since the cause of blackness in bodies is the smallness of their *transparent* parts, which renders them incapable of reflecting any colour; how can black bodies . . . be at the same time opaque?" ("Observations on light and colours" [1756], p. 82; italics added).

certain size must be transparent, just as the central spot in Newton's rings is. In contrast, the existence of opaque and black bodies led to the inescapable conclusion that ultimately the parts must absorb light. The source of Newton's problem was the method of transduction, which he had carried to too deep a level. Transparency for Newton was a property of pores and therefore an attribute of a compound body that consisted of parts and pores. Transduction with a structural property, one that depends on the composition or arrangement of particles, can be carried out at most only as far as that composition still occurs, unlike a universal or primary quality like impenetrability that extends as far as the primordials. If a bottle made out of glass can hold water, it cannot be argued that the corpuscles of glass are also capable of holding water.[77] Newton thus found that he could not construct opaque bodies solely from transparent primordials, and he ended up with a different model than the one he began with. All bodies, transparent and opaque, now consisted of absorbent primordial particles and pores, which are assumed to be filled with aether. If microscopes were sufficiently improved, the primordials would thus not be invisible but would appear like little black holes. Transparency first emerges in corpuscles composed of these particles.

There is no indication that Newton recognized this problem with transduction until shortly before the publication of the *Opticks*. However, in revising the "Observations" for the *Opticks* in 1691, he once again became aware of how different properties – most notably color – emerge at different levels of composition. In this same period he was also revising Rule III of the *Principia,* which justified the attribution of universal qualities. Recall (§1.3) that in the early drafts Rule III was a general justification of transduction and applied to "laws and properties." I would suggest that it was possibly his reencounter

[77] Rudjer Josip Bošković clearly recognized the invalidity of transduction with structural properties: "if any property depends on an argument referring to an aggregate, or a whole, in such a way that it cannot be considered apart from the whole, or the aggregate; then, neither must it (that is to say, by the same argument), be transferred from the whole, or the aggregate, to parts of it. . . . Hence those properties, although they are found in any aggregate of particles of matter, or in any sensible mass, must not however be transferred by the power of induction to each & every particle" (*A Theory of Natural Philosophy*, trans. of 1763 ed., §40, p. 31). Bošković was quoting his own *De continuitatis lege* (Rome, 1754). Yet so seductive and natural was reasoning by transduction, that Bošković accepted Newton's explanation of transparency and opacity as being caused by differences in the density of the parts; *Theory,* §483, p. 172. For a useful guide to eighteenth-century analyses of Newton's third Rule of Reasoning and transduction, see McGuire, "Atoms and the 'analogy of nature,' " n. 8.

with the subtleties of transduction in his theory of colored bodies that led him to restrict Rule III to "qualities." When in an intermediate draft of the rule Newton introduced the requirement of intensification and remission to restrict it to universal qualities, he explicitly mentioned color and blackness and transparency and opacity among those properties that do not apply to all bodies. There is no doubt that Newton continued to believe in the validity of transduction, although he could not justify it by Rule III except for universal qualities. Some criterion was otherwise required to limit transduction to a certain level, but he never formulated one.

When Newton was putting the finishing touches to the *Opticks* in 1703, he drafted a preface, which, as he so often decided, he did not publish. He considered the possibility of deducing all phenomena from just four "general suppositions" or "Principles," because "if for every new Phaenomenon you make a new Hypothesis . . . what certainty can there be in a Philosophy wch consists in as many Hypotheses as there are Phaenomena to be explained." These "principles" were not hypotheses but derived by induction, for "there is no other way of doing any thing with certainty then by drawing conclusions from experiments & phaenomena untill you come at general Principles."[78] The first three principles are, from our perspective, an odd assortment: "ye being of a God," the impenetrability of matter, and the law of gravitational attraction. In describing the fourth principle, he announced that he intended to derive the theory of the colors of natural bodies from his hierarchical, corpuscular theory of matter:

> A fourth Principle is that all sensible bodies are aggregated of particles laid together wth many interstices or pores between them. . . . As by the third Principle we gave an account heretofore of ye motions of the Planets & of ye flux & reflux of ye sea, so by this Principle we shall in ye following treatise give an acc^t^ of ye permanent colours of natural bodies, nothing further being requisite for ye production of those colours then that ye coloured bodies abound with pellucid particles of a certain size & density. This is to be understood of the largest particles or particles of ye last composition. For as bodies are composed of these larger particles with larger pores between them so it is to be conceived that these larger particles are composed of smaller particles with smaller pores between them.[79]

Newton here explicitly corrects his assertion in Prop. 2 that the "least" particles are transparent ("pellucid") and now attributes trans-

[78] Add. 39790, 479^r^; McGuire, "Newton's 'principles of philosophy,' " pp. 182–3, who published two versions of this draft.
[79] Add. 3970, f. 479^v^; McGuire, "Newton's 'principles of philosophy,' " p. 184.

parency to the "largest" particles of the highest composition. I cannot offer any convincing reasons to explain why he did not go back and emend the text of the *Opticks* other than that he simply overlooked it or it was too late to change the text. All that was required was for him to change the "least parts" back to "small parts," as the 1672 text had.[80]

The corpuscular theory of matter was thus for Newton not a hypothesis but a demonstrated principle established with as much certainty as the existence of God or the theory of gravitation. He cites two principal sorts of evidence in its support: Various substances penetrate the pores of bodies, like water into vegetable and animal matter, and quicksilver into metals; and transparency, which shows that light passes through the pores of a great variety of bodies (which, to be sure, assumes an emission theory of light). The theory of colored bodies was not only founded on the corpuscularity of matter; it was a theory of matter attributing specific properties and arrangements to the corpuscles that cause the transparency or opacity and colors of bodies. For Newton to have considered corpuscularity to be a hypothesis or a working assumption would have been to violate one of his most fundamental methodological principles.

Because Newton adopted an emission theory of light, we should not ignore the significance of his second principle – "that matter is impenitrable by matter" – for it is assumed in his concepts of transparency, opacity, and absorption. The impenetrability of matter underlies his idea that if light passes through a body, it must pass through its pores; and if a body is opaque or absorptive, light must run into its corpuscles.

About a decade earlier, when Newton was revising the "Observations" for the *Opticks*, he added a paragraph on matter theory at the end of Prop. 8, immediately after the new paragraph that incorporated the former Prop. 9 on the opacity of the least parts. In this well-known passage Newton wanted us to "understand that Bodies are much more rare and porous than is commonly believed." He reasoned that

> water is 19 times lighter, and by consequence 19 times rarer than Gold, and Gold is so rare as very readily and without the least opposition to transmit the magnetick Effluvia, and easily to admit Quick-silver into its pores, and to let Water pass through it. . . . From all which we may conclude, that Gold

[80] To my knowledge, only 's Gravesande, in the eighteenth century, noticed and corrected this problem; see Ch. 5, note 36.

has more pores than solid parts, and by consequence that Water has above forty-times more pores than parts. And he that shall find out an Hypothesis, by which Water may be so rare, and yet not be capable of compression by force, may doubtless by the same Hypothesis make Gold and Water, and all other Bodies as much rarer as he pleases, so that Light may find a ready passage through transparent substances.[81]

The doctrine of the porosity of bodies was a necessary consequence of the compositional theory of matter, which Newton held since his *anni mirabiles* but still forebore from introducing to the *Opticks* (§2.3). The doctrine was neither controversial nor in itself hypothetical, because it was deduced from a variety of phenomena. Indeed, in explaining the nature of transparency by his wave theory of light, Huygens had arrived at the identical conclusion: "The interstices or pores of transparent bodies . . . occupy much more space than the coherent particles which constitute the bodies."[82]

In the Latin translation of the *Opticks*, Newton added yet another paragraph to the end of Prop. 8 in which he finally set forth his hierarchical matter theory to explain the great porosity of bodies.[83] He offered his theory only as a hypothesis and did not insist on it, for "there are other ways of conceiving how Bodies may be exceeding porous. But what is really their inward Frame is not yet known to us."[84] Once Newton's hierarchical theory was published and known beyond the coterie of Newtonians, its influence was largely on matter theory itself. The interpretation of his theory of colored bodies was scarcely altered. The underlying role of the hierarchical theory in his theory of colored bodies was subtle and could be uncovered only by careful analysis.

3.4. NO RATIONAL DOUBTS

In concluding the theory of the colors of natural bodies, Newton made a rare public concession about doubts that may be legitimately entertained about his theory: "I see nothing materiall that rationally may

[81] *Opticks*, pp. $_2$69–70. Newton already alluded to this problem in a suppressed conclusion to the first edition of the *Principia; Unpublished Scientific Papers*, pp. 332, 345.

[82] Huygens, *Treatise on Light*, p. 31; *Oeuvres complètes*, 19:482. Newton's argument closely parallels Huygens's in the *Traité de la lumière*, a similarity that has not previously been noted; *Oeuvres complètes*, 19:482–3; *Treatise*, pp. 31–2. Huygens sent Newton a copy of the *Traité* in February 1690, when he was in the midst of composing the *Opticks*; see Harrison, *Newton's Library*, no. 822.

[83] See Ch. 2, note 95.

[84] *Opticks* (Dover), p. 269.

be doubted of excepting this Position, That transparent corpuscles of the same thicknesse & density with a Plate, do exhibit the same colour."[85] This "position" is of course the essence of his theory. Newton himself concedes only two possible problems: "And this I would have understood not without some latitude, as well because those corpuscles may be of irregular figures, & many rays must be obliquely incident, & so have a shorter way through them then the length of their diameters," as also because the narrowness of the aether within the body may alter the reflections from the corpuscle; he immediately rejects the latter.[86] Nonetheless, this concession hints at some underlying unease with the theory, as does the sketch of an alternative proof in the draft of a preface to the *Opticks*. This attitude is in strong contrast to his extreme confidence in the theory of color.

Why, we may ask, was Newton so sufficiently convinced of the truth of his theory of colored bodies that despite various problems he considered it to be a theory rather than a hypothesis? The method of transduction, far from discrediting or rendering the theory suspect, as it does from our vantage point, was inherent to the mechanical philosophy and widely utilized. For Newton, it was a variety of induction. The principal conclusions established by transduction ranged from the very reasonable to the totally convincing: Reflection is caused by a difference of optical density, all corpuscles of matter are transparent, opacity arises from reflection within the parts of bodies, and the corpuscles of bodies exhibit the same color as a thin film of the same size and density. Newton proved the first claim by a large number of experiments and observations; the second and third were already demonstrated by Hooke; and the fourth was supported with a variety of phenomena, which, Newton argued, were effects of the same kind. Thus, in Newton's eyes the theory was derived from experiment and observation.

Though not a mathematical theory, it was founded on a mathematical description of Newton's rings and could even predict the size of the corpuscles of bodies of different color. With improvements in the microscope, Newton expected that his predictions could be tested. The quantitative and predictive aspects of his theory would have reinforced his confidence in it. The theory's comprehensiveness would have been another favorable feature. Newton began his account of colored bodies in the "Observations" by proclaiming that he

[85] Add. 3970, f. 515^v; Birch, 3:303.
[86] Ibid.

would "consider how the Phaenomena of thin transparent Plates stand related to those of *all other naturall bodies.*"[87] This, as we have seen, was not an idle boast, for no instance of the coloration of bodies then known escaped the reach of his theory. Of course Newton recognized there were some problems with the theory, such as transparent colored bodies that were the same color by reflection and transmission, a phenomenon that never occurs in thin films. But, as he well knew after wrestling with the moon's motions for so many years, no scientific theory is without its unsolved problems.

In proper Newtonian procedure, having presented an analysis of his analogical theory of colored bodies, I will now briefly synthesize its various elements so that we can get a coherent view of it and more clearly distinguish between what he considered to be demonstrated and hypothetical. It should be borne in mind that Newton was careful not to describe conjectured details of the model, so that some aspects of my account are only implicit in his account or even arbitrary. Newton builds colored bodies from optically absorbing primordial particles and pores. In Figure 3.2 the solid circles are the primordials, which are gathered into small corpuscles of the second order; these are in turn compounded into still larger corpuscles, and then compounded once again.[88] The corpuscles of the highest order composition form macroscopic bodies. Newton's theory explains how the color of a body is produced by the highest order corpuscles just like a fragment of a thin film. All the specific features of my model, such as the number of compositions and the relative size, shape, and arrangements of the particles and pores, are arbitrary or hypothetical, for Newton says nothing about them. Transduction does not allow him to go further. All the particles and clusters within the largest corpuscles represent "the more secret & noble works of Nature." We can see directly from the figure why Newton asserted there was very little matter in bodies.

When incident light rays *I* are incident upon the body, they enter the highest order corpuscles and proceed to *r* at the second surface. Depending on the size of the corpuscle, rays of certain colors *R* are reflected back out of the corpuscle and rays of all other colors – the complement – *T* are transmitted. The transmitted rays continue to

[87] Add. 3970, f. 512^r; Birch, 3:296; italics added.

[88] In describing his hierarchical theory in a suppressed conclusion to the first edition of the *Principia,* Newton described the least particles as forming "very long and elastic rods [*virgas praelongas et elasticas*]"; *Unpublished Scientific Papers,* pp. 328, 341. Nor are the higher order corpuscles necessarily spherical, for they "may be of irregular figures"; see note 86.

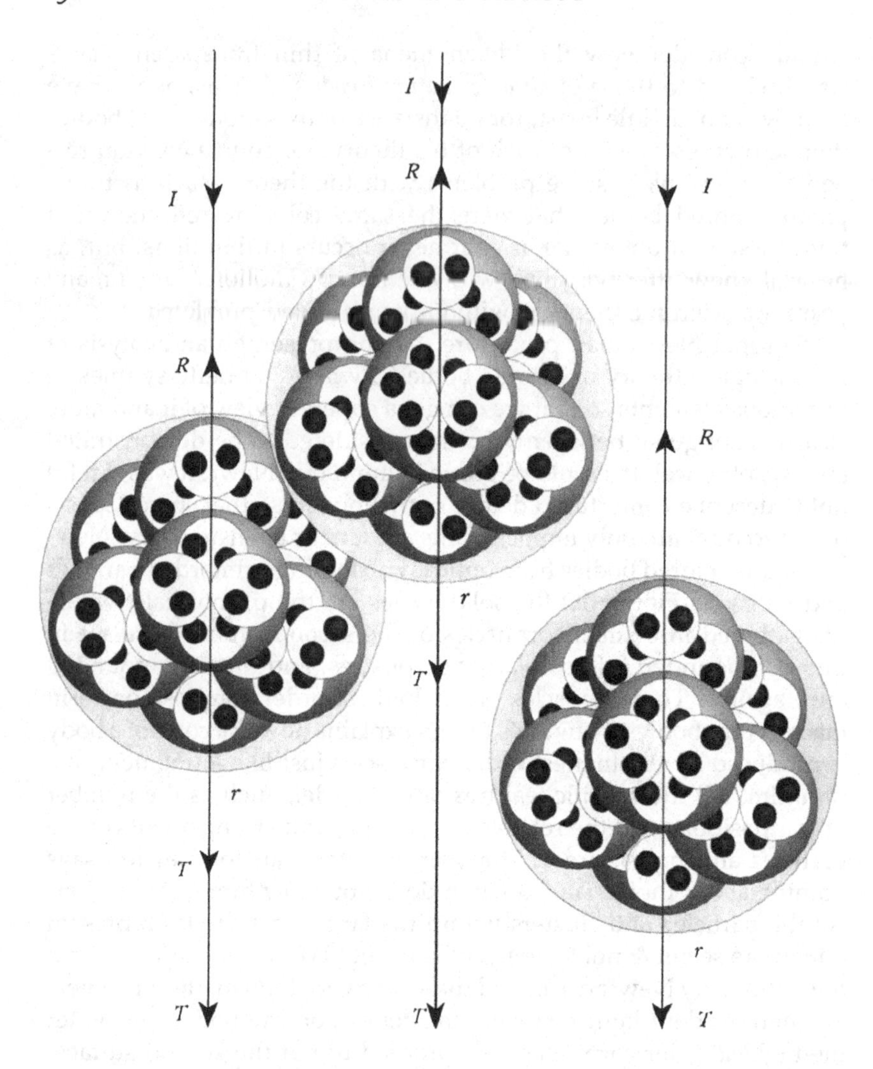

Figure 3.2. Newton's hierarchical structure of matter and his theory of colored bodies. The black spheres represent the lowest order corpuscles, which are compounded three times to form the large corpuscles that make up colored and transparent bodies. The shape, size, arrangement, and number of compositions of the corpuscles are all chosen for the sake of illustration.

move through the body. If the body is opaque, they will be refracted in the pores from corpuscle to corpuscle until they eventually collide with a primordial and are absorbed. But if the body is transparent, the rays proceed in a straight line through its pores and corpuscles until they emerge (except for those which chance upon a primordial) and produce the body's color by transmission *T*. By comparing the color of a body to the colors in Newton's rings, its order can be found; and if the optical density of the corpuscles can also be discovered, then the size of the corpuscles is readily determined from the table of colors (Figure 2.8). The reflected rays *R* are responsible for the body's reflected color.

Although the complementary nature of the reflected and transmitted colors, which is inherent in the model, is fine for explaining Newton's archetypal phenomena, it cannot explain the most common sort of colored transparent body, those that are the same color all around. If the corpuscles are less than a certain size (corresponding to the central spot in Newton's rings), they reflect no light and are therefore in principle invisible. Thus the inner works remain "secret," except, we may add, for the primordials. If the magnification were great enough, the primordials would be black, though Newton never ackknowledged this consequence.

Newton's model assumes that there are primordial particles, which have the property of absorbing light, and that these particles, together with the transparent pores, compose larger corpuscles, which possess the properties of reflecting and refracting light. He does not in his analogical theory attempt to explain the origin of these properties. That, he insisted, requires an hypothesis and his aim was only to demonstrate from experiment that those properties exist. We can now recognize another reason why Newton did not consider his analogical theory to be hypothetical. He did not invoke any conjectured aethers, spirits, or microscopic forces that were not deduced from experiment to explain the colors of bodies. These were left to his "Hypothesis" and queries.

How, then, would I explain the tacit historical judgment that Newton somehow went wrong, despite his apparent care in establishing the theory? It is only from a modern perspective that we can consider that he actually went wrong. Basing our judgment on the theory of white light and color and especially the theory of motion and gravitation, we expect him to be nearly infallible. The theory of colored bodies was, we have seen, a perfectly good one for its day. In its weak points Newton simply shows himself to be a more typical scientist of the seventeenth century. In the first, and perhaps most important, place, he had attempted in his phenomenological theory to encompass too broad a range of instances of coloration within one explanation. He then had to force some phenomena to fit his misguided generalization that the transmitted colors are the complement of the reflected ones. This then served as the observational basis for the analogical theory. We should not, however, forget that Newton's desire to formulate a universal theory, which failed him here, served him so well in the *Principia*.

In the second place, the method of transduction, which was so widely adopted by the mechanical philosophers, was simply not up to the task of treating the colors of natural bodies. Newton had in effect reduced optical properties and processes to mechanical ones, and he formulated his theory with a shrewd awareness of the optical phenomena and possible mechanical explanations. His experiments and observations really could tell him little or nothing about the invisible structure of matter responsible for the optical phenomena. Here he was guided by his mechanical theory, and this led to his problems with assigning particular properties to particular structures or parts. Perhaps the oddest feature of Newton's theory in our eyes is the near circularity of its analogical formulation: Colored bodies are composed of parts that are macroscopic pieces of colorless, transparent bodies (fragments of thin films of glass or water, and thus even of themselves), while these pieces already possess a microscopic structure of parts and pores. By further transduction he then deduced the microscopic structure of transparent and colored bodies. It is no wonder that the properties of the parts and the whole are so intertwined. If to us his theory looks something like a snake swallowing its tail, to the mechanical philosophers transduction was a fruitful method (as Newton himself proved in mechanics) intrinsic to their philosophy. Descartes, we should recall (§1.3), asserted that the properties of perceptible and imperceptible bodies are the same, and Boyle that the same mechanical processes occur "not only in the great masses, and the middle sized lumps, but in the smallest fragments of matter." Newton's theory of the structure of matter, its chemical transformations, and its interactions with light was mechanical, and so the method of transduction – where the whole and its parts followed the same laws – was unquestionably applicable. One could pass freely from the great masses of prisms and the middle-sized lumps of shattered glass to their smallest fragments.

In the eighteenth century Newton's methodology in establishing his theory of colored bodies went unquestioned, except for the one critique by Jordan. This should not be surprising, for scientific theories are rarely, if ever, overturned on methodological grounds. Moreover, most eighteenth-century scientists, both critics and supporters of Newton's theory, still utilized transduction, even if sometimes more cautiously and within a different natural philosophical framework. By the last third of the eighteenth century, when the strict mechanical interpretation embedded in the theory of colored bodies was widely rejected, Newton's theory was less persuasive to many. Newton's experiments and observations, his interpretations of them, and his principal conclusions were subject to criticism, but not his method. The subtle points most affected by transduction, such as the compositional theory of matter and the level of structure where reflection and absorption occurred, were of no great concern, probably because they were perceived as irresolvable. Subsequent generations quite fairly took the following as the principal tenets of the theory: Transparency and opacity were caused by differences in the size and density of the parts and pores, the colors of bodies were caused by their parts just as thin films produced colors, and some colors are reflected and others (their complement) are transmitted. Only the process of absorption and the compound nature of the colors of thin films and of bodies were not properly appreciated. These principles and their consequences were put to the test in the late eighteenth and early nineteenth centuries. Despite many negative results, the theory turned out to be extraordinarily robust, especially after a new generation of quantitative physicists succeeded in reinterpreting Newton's theory of colored bodies within the framework of Newtonian forces and fits.

4

THE THEORY OF FITS

Newton's theory of fits of easy reflection and transmission – his description of the periodic properties of light exhibited in thin films and thick plates – bears the unique stamp of its creator. In that theory Newton teaches that light rays are put into periodically recurring states (fits) that dispose them to be easily reflected or transmitted. From the time that he first proposed the theory in the *Opticks,* it has found few empathetic readers, let alone adherents. It has been judged by many to be obscure and by many others to be a physical speculation – "a strange hypothesis" – masquerading as a series of abstract propositions.[1] Very few have declared as enthusiastically as David Brewster that it "will ever be considered as one of the finest specimens of generalization," but even he had to acknowledge that "it has often been ridiculed by those who could not follow the train of his reasoning."[2] Thomas Young, who certainly was no supporter of the theory of fits, also recognized that it has "never been very generally understood."[3] By tracing the development of Newton's theory and placing it within the context of his methodology, we will be able to gain a number of insights into why it was formulated in a way so difficult to comprehend.

The theory of fits emerged out of Newton's investigation of the periodic colors of thick plates, which is set forth in Book II, Part IV, of the *Opticks* and now read by no one. The colors of thick plates were an entirely new phenomenon that Newton had discovered most probably in the early 1690s. Utilizing the concept of vibrations that he had

[1] Tiberius Cavallo, *The Elements of Natural or Experimental Philosophy* (1803), 3:207.
[2] Brewster, "Optics" (1830), p. 477.
[3] Young, *Course of Lectures,* 1:457.

devised earlier to explain the colors of thin films, he was able to account precisely for the observed appearances of the colored rings of thick plates. This not only confirmed the properties of vibrations by "general induction" but showed that they were a more universal property of light than he could previously demonstrate. By the fall of 1691, Newton had completed and written up the first phase of his investigation of thick plates, which, together with his research on diffraction, was to form the concluding book of the *Opticks*. Because of his successful explanation of thick plates by vibrations, he felt justified in elevating the status of the hypothesis of vibrations to a theory. At this time he drafted some propositions on the vibrations and their properties, though having abandoned the aether after the *Principia*, he now attributed the vibrations to the corpuscles composing reflecting and refracting bodies. Newton, however, soon suppressed these propositions, apparently because he recognized that corpuscular vibrations were as hypothetical as aethereal ones. He then completed the *Opticks* without any propositions on either vibrations or periodicity. Before the winter of 1691–2 ended, however, he had devised the theory of fits, drafted a new series of propositions on it, and revised the previously completed *Opticks*, which entailed rewriting the part on the colors of thick plates and explaining them in terms of fits.

With the theory of fits, Newton had succeeded in carrying out the dictates of his methodology by distinguishing the properties of periodicity from the hypothetical nature of the vibrating medium. He now declared periodicity to be a general property of light and not, as he had held for twenty years, one exhibited only in thin films and thick plates. Newton's "abstracting" of all hypothetical elements from the theory of fits (or so he believed) is largely responsible for the oft-commented-upon unintelligibility of the theory. Another factor contributing to its obscurity is that the theory was formulated after most of the *Opticks* was composed and then was inserted into the previously completed work. Thus the theory was neither well integrated nor fully motivated. In particular, its fundamental dependence on the colors of thick plates was apparent only to a diligent reader.

The key to appreciating the explanatory power of the theory of fits is Newton's calculation of the diameters of the colored rings of thick plates. It is a sophisticated piece of mathematical physics in which he develops a model capable of accurately predicting all the observed phenomena. Equally important for understanding the theory, one must recognize that Newton developed his explanation and carried out this calculation by means of corpuscular vibrations. Only afterward did he purge the hypothetical vibrations and recast it in terms of fits. Because the colors of thick plates and Newton's calculation play such a central role in the theory of fits – and this was recognized by the few supporters that the theory garnered – we shall follow it in some detail. In order to appreciate the significance of the colors of thick plates for the development of the theory of fits, we must first briefly look at Newton's efforts to conclude the *Opticks*.

4.1. COMPLETING THE *OPTICKS*

Newton wrote the books and parts of the *Opticks* at different times over about sixteen years. He began the first book in 1687 or so and did not complete the last book and its queries until 1703, shortly before it appeared in early 1704. Deducing the order and especially the date of composition of the extant worksheets, outlines, drafts, and final printer's manuscript requires meticulous textual and historical analysis.[4] Although I am confident of the essential chronology that I will set forth, it will no doubt require refinements as scholarly study of the *Opticks* continues. I will focus on the development of the colors of thick plates and theory of fits and treat the other portions of the *Opticks* only insofar as they illuminate the former.

The central document in my analysis is entitled "The fourth Book of Opticks. The first Part. Observations concerning the reflexions & colours of thick transparent polished plates" (henceforth cited as Book IV, Part I, to distinguish it from a projected "Fourth Book").[5] This, together with a second part on diffraction, formed the concluding book of the *Opticks* in its first complete state. Within a few months, however, Newton grew dissatisfied. He removed the part on diffraction, revised the part on thick plates, and put it into very nearly its final state as Book III, Part IV. For us, the significance of this document is that it captures the transition from the theory of corpuscular vibrations to the theory of fits. In its original state, Newton used vibrations in the corpuscles of bodies to explain the colors of thick plates,

[4] The manuscript from which the *Opticks* was printed was returned to Newton and is Add. 3970, ff. 17–78, 91–233, 339; despite the gaps in foliation it is complete. The composition of the *Opticks* is treated more fully in my "Beyond the dating game." See Westfall, *Never at Rest*, pp. 520–4; and Bechler, "Newton's search for a mechanistic model," pp. 20–3, for brief descriptions of the composition of the *Opticks*.

[5] The now scattered sheets that composed Bk. IV, Pt. I, form one continuous text, with the handwriting, catchwords, and watermarks matching perfectly; Add. 3970, ff. 202–5, 381–4, 344–5, 385–6, 306–7.

though he did not formally expound that theory. In revision he substituted an explanation based on his newly devised theory of fits. I will be able to demonstrate that in the six months between 27 August 1691 and 24 February 1692 Newton developed his theory of fits and composed the new propositions on it for the end of Book III, Part III, and subsequently revised Book IV, Part I, to make it Book III, Part IV.

From the manuscript of the *Opticks* and related papers, it is clear that until shortly before publication the two parts of the published Book I were separate Books I and II and the published Books II and III were then Books III and IV. In order to follow the successive stages in the composition of the *Opticks,* I adopt the manuscript's original division into books. This should create little confusion, as the published Book III is not divided into parts, and it is a simple matter to convert to the familiar arrangement of the published work with its one less book.

In about 1687 or 1688, Newton wrote the "Fundamentum Opticae" ("The Foundation of Optics" or "The Foundation of the *Opticks*"), which is the first draft of Book I of the published *Opticks.*[6] He started to revise this, but after writing just one page he abandoned the Latin and started anew in English, the language in which it was published.[7] In the published work this forms the two parts of Book I (recall that they were then still separate Books I and II). This essentially presents the theory of white light and color that he had developed in his *Optical Lectures* and optical correspondence in the early 1670s. Even though he took these writings as his foundation, this was a completely new exposition.

The composition of Book III (II in the published work) on the colors of thin films and natural bodies was at first much easier, for he simply lightly revised the three parts of the "Observations" of 1675, which then made up Book III, Part I through Prop. VIII of Part III.[8] Newton now had to decide how to conclude the *Opticks,* and this turned out to be no easy matter. He ultimately added twelve new propositions to Part III (of which the last ten presented the theory of fits), the entirely new Part IV on thick plates, and a brief final book on diffraction together with sixteen appended queries. Book III was essen-

[6] Add. 3970, ff. 409–10, 415–16, 394–8, 583–4, 425–6, 647–8, 407–8, 405–6, 403–4, 401–2, 399–400, 419, 422, 420–1, 411–14, 423–4, 417–18.

[7] Ibid., f. 302.

[8] Newton made so few changes in the text that he was able to mark up the manuscript from 1675 for his amanuenses to transcribe for the *Opticks.*

tially complete by February 1692, but diffraction continued to elude Newton, and he did not write the last book until 1703, shortly before publication.

Newton's own account of the composition of the *Opticks* in his "Advertisement" is somewhat vague and not entirely consistent, but we must bear in mind that it is only a rough recollection written about sixteen years after he began the work. In the Advertisement to the *Opticks*, he recounts:

> Part of the ensuing Discourse about Light was written at the desire of some Gentlemen of the *Royal Society*, in the Year 1675. and then sent to their Secretary, and read at their Meetings, and the rest was added about Twelve Years after to complete the Theory; except the Third Book, and the last Proposition[s] of the Second, which were since put together out of scattered Papers. To avoid being engaged in Disputes about these Matters, I have hitherto delayed the Printing, and should still have delayed it, had not the importunity of Friends prevailed upon me.[9]

The "part" written in 1675 refers to the "Observations," which was sent to the Royal Society on 7 December 1675 and read at its meetings in January and February 1676 (§2.2). This formed most of Book II (and throughout this and the next paragraph I will follow Newton's references to the books of the published work), so that "the rest" refers to Book I, which was written "about" 1687–8. In recalling that Book I was written "about" twelve years after the "Observations," Newton does not enter into such a fine distinction as that between the first draft (the "Fundamentum") and the text of the *Opticks* itself. Surely the enterprise was under way by June 1689, when Huygens met with Newton, for he then told Huygens that a work on color by him would appear.[10]

Turning now to Book II, we find that his recollections in the Advertisement are similarly incomplete: He "since" added twelve new propositions to Part III of Book II, whence I have added the "s" to "Proposition"; and he altogether ignores the new Part IV of Book II. The manuscripts unambiguously show that he added the last ten propositions to the end of Part III all at once. Nonetheless, all the evidence supports the thrust of the account in the Advertisement, namely, that he wrote Book III and the new material for Book II after he had completed Book I and the earlier portion of Book II. One significant omission from Newton's account is that he does not say

[9] *Opticks*, p. [iii].
[10] Huygens to Leibniz, 24 August 1690; Huygens, *Oeuvres complètes*, 9:471.

when he revised the "Observations" for the *Opticks* (Bk. II, Pts. I, II, and much of III), perhaps because it went so quickly. Let us now attempt to add some refinements to Newton's recollections.

In 1675, the "Observations" was supplemented by the accompanying "Hypothesis," which (as we saw in §2.3) presented his speculations on the nature of light and aether and the causes of reflection, refraction, color, and the newly discovered phenomenon of diffraction (called "inflexion" by Newton). Initially he planned to follow the end of Part III with new research on diffraction,[11] but he soon conceived of a more comprehensive scheme, a projected "Fourth Book" that was modeled on the "Hypothesis." In fact, the first outline or sketch for the "Fourth Book" (although not titled as such) appears to be a summary of the "Hypothesis" put in propositional form. This relation, I believe, accounts for the fact that this is the only manuscript related to the first edition of the *Opticks* that utilizes an aether (a "certain medium" or "Medium of activity") to account for optical processes and sensation.[12] The actual drafts of the "Fourth Book" invoke fundamentally different physical causes than the "Hypothesis," for the aethereal density gradients and vibrations are replaced by short-range forces ("a certain power whereby the parts of bodies act . . . at some little distance") and vibrations excited in the corpuscles of bodies.[13] The "Fourth Book" also assigns a more prominent role than the "Hypothesis" to diffraction, which Newton had learned about just as he was composing the "Hypothesis."

There are three successive drafts of "The Fourth book concerning the nature of light & ye power of bodies to refract & reflect it" – to take the title of the intermediate version.[14] The final draft remains incomplete. It is important to recognize that these papers are an outline of a projected "Fourth Book" and not, as has sometimes been misunderstood, drafts of the *Opticks* itself. Newton was sketching out, through a series of propositions without proof, what he hoped to establish about the relation between various phenomena and their causes. For instance, he hoped to be able to explain diffraction by the same mechanism (namely, vibrations in the corpuscles of bodies) that he used to explain the colors of thin films, as can be seen in Prop. 13; but he abandoned this approach. Much has been made of the "hypotheses," as Newton called them, which conclude the latter two

[11] See, for example, Add. 3970, ff. 371–2, 377–8.
[12] Add. 3970, f. 374^{v}.
[13] Ibid., f. 342^{r}.
[14] In order of composition, these are ibid., ff. 342–1, 337–8, 335–6.

versions of these sketches.[15] These, as the entire sketch, reveal a great deal about the speculative schemes that Newton was then employing to explore and explain nature, such as "All solid bodies are atoms." We cannot, however, seriously use these documents to examine his methodology, for he had no intention of publishing them as they stood and made no attempt to demonstrate the propositions.

The propositions of the projected book are an odd assortment. Some Newton judged to have been demonstrated by experiment and, with some minor differences, were propositions in the "Observations" and the *Opticks*, such as the following:

Prop. 5. Refraction & Reflexion as well as Inflexion are performed wthout the rays impinging upon the solid parts of the body.
Prop. 6. Pellucid bodies are sufficiently porous to transmit a competent quantity of light every way in right lines.

A good number of the propositions dealt with the vibrations excited by light particles in matter and were like the speculations in the "Hypothesis":

Prop. 10. The rays of light in being refracted or reflected do mutually cause an agitation in ye parts of the refracting or reflecting bodies.

. . . .

Prop 12 The motion excited in pellucid bodies by the impulses of the rays of light is a vibrating one & the vibrations are propagated every way in concentric circles from the points of incidence through ye bodies.
Prop 13 The like vibrations are excited by ye inflected rays of light in their passage by ye sharp edges of dense bodies: & these vibrations being oblique to the rays do agitate them obliquely so as to cause them to bend forwards & backwards wth an undulating motion like that of an Eele.
Prop 14 As the oblique vibrations excited by inflected rays do agitate the rays sideways wth a reciprocal motion so the perpendicular vibrations excited by the refracted rays do agitate ye rays directly wth a reciprocal motion so as to accelerate & retard them alternately & thereby cause them to be alternately refracted & reflected by thin plates of transparent substances for making those many rings of colours described in ye Observations of the third book.[16]

[15] Portions of these "hypotheses" are presented in Cohen, "Hypotheses in Newton's philosophy," pp. 179–81, and Westfall, *Never at Rest*, pp. 521–2, but they conflate two different drafts.

[16] Add. 3970, f. 335^{r,v}. An earlier version contained a deleted proposition that anticipates the chemistry of light (described in Part II) that was developed about a century later based on short-range forces or affinities: "Prop 17 If ye rays of light be bodies they are refracted by attraction [of] the parts of refracting bodies by some such principle as the parts of acids & alcalies rush towards one another & coalesce" (f. 338^{r}).

All of these propositions with the exception of 13 were included in the *Opticks*, where they were either labeled a hypothesis or relegated to the queries. Newton no doubt suspected that they were true, but it is equally true that he never considered them to be proved by experiment.

After Newton penned these speculations, he set them aside and turned to more empirical investigations. He continued his experiments on diffraction and also instituted an extensive series of experiments and calculations on the newly discovered phenomenon of the colors of thick plates. Because the colors of thick plates was not mentioned in any version of the "Fourth Book," despite numerous opportunities to introduce it, we may conclude that it was most likely discovered, or at least investigated, afterward. Having carried out these new investigations, Newton decided to conclude the *Opticks* with them and composed a Book IV with Part I on the colors of thick plates and Part II on diffraction.[17] He also worked out an ending for Book III, Part III: Several paragraphs were added to the end of Prop. VIII, and two new propositions, IX and X, on the nature of reflection and refraction, were added. Books I and II were already complete except for one proposition, and a fair copy had already been made. Newton arranged for a fair copy of Book III to be made by amanuenses, and he wrote out Book IV himself. This initial state of the *Opticks* with its Book IV has hitherto been unknown to scholars.

In about the fall of 1691 the *Opticks* was complete. If Newton intended to append his queries at this time, there is no evidence of it. He did not long remain satisfied with the work as it stood. By February 1692, Book IV was completely dismembered. Book IV, Part II, on diffraction, was suppressed and removed, while Part I was revised and shifted to Book III, which was then essentially put into its published state. To accomplish this revision, Newton composed ten new propositions, XI to XX, on his theory of fits and tacked them onto his theory of colored bodies in Book III, Part III. He also revised Book IV, Part I, on the colors of thick plates – now Part IV of Book III because of their affinity to the colors of thin films – and explained them in terms of fits while redoing the calculations. By carrying out only a minimal revision – adding the propositions on fits to the others already in Part III and rewriting only portions of Part IV – the nature of the theory of fits and its relation to the colors of thick plates was

[17] Bk. IV, Pt. II (ibid., ff. 79–90), was entitled "The Second Part. Observations concerning the inflexions of the rays of light in their passage by the surfaces of bodies at a distance." For Pt. I, see note 5.

Table 4.1. *Newton's correspondence on eclipse colors and successive conclusions for Book III, Part III,* Opticks

Document	Contents
The Fourth Book	Speed of different colors varies
10 Aug. 1691, Newton to Flamsteed	Inquiry about eclipse colors
27 Aug. 1691, Gregory to Newton	No eclipse colors
ff. 348–9	Thick plates, corpuscular vibrations, no eclipse colors
7 Nov. 1691, Gregory to Newton	Cassini report on eclipse colors, red slower
ff. 339–40, 363–4*	Draft propositions on fits
ff. 365, 370	Fits; red slower
24 Feb. 1692, Flamsteed to Newton	No eclipse colors

*Ff. 339–40 and 363–4 fall between ff. 348–9 and 365, 370, but there is no evidence whether they were written before or after 7 Nov. 1691.

obscured. This decision probably contributed to the theory's poor reception.

It is possible to establish a chronology for Newton's completion of the research on the colors of thick plates, his creation of the theory of fits, and his final revision of the end of Book III. Though it is always exciting for an historian to establish absolute dates, as I do here, in this case it is the relative dates of the various documents that are important and allow us to follow Newton in the formulation of his new theory. In any case, my dating of the composition of the *Opticks* only serves to confirm what other scholars have estimated, "the early 1690s." In order to show that the revision of Book III and the development of the theory of fits took place between 27 August 1691 and 24 February 1692, I must take somewhat of a detour and consider Newton's ideas on the velocity of light rays of different color (my argument is summarized in Table 4.1).

As early as 1666, Newton had developed a mechanical model of refraction that allowed him to explain chromatic dispersion and calculate its magnitude by assuming that the particles of red light move faster than those of violet. To develop new mechanical models of optical phenomena, these fundamental parameters continued to interest him in the "Fourth Book." He assumed there that the vibrations excited by the particles or "rays" of each color propagate with the same velocity, but that the "rays of light differ in swiftnesse."[18] In August 1691, Newton recognized that this assumption could be tested by observations of eclipses of the moons of Jupiter, for when a satellite disappears behind a planet the slowest color should be seen last. On 10 August, he wrote John Flamsteed, astronomer royal and his long-time correspondent, to ask if Flamsteed had observed any change of color (either to red or blue) in Jupiter's satellites immediately before they disappeared.[19] Flamsteed's initial reply came through David Gregory, who, after visiting Flamsteed in Greenwich, wrote Newton on 27 August that "he said that he never observed any change of Colour in the light of the appearing or disappearing Satellits of Jupiter. But that hereafter he shall advert if ther be any such thing."[20]

Newton took heed of Flamsteed's observation in the earliest extant draft of the new propositions added at the end of Book III, Part III. This draft (ff. 348–9) is written over a letter from 9 January 1690 to Newton and contains the first version of some additions to Prop. 8, the new Prop. 9 (here number 10), and two propositions (11 and 12, soon to be abandoned) on the vibrations excited in the corpuscles of bodies by light rays.[21] In Prop. 12 he states that the times of the vibrations excited by rays of different colors would follow the same ratio as the tones of an octave, "provided that all the rays of light be equally swift. For if they differed much in swiftness that difference would be discoverable by the eclipses of Jupiters satellites."[22] This draft, then, must be some time after Gregory's report of 27 August 1691. Moreover, in his draft of Prop. 10 – "There is one common cause of Reflexions & Refractions" – he appeals to the colors of thick plates. He first invokes the colors of thin films, for "tis doubt[1]ess the same cause wch at one thickness of the plate makes it to be reflected & at another to be transmitted & refracted." Then, in a passage that has no equivalent in the published *Opticks*, he invokes the evidence of thick plates: "And as in thin transparent plates so in thicker transparent bodies it may depend upon the thickness of ye body whether any ray shall be reflected or transmitted after it has past through y^t^ thickness."[23] Thus some time before this draft was composed, New-

[18] Prop. 19, ibid., f. 335^v^; Prop. 16, f. 337^v^ is an earlier version.
[19] *Correspondence*, 3:164. Bechler, "Newton's search for a mechanistic model," p. 22, discusses this eclipse test, but by considering only some of the relevant drafts and correspondence he arrives at a different conclusion.
[20] *Correspondence*, 3:165.
[21] Newton already used this large sheet with its brief letter to him (ibid., 3:66) twice, for jottings on theology and then on geometrical optics.
[22] Add. 3970, f. 349^r^. Prop. 12 is discussed in §4.3.
[23] Ibid., f. 349^v^–8^v^.

ton's research on thick plates was sufficiently advanced for him to know that the phenomena of thin films and thick plates were analogous and had the same cause. Book IV, Part I, must have been finished or very nearly so, as he was trying to end Part III with material in part derived from that research.

Newton ends another draft (ff. 365, 370) of the new propositions for Part III, with an incomplete proposition:

> Prop. 12. The most refrangible rays are swiftest.
> For the light of Jupiters Satellites is red at their immersion[24]

In this manuscript the theory of fits had already been formulated. It also contains drafts of Props. 10 and 11, and on textual and conceptual grounds there can be little doubt that it is later than folios 348–9, which we just considered. How, then, can we explain the reversion to the view – supported, no less, by eclipse observations – that the velocities of different colored light rays differ, though now he holds violet rays are fastest, contrary to his earlier view?[25] David Gregory, ever devoted to serving Newton, had not forgotten the question of eclipse colors, for on 7 November 1691, he reported to Newton: "Mr Cassini assures a gentleman of my acquaintance at Paris that befor the immersion the Satellites appear red and contrariwise at the emersion."[26] Newton had evidently accepted, at least temporarily, the reliability of Cassini's observation.

More news on eclipse colors and another change of view were to follow. On 24 February 1692, Flamsteed, who had promised to observe eclipse colors more carefully, informed Newton that "I cannot say that I ever saw any change to a blewish colour or red."[27] Newton always justly respected Flamsteed's skill as an observer, and he took this as the definitive judgment, for he never again adopted the assumption of different velocities for different colors.

Now let us return to the task at hand and apply this sequence of events and documents to dating the various stages in the composition of Book III. The earliest of three attempts to end Book III, which is

[24] Ibid., f. 370^v. The proposition ends abruptly at the bottom of the sheet with "immersion" as the catchword.
[25] In one earlier passage, in the first of the drafts of the "Fourth Book" (ibid., ff. 342^v–1^v), Newton tentatively assumed that violet rays are faster than red in the ratio of 81 to 80. I do not understand this passage or why he made this unusual choice of velocities, but he deleted it and did not carry it (or an equivalent) over into the later drafts.
[26] *Correspondence*, 3:171.
[27] Ibid., 3:202.

represented by the various drafts of the "Fourth Book," was undertaken most likely before 27 August 1691, when Gregory wrote Newton about his visit with Flamsteed. The next attempt, when he was using corpuscular vibrations in folios 348–9, was written between Gregory's letter of 27 August and his next letter with Cassini's observations on 7 November. This dates an important intermediate stage in the development of the theory of fits. The last of his endings for Part III, which contains the draft of the theory of fits (ff. 339–40, 363–4), was composed by 24 February 1692, when Flamsteed himself wrote Newton on eclipse colors, because it can be shown on textual grounds to precede folios 365, 370, which were written between 7 November and 24 February.

Because the draft propositions on fits initially cite the portion on the colors of thick plates as being in the "next" or "following" book, whereas folios 365 and 370 refer to it as being in "the next Part of this Book," it was during this same period that Book IV, Part I, was absorbed into Book III to become Part IV.[28] In fact, because the earlier revisions of Part IV refer back to these propositions as in "ye 3^{d} Book," whereas the later ones refer to "the third part of this book," Newton decided to combine the colors of thin films and thick plates into a single book while he was in the midst of revising Book IV, Part I, and formulating the theory of fits.[29] Thus we can confidently date the conception of the theory of fits, the final revision of Book III, Part III, and Book IV, Part I, and the latter's incorporation into Book III to the six months between 27 August 1691 and 24 February 1692.

The date of the first complete state of the *Opticks,* and in particular that of Book IV, Parts I and II, cannot be established as decisively. It was probably carried out around the time folios 348–9 were written, that is, in the summer or fall of 1691, for on this sheet Newton was attempting to compose an ending for Part III utilizing, in part, knowledge of the cause of the colors of thick plates. When did Newton first begin his attempt to end Book III and the *Opticks,* which culminated in his work on diffraction and thick plates? I would conjecture some time earlier in 1691, in the winter or spring, so that his last active period of optical research lasted six months to a year.[30]

[28] Compare, for example, the paragraph introducing the theory of fits, Add. 3970, f. 339^{r}, with its revision on f. 370^{v}.

[29] See the additions to Obs. 7 and 9 cited in notes 53 and 74.

[30] There is evidence that helps to establish an earlier possible date. One version of the "Fourth Book" contains a proposition that alludes to the "sides" of light rays that Newton used to explain polarization in Iceland crystal: "Prop 17 Every ray hath length breadth & thickness distinguished from one another by different quali-

Thus far I have established that by February 1692 Books I and II and those portions of Book III that corresponded to the "Observations" were in final form, and the revision of the remainder of Book III had only been drafted. The final touches to the revision – in particular, the elimination of Prop. 12, which asserted that violet rays move fastest – and the preparation of the fair copy occurred some time after. Certainly the manuscript of the entire *Opticks*, except for the last book, was in publishable form by May 1694, when David Gregory spent several days with Newton in Cambridge. In a contemporaneous memorandum dated 5, 6, 7 May, Gregory jotted down a perfunctory description of the *Opticks*, but shortly afterward in his *Notes on Newton's Principia* he prepared a more elaborate and unambiguous account of Newton's manuscript. He inserted a description of the *Opticks* in his commentary on Prop. 96 in Book I, Section XIV, where a number of optical theorems are derived:

I saw *Three Books of Opticks*, written in the English language, which if printed would equal the *Principia mathematica*. The fourth [book], about what happens to rays in passing near the corners of bodies, is not yet complete. Nonetheless, the first three make a most complete work. The author observes that the progress of light rays occurs by jumps, and they possess fits [*paroxismos*], indeed indeterminate, of being sometimes refracted and sometimes reflected.[31]

Inasmuch as the *Opticks* was then still divided into four books, and not the three of the published editions, Book IV on diffraction was not yet complete.

In my account of the composition of the *Opticks*, I have focused on Book IV, Part I, and largely ignored its companion on diffraction.

ties. Prove this by the experiment of Island glasse" (Add. 3970, f. 338r). The polarization of light in Iceland crystal was discovered by Huygens and announced in his *Traité*, which Newton received in February 1690 or a little afterward. Perhaps Huygens had described it to him when they met during his visit to England in the summer of 1689; Huygens, *Oeuvres complètes*, 9:333, n. 1.

[31] "Vidi ego *tres Optici libros* anglico idiomate conscriptos, qui impressi princip: Math: aequarent. quartum de ijs quae radijs accidunt in transitu prope corporum angulos nondum absolvit. priores tamen tres opus absolutissimum constituunt. Observat auctor, Radiorum lucis progressum subsultim fieri. eosque nunc refringendi nunc reflectendi paroxismos obtinere, et quidem incertos" (Gregory, *Notae in Newtoni Principia*, Royal Society, MS 210, insert between pp. 55, 56). Gregory completed his *Notae* by January 1694, so that inserts were added afterward; see Cohen, *Introduction to Newton's 'Principia,'* pp. 189–90. In Gregory's briefer, contemporaneous memorandum, he simply notes: "I also saw 'Three Books of Opticks' "; *Correspondence*, 3:336, 338. This has hitherto been interpreted by historians to mean that the *Opticks* was then complete, but this interpretation is based on the mistaken identification of the three books with those of the published work.

Newton evidently long remained dissatisfied with it. When he shifted Book IV, Part I, to Book III, he also removed Part II on diffraction from the manuscript with the intention of revising it afterward. In May 1694, when Gregory visited him, he had still not carried out the revision. Word of the *Opticks* spread in scientific circles. In 1695, John Wallis, the Savillian Professor of mathematics at Oxford and doyen of English mathematics, undertook a campaign to persuade Newton to publish it. Although Newton's letters to Wallis do not survive, it is apparent from Wallis's letters to him that, besides his abiding fear of controversy, Newton was withholding his book because he judged the last part on diffraction to be still incomplete.[32]

In 1696, after spending most of his life at the University of Cambridge, Newton moved to London to assume the duties of warden, and then master, of the mint. He no longer had the time and, more importantly, the interest to carry out further experiments on diffraction. By November 1702, according to another memorandum by Gregory, he finally committed himself to publishing the *Opticks*.[33] He retrieved Book IV, Part II, written more than a decade earlier, from his papers and revised it for publication. At the conclusion of the last of the eleven observations of Book IV (which was no longer divided into parts) he explained:

> When I made the foregoing Observations, I designed to repeat most of them with more care and exactness, and to make some new ones for determining the manner how the rays of Light are bent in their passage by Bodies for making the fringes of Colours with the dark lines between them. But I was then interrupted, and cannot now think of taking these things into further consideration. And since I have not finished this part of my Design, I shall conclude, with proposing only some Queries in order to a further search to be made by others.[34]

The *Opticks* was now complete. Newton decided some time after Book IV was finished to combine Books I and II into one book with two parts, and then went through the manuscript and renumbered

[32] Wallis to Newton, 30 April 1695, *Correspondence*, 4:116–17; and also pp. 100, 115, 186, 188. See Newton's comment on disputes in the Advertisement to the *Opticks* at note 9.

[33] David Gregory recorded in his notebook that "On Sunday 15 Nov. 1702 He [Newton] promised Mr. Robarts, Mr. Fatio, Capt. Hally & me to publish his Quadratures, his treatise of Light, & his treatise of the Curves of the 2^{d} Genre" (Gregory, *David Gregory, Isaac Newton and Their Circle*, p. 14).

[34] *Opticks*, p. $_{2}$132.

Books III and IV to II and III.[35] The *Opticks* was published by 16 February 1704, when Newton presented a copy to the Royal Society.[36]

4.2. THE COLORS OF THICK PLATES

Newton's work on the colors of thick plates, undertaken when he was in his late forties, was his last major, successful optical investigation. It is as fascinating as his earlier investigation of Newton's rings as we follow him refining his measurements, deducing and testing laws, and formulating his theory. The surviving papers are perhaps unique among his optical corpus in the view they provide of his working out a theoretical explanation of a class of phenomena.

The heart of Newton's explanation of the colors of thick plates was a complex calculation that could predict the observed dimensions of the rings in all circumstances. He was able to accomplish this by extending to this new phenomenon the mathematical-physical concepts – most notably, periodicity – that he had formulated earlier for the thin films of Newton's rings and by introducing the concept of a path difference between two rays. However, he had to grapple with two fundamental differences between the phenomena of Newton's rings and thick plates: The films in the former were of varying thickness, whereas in the latter they were of constant thickness; and the circles vanished in the former after a thickness of about 1/10,000 inch, whereas in the latter they first appeared at ¼ inch. After resolving these (and other) problems, he developed sufficient confidence in his calculation that a small discrepancy between the observed and predicted values led him to redetermine the vibration lengths or intervals of the fits, the most fundamental parameter of his theory. The new value, close to the modern one, was significantly smaller than that which he had been using for twenty years and necessitated redoing not only his initial calculations in Book IV, Part I, but also his earlier work on thin films.

There was, however, a more profound reason for Newton to revise Book IV, Part I than simply to redo the calculations. He had, in the interim, devised the theory of fits to replace the hypothetical corporeal vibrations that he had used to solve the problem of thick plates and had at one time considered introducing in Prop. 12 to conclude Book

[35] The change of book numbers was made close to the time of publication because the last book was initially headed "The fourth Book of Opticks" and only afterward was "third" substituted; Add. 3970, f. 219.

[36] Newton, *Correspondence of Sir Isaac Newton and Professor Cotes* (1850), p. lxxi, n. 147.

III, Part III. My aim in this section will be primarily to present Newton's mathematical and physical explanations of the colors of thick plates and the changes it underwent in the transition from vibrations in Book IV, Part I, to fits in Book III, Part IV. It is important to recall the sequence of documents that I established in the preceding section: After completing the first state of his account of thick plates in Book IV, Part I, Newton devised the theory of fits in the draft folios 339–40 and 363–4, and then in the midst of revising Book IV, Part I, he incorporated that part into Book III. The manuscript of Book IV, Part I, literally served as the draft for Book III, Part IV, as Newton made virtually all his changes on it – hence its importance for my account. My presentation will be based on the published Book III, Part IV, and then I will turn to the draft to describe the development of his ideas.

In our century the colors of thick plates has vanished entirely from the historical literature and nearly so from the scientific. A complete explanation of the phenomenon according to the wave theory of light was given by George Gabriel Stokes in 1851, although Thomas Young and John Herschel already had some understanding of its production.[37] Because the reader is unlikely to be familiar with the phenomenon or its explanation, I will present a brief physical description. In Figure 4.1 a white screen *LCQ* is placed at the center *C* and perpendicular to the axis *CN* of a spherical glass mirror *MNVT*, which is silvered on the back side *NV*. If sunlight is admitted through a small hole in a window shutter (not illustrated) and then passes through a small hole *C* in the screen and is reflected from the mirror back to the screen, there will be seen four or five colored circles around center *C* like Newton's rings, but larger and fainter. The circles are in fact seen when the screen and window, or light source, are at any distance from the mirror, though they are most distinct when the screen and window are coincident at the center of the mirror.

The rings are produced by the interference of two beams of light

[37] Stokes, "On the colours of thick plates" (1856), *Mathematical and Physical Papers*, 3:155–96; Young, "On the theory of light," Prop. 8, Cor. 3, pp. 41–2; and Herschel, "Light," §§676–87, pp. 473–5. Because of the development in the 1950s of a simple interferometer based on a flat scatter plate, there has been a resurgence of interest in interference in scattered light. See Eugence Hecht and Alfred Zajac, *Optics*, pp. 378–81; and the review article, which covers research since Young, by A. J. de Witte, "Interference in scattered light." These recent works consider only the simple case of a flat plate and not the curved one that Newton and his successors used, so that Thomas Preston's *Theory of Light* (1901), pp. 203–7, is still useful. For historical accounts one must consult such older works as Joseph Priestley, *The History and Present State of Discoveries Relating to Vision, Light, and Colours* (1772), pp. 311–16, 515–20, but these are not particularly helpful; see note 150.

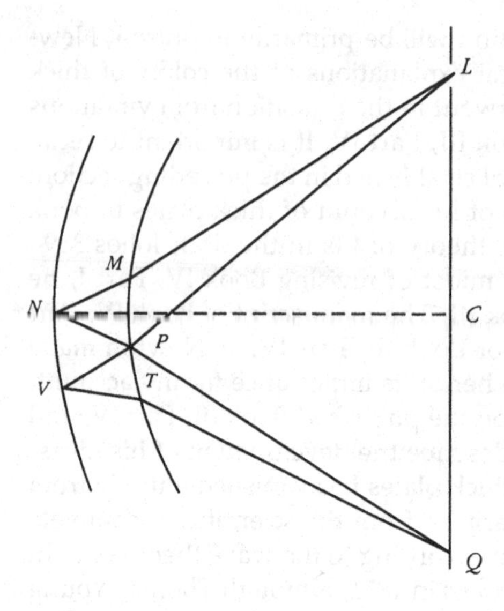

Figure 4.1. Explanation according to the wave theory and principle of interference of the colored rings produced by the spherical mirror *MNVT*. The two interfering rays, *LMNPQ* and *LPVTQ* are scattered at point *P*.

that are reflected from the rear surface and scattered by some imperfection on the first surface. Let us consider the more general case (eventually considered by Newton) of light incident obliquely on the mirror from a hole *L* a small distance away from the axis. One ray *LM* is refracted at *M*, reflected at *N*, and on its emergence at *P*, some portion of it is scattered to *Q*. Another ray *LP* is also scattered at *P*, and some portion of it enters the mirror at *P*, is reflected at *V*, and after refraction at *T* it passes to *Q*. The two rays will interfere because they are scattered from the same point and the path difference of the rays *LMNPQ* and *LPVTQ* is small. The calculation of the path difference is straightforward, and I will only state the resultant equation for the diameter *D* of the rings.

Let d be the mirror's thickness, S the distance of the hole from its inner surface *MP* (in general, S will be the mirror's radius R, for Newton almost always placed the hole in the screen at C), n the index of refraction, u the distance *LC* of the hole or source from the mirror's center C, and λ the wavelength. Then the diameters of the rings will satisfy the equation

$$\frac{d}{nS^2}\left(\frac{D^2}{4} - u^2\right) = \frac{m\lambda}{2},$$

where the rings will be bright for even values of the integer m and dark for odd values. For the simplest case, on which Newton founded his investigation, when the hole admitting the light is at the center ($u = 0$ and $S = R$), the preceding becomes

$$D^2 = \frac{4nR^2}{d} \cdot \frac{m\lambda}{2} = \frac{8nR^2}{d} \cdot \frac{mI}{2}; \qquad (4.1)$$

where I have also expressed the result in terms of Newton's vibration lengths $I = \lambda/2$.

In a virtuoso display of experimental skill and analogical, physical reasoning, Newton arrived at a result equivalent to this. While he conceived of the colors of thick plates by "the Analogy . . . with those of the like Rings of Colours" of thin films, it should not at all be imagined that this was a close analogy.[38] The equation describing the diameters of Newton's rings for perpendicularly incident light from Eq. 2.2 is

$$D^2 = \frac{4R}{n} \cdot \frac{m\lambda}{2} = \frac{8R}{n} \cdot \frac{mI}{2}.$$

Both kinds of rings obey the law of arithmetic progression, though, as Newton observed, the rings of thick plates are like the transmitted, and not the reflected, rings of thin films; that is, for a given integer m, the light and dark rings are interchanged in the two phenomena.[39] Otherwise, the relations are altogether different. The law governing thick plates could be derived only by an independent mathematical-physical analysis. For example, in thin films the square of the diameter of the rings varies linearly with the radius of the lens, whereas in thick plates it varies as the square of the radius of the mirror. It is important to recognize that Newton did not derive his result for the diameters of the rings in analytic form. Rather, by introducing the concept of path difference, he was able to transform the problem of thick plates to an equivalent one of thin films and calculate the diameters of the rings. Newton's calculation is difficult, but less so than many sections of the *Principia*. To appreciate the development of the theory of fits and its subsequent reception, even those readers who are not enamored of things mathematical should at least skim the following account. The theory of fits is, after all, a mathematical theory.

The numerous surviving drafts of Part IV, but especially Book IV,

[38] *Opticks*, Bk. II, Pt. IV, Obs. 7, p. $_2$95.

[39] Because the two interfering rays that produce the rings in thick plates traverse the same sequence of media, no change of phase has to be introduced as with thin films.

Part I, allow us to watch Newton develop the theory of fits and refine his measurements, but the earliest papers in which he conceived of the explanation by vibrations and scattering are not extant. Thus there is no evidence indicating how he arrived at the crucial insight that the rings are caused by scattered light. As he explained in his introduction to this part:

> There is no Glass or Speculum how well soever polished, but, besides the Light which it refracts or reflects regularly, scatters every way irregularly a faint Light, by means of which the polished surface, when illuminated in a dark Room by a beam of the Sun's Light, may be easily seen in all positions of the Eye. There are certain Phaenomena of this scattered Light, which when I first observed them, seemed very strange and surprising to me.[40]

This passage has misled those who have not carefully worked through the entire part into incorrectly claiming that Newton held that the phenomenon is caused by light scattered at the first surface. Although Newton did believe that some scattering occurs at the first surface, in his single-ray theory only light scattered at the second surface produces the rings. It was quite natural for Young and other nineteenth-century wave theorists to misinterpret Newton's explanation in this way, especially after the duc de Chaulnes observed that the rings become much brighter and more distinct when the first surface is fogged up or coated with a dry powder (§4.5). Nonetheless, this interpretation altogether misrepresents Newton.

The earliest papers with Newton's observations on thick plates no longer survive, so it is difficult to date his discovery of the phenomenon. It has been quite reasonably suggested that he discovered and investigated the colors of thick plates in the early 1680s, when he unsuccessfully attempted to construct a reflecting telescope with a glass mirror silvered on the back side in place of his earlier metal ones.[41] However, as Allen Simpson has noted, the mirror of the 4-foot telescope that Newton describes in the *Opticks* would have had a radius of 8 feet, whereas that used for his research on thick plates

[40] Ibid., Bk. II, Pt. IV, p. $_2$87. Most likely Newton arrived at his insight by recognizing that by ordinary specular reflection all light that enters the hole at the center should be reflected back through the hole, so that some other mechanism – namely, diffuse reflection or scattering – was required to reflect light to rings away from the center. David Gregory, in what is otherwise a puzzling remark in one of his memoranda, was no doubt referring to Newton's use of diffuse reflection to explain the colors of thick plates: "The angle of reflection is generally but not always equal to the angle of incidence [*Ang Reflectionis est plerumque non semper, aequalis angulo incidentiae*]" (Cambridge, 5, 6, 7 May, *Correspondence*, 3:335).

[41] Westfall, *Never at Rest*, p. 392.

had a radius of 6 feet or 5 feet 11 inches.[42] Though it is possible that he first observed the colored rings in the early 1680s, it would then appear that he postponed seriously examining them for a decade. None of the extant manuscripts on thick plates precedes the early 1690s, if we may judge by handwriting and watermarks, nor does he mention the phenomenon before 1691.

The thrust of the first seven observations is to establish the analogy of the colors of thick plates with those of thin films, and consequently that they are produced "much after the manner that those were produced by very thin plates."[43] In particular, his aim is to show that the colors of thick plates are periodic, with the same periods, or vibration lengths, as found for thin films. In this series of observations the screen, hole, and window are generally placed at the center of the mirror where the rings are most distinct. The sequence of colors is found to be the same as the transmitted rather than the reflected colors of Newton's rings; that is, the central spot is white (Obs. 2). Measurements of the diameters of the first four bright and dark rings established that the squares of their diameters follow the same arithmetic progression as those of thin films (Obs. 3). The diameters of the rings, it should be noted, are an order of magnitude larger than those of Newton's rings, so that they may be more precisely measured. He also reported here: "If the distance of the Chart from the Speculum was increased or diminished, the Diameters of the Circles were increased or diminished proportionally."[44] This simple result, which is a later addition to the manuscript, followed immediately from his theory that the rings are produced by rays that emerge from the first surface and proceed rectilinearly to the screen (Obs. 8).

Pursuing his demonstration of the analogy, he observed the rings in monochromatic light and found them to behave as those in thin films (Obs. 5). From some measurements of the rings in monochromatic light (which, as ever, he considered "hard to determine . . . accurately") and an awkward calculation, he determined the ratio of the squares of the diameters of the extreme red and violet rings, that is, the ratio of their vibration lengths.[45] The result, "3 to 2 very nearly,"

[42] Simpson, "The early development of the reflecting telescope," p. 284. Newton describes this telescope, which he tried to make "about five or six Years ago," in the *Opticks*, Bk. I, Pt. I, Prop. VII, pp. 77–8. Inasmuch as this passage can be dated no closer than to 1686 or 1687, he undertook this between 1680 and 1682.

[43] *Opticks*, Bk. II, Pt. IV, Obs. 7, p. $_2$95.

[44] Ibid., Obs. 3, pp. $_2$90–1; Add. 3970, f. 204^r.

[45] *Opticks*, Obs. 5, pp. $_2$92–3. The musical scale utilized for this calculation differs from all others in the *Opticks*. It is, however, only a hypothetical scale introduced to

he judged "differs not much" (less than 4%) from the 14 to 9 determined in Newton's rings.[46] As tenuous as this conclusion may be, it was crucial, for it established that the previously determined vibration lengths could be applied to thick plates.

The aim of Observation 7 is to establish the exact nature of the analogy and to show how the explanation of the colors of thin films may be extended to those of thick plates. To demonstrate that the colors of thick plates, like those of thin films, "depend upon the two surfaces of the plate of Glass whereof the Speculum was made, and upon the thickness of the Glass," Newton carried out a pair of simple experiments.[47] He rubbed off the quicksilver from the back of the mirror and saw the rings formed as before, except fainter, so that the quicksilver only increases the brightness of the rings and does not cause them. He also found that the rings were not produced in the single surface of a metal mirror. His observation that the color of the rings changed as the obliquity of the incident light varied was essential for the formulation of his explanation, for it informed him that these rings were like those of thin films of constant thickness rather than Newton's rings with varying thickness.

The reason why the colors of a thin film of constant thickness varied was that at any obliquity only rays of a particular kind are reflected while the others are transmitted:

> So the reason why the thick plate of Glass whereof the Speculum was made did appear of various Colours in various obliquities, and in those obliquities propagated those Colours to the Chart, was, that the rays of one and the same sort did at one obliquity emerge out of the Glass, at another did not emerge but were reflected back towards the Quick-silver by the hither surface of the Glass, and accordingly as the obliquity became greater and greater emerged and were reflected alternately for many successions, and that in one and the same obliquity the rays of one sort were reflected, and those of another transmitted.[48]

At this key point, Newton made a series of significant changes in the initial version, Book IV, Part I, to clarify his argument and to accommodate the newly developed theory of fits. The fundamental question that his explanation had to address was, as he put it in the

estimate the ratio of the vibration lengths of the extreme colors. When Newton later came upon the 2/3 power law for the intervals (Bk. II, Pt. III, Prop. XVI), he did not return to this observation and recompute the ratios.

[46] Ibid., p. $_2$93.

[47] Ibid., p. $_2$95.

[48] Ibid., p. $_2$96.

first version, "how these rings wch in thin plates by encreasing the thickness of the plates mixed with one another & thereby soon grew dilute & vanished, do now again after y^{e} glass is become $\frac{1}{4}$ of an inch thick appeare distinctly."[49] In the earlier version Newton answered this question in Observation 9, only after he had carried out a detailed calculation of the rings' size according to his theory. He then recognized that a more complete physical explanation was called for earlier, especially because there is a major difference, or disanalogy, in the production of colors in the two cases. Subsequently, he thoroughly rewrote his earlier account, couching it now in terms of fits instead of vibrations (or "agitations," as we shall soon see) and inserting it at the end of Observation 7. He began by explaining how the two sorts of rings differed:

> It seemed to me therefore that these Rings were of one and the same original with those of thin plates, but yet with this difference that those of thin plates are made by the alternate reflexions and transmissions of the rays at the second surface of the plate after one passage through it: But here the rays go twice through the plate before they are alternately reflected and transmitted; first, they go through it from the first surface to the Quick-silver, and then return through it from the Quick-silver to the first surface, and there are either transmitted to the Chart or reflected back to the Quick-silver, accordingly as they are in their fits of easie reflexion or transmission when they arrive at that surface.[50]

In Newton's explanation of thin films, the reflection and transmission of the rays occur at the second surface. Here they occur at the first surface, after the rays pass back and forth through the mirror.

The essence of Newton's explanation is straightforward, even though there are subtleties in interpreting the theory of fits and computing the rings' magnitude. In Figure 4.2, *LCQ* is a white screen with a small hole *C* placed at the center of the mirror *PP′VN* and perpendicular to the axis *CN*. Ray *CP* falls on the mirror at *P* and proceeds through it to the second surface, where it is reflected at *N*. The perpendicularly reflected rays pass through the first surface again at *P* and form the central white spot *C* on the screen *CQ*. Some rays, however, at *N* are scattered back to the first surface at *P′* and after refraction proceed to the screen, where they form colored rings at *Q*.

[49] Add. 3970, f. 345^{r}.

[50] *Opticks*, Obs. 7, p. $_{2}$97. Beginning with this passage, the remainder of Obs. 7 was added during revision of Bk. IV, Pt. I; Add. 3970, ff. 383^{v}, 384^{v}.

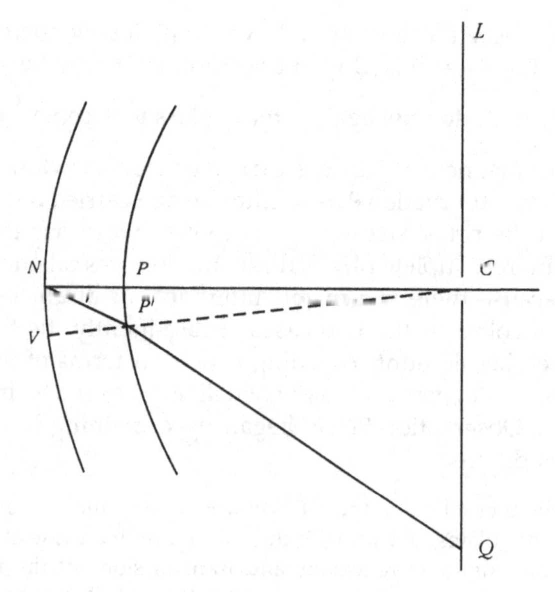

Figure 4.2. Newton's explanation of the colored rings produced by the spherical mirror *PP'VN*. The incident rays *CN* are scattered at point *N*.

The scattered rays depict colored rings, because with a longer, oblique path *NP'* than the perpendicular rays, some sorts will be in fits of easy transmission when they return to the first surface and so will emerge from the mirror, whereas others will be in fits of reflection and be turned back into the mirror. With this consideration, Newton has introduced the concept of path difference. When *NP'* differs from *NP* by half a vibration length, no rays will be transmitted, and there will be a dark ring. When that difference increases to a whole vibration, some rays will again be in fits of easy transmission and depict colored rings, and so on for a number of successions, until the rings of different color overlap and blend into white.

Qualitatively the concept of fits is equivalent to that of vibrations. The fits of easy reflection and transmission are periodically recurring dispositions of the rays that correspond to the rarefaction and compression of the vibrations, and the intervals of the fits correspond to the vibration lengths. Because Newton had already established that the same intervals apply to both thin and thick plates, he was able to show that the appearances predicted by his explanation agree with observation because the intervals of the fits at equal obliquities are greater and fewer in the less refrangible rays, and less and more numerous in the more refrangible, therefore the less refrangible at equal obliquities shall make fewer Rings than the more refrangible, and the Rings made by those shall be larger than the like number of Rings made by these; that is, the red Rings shall be larger than the yellow, the yellow than the green, the green than the blue, and the blue than the violet, as they were really found to be.... And therefore the first Ring of all Colours incompassing the white Spot of Light shall be red without and violet within, and yellow, and green, and blue in the middle, as it was found.[51]

These rings, then, are formed much like those of thin films of constant thickness, but let us examine the difference more carefully.

In the earlier version (Bk. IV, Pt. I), Newton directly replied to his rhetorical question, "If you now ask" how rings could so distinctly reappear in a quarter-inch plate when they vanish so quickly in a thin film:

I answer that the phaenomena of thin plates depend ... only on the length & obliquity of but one passage through ye plate: here it depends on the proportion of ye lengths & obliquities of the two passages. If the obliquities of ye two passages to ye reflecting surface be equal the ray returns back from ye reflecting surface after the same manner that it came to it, so as every where at equal distances from that surface to be alike affected in returning from that surface as in coming to it. And therefore if the length of the passage between the surfaces after reflexion be equal to the length of the passage between them before reflexion *the ray shall go out of the glass after the same manner & wth the same kind of agitation & degrees [of] velocity that it entred.* And this is the reason why all the rays of what sort of colour soever which enter ye glass perpendicularly & return back by the same lines to ye places where they entred, do there come out again as they entred, & go to ye white spot in the common center of the rings of colours on the chart & by their mixture compound ye white colour of that spot.[52]

Thus, the reason why rings can appear at such great thicknesses is that their appearance depends on the "proportion" or difference of two lengths. As we shall see in his calculation of the rings' diameter, Newton's concept here is closely akin to that of a path difference in the wave theory. The difference of the two path lengths *NP*, *NP'* determines whether the rays are transmitted or reflected, and not the

[51] *Opticks*, Obs. 7, p. $_2$98. By the "greater and fewer" intervals of the fits of red than violet rays, Newton means that the wavelength is larger and the frequency lower, so that fewer intervals will occupy a given distance.

[52] Add. 3970, ff. 345^r, 385^r; italics added.

absolute length of the path traversed by the rays, as in thin films. Although the paths are very large (approximately 35,000 intervals), the first ring appears when their difference is just one interval, and in this way the analogy to thin films is restored. Nonetheless, the analogy between the two phenomena is a limited one: The underlying theory of fits or aethereal vibrations is the same in both explanations, but the effective path that governs the appearance of the rings is altogether different. In the first phase of his solution, when he was still working with vibrations, he based his explanation of thick plates on symmetry considerations: If the lengths and obliquities of the rays before and after reflection at the second surface are equal, then the rays will behave identically at the same distance from the second surface. Thus perpendicularly incident rays that are in a state that allows them to enter the mirror freely will, after reflection, be able to leave it just as readily, and they will therefore form the central white spot. Those rays scattered at the second surface will return by a more oblique path; and when that path is longer by ½, 3/2, 5/2, and so on, of a vibration length, they will be unable to leave the first surface.

Newton's explanation of the colored rings of thick plates gained its true support from his calculation of their magnitude and not from his physical interpretation. Yet his qualitative account of the formation of the central white spot by the reflection of perpendicularly incident rays entailed a change in his concept of the periodic disturbances governing the transmission and reflection of light rays. Because his symmetry argument applies to rays of every color, and reflection occurs at a compression of the waves, then whatever the mirror's thickness and the vibration length, the vibrations must so precisely fit into that thickness that they will terminate at the second surface in the midst of a compression. It is simply impossible that all vibration lengths exactly divide any thickness in this way. Consequently, it is impossible for the state or phase of all vibrations before and after reflection to be perfectly symmetrical or identical at equal distances from the reflecting surface. Newton resolved, or rather eluded, the problem by the theory of fits, which invokes only a disposition to be reflected at some phase of a fit of easy reflection. In adopting this approach, however, he abandoned the strict relation that he had established in thin films between the path traversed, vibration length, and reflection or transmission. In thin films he had calculated the reflection and transmission of rays according to the strict mathematical law of arithmetic progression, where they are reflected or transmitted at integral multiples of half the vibration length.

When Newton revised his account of the formation of the central white spot in terms of fits for Observation 7, he eliminated the argument of the perfect symmetry of the two paths:

For the intervals of the fits of the rays which fall perpendicularly on the Speculum, and are reflected back in the same perpendicular Lines, by reason of the equality of these Angles and Lines, are of the same length and number within the Glass after reflexion as before by the 19th Proposition of the third Part of this Book. And therefore since all the rays that enter through the first surface are in their fits of easy transmission at their entrance, and as many of these as are reflected by the second are in their fits of easy reflexion there, all these must be again in their fits of easy transmission at their return to the first, and by consequence there go out of the Glass to the Chart, and form upon it the white Spot of Light in the center of the Rings. For the reason holds good in all sorts of rays, and therefore all sorts must go out promiscuously to that Spot, and by their mixture cause it to be white.[53]

Newton no longer requires, as in thin films, that at a given thickness all the rays of a particular color are reflected; he now refers only to "as many of these as are reflected." If he abandoned the strict law of arithmetic progression for his qualitative treatment of perpendicular rays, he still rigorously applies it to determine the path difference of the oblique rays that produce the colored rings, which is in fact the phenomenon under investigation. "These seem to be the reasons of these Rings in general," he concluded Observation 7, "and this put me upon observing the thickness of the Glass, and considering whether the dimensions and proportions of the Rings may be truly derived from it by computation."[54]

Despite the difference in the physical cause of the formation of the rings in thin and thick plates, Newton succeeded in drawing a formal analogy between them to calculate the rings' diameters in thick plates. He reduced the two passages of the ray through the mirror to the equivalent case of a single path, as in a thin film of uniform thickness, by considering only the ray *NP′* (Figure 4.2) scattered from the rear

[53] *Opticks*, pp. $_{2}$97–8. The reference to Prop. 19 should be to 20, for after Newton composed this passage and the propositions on fits, he added a new proposition 14. In the manuscript Newton initially began the first sentence of this quotation: "For the rays wch fall perpendicularly on y^{e} speculum & return back in [the] same lines, since they are in their fits of transmission at their entrance into ye glass & in their fits of reflexion at the points of their reflexion where they are reflected by the back side & return back in angles & lines equall to those of their entrance, they must" (Add. 3970, f. 383^{v}). Moreover, in the conclusion of this sentence in the manuscript Newton originally cited the proposition in "y^{e} 3^{d} Book," and only in further revision did he add "part of this."

[54] *Opticks*, p. $_{2}$99.

of the mirror, although it is the difference or ratio of the path lengths NP', NP that enters into his calculation. In turn, he cleverly transformed the case of a film of uniform thickness to a variable one as in Newton's rings.

In order to follow Newton's calculation and the transformation from a film of constant thickness to one of varying thickness, let us see how this could be explained by the wave theory. Although I just showed that the colors of thick plates are not at all explained in this way by the wave theory, it is a useful exercise that offers insight into Newton's calculation and shows more precisely how it is based on an analogy to thin films. The reason that his calculation can be interpreted or "justified" by a wave theoretical explanation of thin films is that the concept of path difference underlies both. That is the key concept. His specific device of increasing the thickness of the film by steps is simply a technique that allows him to carry out his computation as if he were dealing with a film of varying thickness. According to the wave theory (Eq. 2.3), a bright ring is transmitted in a thin film when $2nd_0\cos r = m\lambda$, where m is an integer, λ the true wavelength, d_0 the thickness of the mirror, and n the index of refraction.[55] For a perpendicularly incident ray a ring of order m will be formed such that

$$2nd_0 = m\lambda.$$

When that ray is obliquely incident, the next ring of order $m - 1$ will be formed such that

$$2nd_0\cos r_1 = (m - 1)\lambda,$$

where r_1 is the angle of refraction; and so on at angles $r_2, r_3, \ldots r_p$ for succeeding orders $m - 2, m - 3, \ldots m - p$ (where $p = 1, 2, 3, \ldots$).[56] Therefore for successive rings of decreasing order and increasing obliquity

$$\sec r_1 = \frac{m}{m - 1}, \sec r_2 = \frac{m}{m - 2}, \ldots \sec r_p = \frac{m}{m - p}. \quad (4.2)$$

We can also treat this example, as Newton will, like Newton's rings and consider the thickness to increase with increasing obliquity. If

[55] Newton treats the reflected rings as if they are transmitted ones because of the troublesome central white spot. This entails neglecting $\lambda/2$ in the total path traversed, but because the path is so large, it is of no quantitative significance.

[56] In films of uniform thickness the order of successive rings decreases rather than increases as in Newton's rings.

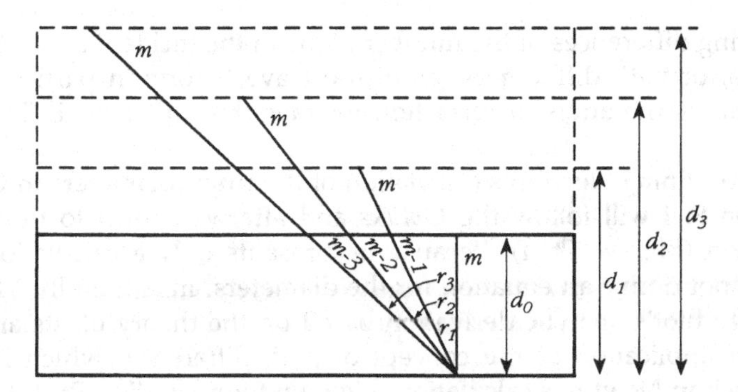

Figure 4.3. Newton's transformation of the appearance of rings in a plate of constant thickness to the equivalent case of a plate of varying thickness. He recognized that a plate of uniform thickness produces rings of decreasing order at exactly the same angles that one of increasing thickness produces rings of constant order.

we increase the plate's thickness d_0 by steps to $d_0/\cos r_1 = d_0\sec r_1 = d_1$, $d_0\sec r_2 = d_2, \ldots d_0\sec r_p = d_p$, then an oblique ray of the mth order would appear in this thicker film at the same angle r_p as the ring of order $m - p$ in the film of constant thickness. We then have the relation

$$\frac{d_1}{d_0} = \sec r_1 = \frac{m}{m-1}, \frac{d_2}{d_0} = \sec r_2 = \frac{m}{m-2}, \ldots,$$
$$\frac{d_p}{d_0} = \sec r_p = \frac{m}{m-p}. \tag{4.3}$$

Thus a plate of uniform thickness will produce rings of decreasing order at exactly the same angles that one of increasing thickness will produce rings of constant order. This is illustrated in Figure 4.3, where identical rings of decreasing order $m - 1$, $m - 2, \ldots$ are formed at thickness d_0 as rings of constant order m in plates d_1, $d_2, \ldots$ Though not a "natural" way to conceive of the generation of colored rings, this formal analogy to the tractable phenomena of Newton's rings allowed Newton to solve the problem of obliquely incident rays in films of constant thickness.[57] By this approach he is essentially cal-

[57] Needless to say, Newton did not work backwards from the wave theory to derive Eq. 4.2 and 4.3, and I will briefly indicate how he may have deduced them. Since I will draw on material to be presented later, my reconstruction may be better appreciated after completing §4.4. For the perpendicular ray, m vibrations (or fits) of length or interval I, occupy the thickness d_0, so that $mI_0 = d_0$; while for an oblique

culating differences of the number of fits in the incident and reflected paths, or path differences, as in the wave theory; in particular, he computes the angle of refraction for rays that differ by 1, 2, 3 . . . p fits.

In outlining Newton's calculation of the rings' diameters in Observation 8, I will follow the *Opticks* and afterward turn to the initial version (Bk. IV, Pt. I). Because he presents only a calculation and does not derive an equation for the diameters, in Appendix 2 I summarize Biot's analytic derivation based on the theory of fits and the direct application of the concept of path difference, which is only implicit in Newton's calculation. Newton found earlier (Pt. I, Obs. 6) that a thin film of air 1/89,000 inch thick transmits the bright yellow light of the first ring,[58] and also (Pt. I, Obs. 10) that in glass this same light will be transmitted at a smaller thickness, 1/137,545 inch, which is inversely proportional to the index of refraction $n = 17/11$.

Because the mirror's thickness is precisely ¼ inch, it will transmit the light of the 34,386th ring. Let us assume (Figure 4.2) that this is the light *NP* that is reflected perpendicularly from the rear at *N* and forms the central spot *C*. Citing Observation 7, Part I, and Props. XV and XX, Part III, on the law of the variation of the interval for obliquely refracted and reflected rays, Newton notes that the thickness d_p required to transmit the same light of the same ring or order will have a ratio to the mirror's thickness d_0 such that

$$\frac{d_p}{d_0} = \sec u,$$

where

$$\sin u = \frac{105n + 1}{106n} \sin i;$$

and u is "an Angle whose sine is the first on an hundred and six arithmetical means between the sines of incidence and refraction,

ray, $(m - p)I = s$, where I is the interval at an angle r_p and s is the oblique path traversed by the ray. Since $s = d_0 \sec r_p$ (a straightforward geometrical relation [see Figure 2.4], which is incorporated in the table in Obs. 19 of Bk. II, Pt. I, of the *Opticks*, and also cited by Newton in a draft Prop. 12, Add. 3970, f. 349^r), then $m/(m - p) = I/(I_0 \sec r_p) = \sec u \approx \sec r_p$. The last step follows from Newton's law for the variation of the interval of fits with obliquity, $I = I_0 \sec u \cdot \sec r$ (ultimately Prop. XV in the theory of fits). The angle u is very nearly equal to r_p and obeys Newton's law of 106-mean-proportionals for the variation of the thickness of a film with obliquity, $d = d_0 \sec u$ (Bk. II, Pt. I, Obs. 7, *Opticks*).

[58] The origin of the new value for the interval, which replaces the former 1/80,000 inch, will be explained shortly.

counted from the sine of incidence when the refraction is made out of any plated Body . . . that is, in this case, out of Glass into Air."[59] (The origin of this formula is explained in §4.4.) For convenience of calculation, Newton chooses to consider the rays to pass from glass to air rather than from air to glass, so that he now takes $n = 11/17$. Thus his angle i becomes that within the glass, or what I have hitherto called – and for consistency will continue to call – the angle of refraction r. Substituting $n = 11/17$ and changing his angle i to my r, we get $\sin u = ([11\frac{6}{106}]/11)\sin r$. Consequently, the angle u will be very nearly equal to r, and only at very large angles will sec u deviate significantly from sec r; for small angles Newton himself took sec u to be equivalent to sec r.[60]

Continuing with the calculation, Newton next considers the thickness of the glass to increase stepwise so that it bears to the mirror's actual thickness

> the proportions which 34386 (the number of fits of the perpendicular rays in going through the Glass towards the white Spot in the center of the Rings,) hath to 34385, 34384, 34383, and 34382 (the numbers of the fits of the oblique rays in going through the glass towards the first, second, third and fourth Rings of Colours,).[61]

Newton has thus deduced the relation $d_p/d_0 = \sec r_p = m/(m - p)$ that we derived in Eq. 4.3. From the ratios of fits, he successively computes sec u, u, sin u, and then sin i, where i is the angle of emergence into air, although Newton calls this the angle of refraction. Sin i is found by the sine law of refraction – $\sin r = (11/17)\sin i$ – and the law of 106-mean-proportionals, which yield $\sin i = (17/11\frac{6}{106})\sin u$.[62] The key steps of the computation thus far are summarized in the first two columns of Table 4.2.

To complete the transformation of the problem of rings of decreasing order in a plate of constant thickness to the equivalent case of

[59] *Opticks*, Bk. II, Pt. IV, Obs. 8, p. $_2$100

[60] See, for example ibid., Bk. II, Pt. I, Obs. 6, p. $_2$10; and the draft Prop. 12, Add. 3970, f. 349^r. Many of the approximations in Newton's calculation are more fully explained in Appendix 2.

[61] *Opticks*, Obs. 8, p. $_2$100; the two parenthetical clarifications were added in proof, as they are lacking in the printer's manuscript of the *Opticks*; Add. 3970, f. 210^r.

[62] The text (*Opticks*, p. $_2$100) mistakenly has sec u, but it is just a slip, for the subsequent computation is carried out using the correct sin u.

Table 4.2. *Summary of Newton's calculation of the diameters of the colored rings of thick plates and his observed values*

$\frac{d_p}{d_0} = \sec u$	$\sin i$*	$D = 2R \sin i$ (inches)	D, Obs. 3 (inches)
Opticks			
$\frac{34386}{34385}$	1172	1.688	$1\frac{11}{16}$ [1.688]
$\frac{34386}{34384}$	1659	2.389	$2\frac{3}{8}$ [2.375]
$\frac{34386}{34383}$	2031	2.925	$2\frac{11}{12}$ [2.917]
$\frac{34386}{34382}$	2345	3.375	$3\frac{3}{8}$ [3.375]
First version, Book IV, Part I			
$\frac{31000}{30999}$	12386	$1\frac{3}{4}$ [1.750]	$1\frac{2}{3}$ [1.667]
$\frac{31000}{30998}$	17518	$2\frac{1}{2}$ [2.500]	$1\frac{4}{11}$ [2.364]
$\frac{31000}{30997}$	21452	$3\frac{1}{15}$ [3.067]	$2\frac{8}{9}$ [2.889]
$\frac{31000}{30996}$	24772	$3\frac{7}{13}$ [3.538]	$3\frac{1}{3}$ [3.333]

*The sines are multiplied by 10^5 for the *Opticks* and 10^6 for Book IV, Part I.

rings of constant order in a plate of variable thickness, Newton indicates:

> And therefore if the thickness in all these cases be $\frac{1}{4}$ of an Inch (as it is in the Glass of which the Speculum was made) the bright Light of the 34385th Ring shall emerge where the sine of refraction is 1172, and that of the 34384th, 34383th and 34382th Ring where the sine is 1659, 2031, and 2345 respectively. And in these Angles of refraction the Light of these Rings shall be propagated from the Speculum to the Chart, and there paint Rings about the white central round Spot of Light which we said was the Light of the 34386th Ring.[63]

When the screen is placed at the center of the mirror, to a very close approximation the rings' diameters $D = 2R\sin i$, where the

[63] Ibid., Obs. 8, p. $_2$101.

mirror's radius R is 6 feet.[64] In columns 3 and 4 of Table 4.2, I summarize the calculated and observed values of the diameter, and in square brackets I give decimal equivalents. Newton concluded:

> Now these Diameters of the bright yellow Rings, thus found by computation are the very same with those found in the third of these Observations by measuring them... and therefore the Theory of deriving these Rings from the thickness of the plate of Glass of which the Speculum was made, and from the obliquity of the emerging rays agrees with the Observation....
>
> And thus I satisfied my self that these Rings were of the same kind and original with those of thin plates, and by consequence that the fits or alternate dispositions of the rays to be reflected and transmitted are propagated to great distances from every reflecting and refracting surface.[65]

Newton was justifiably satisfied with the agreement between theory and observation and the consequent confirmation of his extension of the theory of vibrations to the new phenomenon of thick plates. In demonstrating the theory of fits this would serve as his principal evidence that periodicity was a general property of light, or "propagated to great distances," and not something that was exhibited solely in the tiny realm of thin films. Previously he had established only that periodicity extends a few ten thousandths of an inch in thin films, barely beyond the realm of the microscopic. Now he had demonstrated that the same property continues for at least a quarter of an inch, safely in the macroscopic world. This calculation was the most complex quantitative derivation in all of his work on physical optics. Yet, as his papers show, this agreement was hard won and ultimately necessitated redetermining the vibration length or interval of the fits.

Observation 8 in the first version (Bk. IV, Pt. I) contains the calculation of the published *Opticks* as a set of revisions. Initially Newton performed the computation with the value for the vibration length, 1/80,000 inch. Because at this time he was using $n = 31/20$ for glass, the vibration length in the 1/4-inch glass mirror became 1/124,000 inch, so that the mirror transmits the perpendicularly incident light of the 31,000th ring.[66] Except for the different parameters, the calculation proceeded in exactly the same way as it would in the *Opticks;* the

[64] That the rings' diameters are proportional to their distance from the mirror is the "law" that Newton added to Obs. 3; see note 44.

[65] Ibid., Obs. 8, pp. $_2$101–2. The second paragraph was added during revision of Bk. IV, Pt. I; Add. 3970, f. 344^v.

[66] The change in index of refraction, which is quantitatively insignificant for this calculation, was made for reasons independent of Book II and the theory of fits.

results for the key steps are given in Table 4.2.[67] To calculate the rings' diameters (column 3), Newton took the radius of the mirror to be "six feet wanting $\frac{1}{2}$ an inch," rather than six feet.[68] The calculated values are about 5.5% larger than the initial observations in Observation 3 of what was then Book IV, Part I. Newton was not satisfied with such a large discrepancy and recalculated the diameters, assuming now that the mirror's radius was 5 feet and $8\frac{1}{8}$ inches, or about 5.4% smaller. Inasmuch as the diameters vary directly with the radius, it is not surprising that his newly calculated values – "$1\frac{11}{16}$ [1.688], $2\frac{3}{8}$ [2.375], $2\frac{11}{12}$ [2.917], $3\frac{3}{8}$ [3.375] without any sensible errors" – agreed so well with the observed ones.[69]

For the time being Newton was content with this artifice, for the draft contained, with few differences, the same conclusion that was in the first paragraph of my last quotation from the *Opticks*. Immediately after claiming that this "agrees with the Observation," he attempted to justify his maneuver:

Tis true y^t the distance of the chart from y^e [mirror] is assumed less by about y^e 20th part of y^e whole distance then it was in that Observation, & therefore if you make y^e distances equal the diameters of y^e circles found by calculation will be bigger by about $\frac{1}{20}$th part o[r] in y^e proportion of 21 to 20 then those found by y^e Observation. but this little difference is not worth considering, being no greater then what might easily arise from some small [*del:* & almost insensible] errors in the measures uppon wch y^e computation is grounded.[70]

Newton, nonetheless, was troubled by this transparent device and set about searching for the source of the discrepancy. He determined (correctly) that the vibration lengths that he had found twenty years earlier in thin films were inaccurate and had to be reevaluated. The diameters of the rings in Newton's thick plates were about six times greater than those in his thin films, so that the same error in measurement would have a significantly smaller effect on the former than the latter. It is not now clear exactly how he set about evaluating all

[67] Obs. 8, Add. 3970, ff. 383^r, 384^r, 344^r; Obs. 3, f. 204^r.
[68] Ibid., f. 344^r.
[69] Ibid.
[70] Ibid. Newton inadvertently wrote "Lens" instead of "mirror."

of his previous measurements of the parameters that entered into his calculation, but he did return to Observation 8 in Part I and repeat and thoroughly revise his determination of the vibration lengths in thin films. The new value is a significant 11% smaller and, as best as one can judge, very close to the modern value of the wavelength of yellow light. During this endeavor he undoubtedly repeated his measurements in thick films (Obs. 3), for the manuscript shows that he changed them. These results, however, must have been sufficiently close to those calculated in Observation 8 with the radius diminished to 5 feet $8\frac{1}{8}$ inches that he chose to insert those fractions in the manuscript of the *Opticks* as his new observations. Serious students of Newton's optics have long wondered how after twenty years he came to change his value for the wavelengths by such a large amount and arrive so close to the modern value. Now we know.

In the concluding paragraph that Newton added to Observation 8 during revision, he declared himself satisfied that the colored rings in thin and thick plates had the same cause and consequently that the fits were propagated to great distances. "But yet," he declared, "to put the matter out of doubt I added the following Observation."[71] In Observation 9 he argued that if it is true that these rings depend on the thickness of the mirror, then when they are observed at the center of a mirror of the same radius but different thickness, then their diameters should vary inversely as the square root of the thickness. "And if this proportion be found true by experience it will amount to a demonstration that these Rings (like those formed in thin plates) do depend on the thickness of the Glass."[72] Although this observation was already in the draft, Newton had not yet actually performed the experiment, for the preceding quoted sentence originally continued:

[& by consequence on both its surfaces, suppose by ye mediation of some reciprocating action propagated with the rays from one surface to ye other. And tho I havę not yet tried this in such concavo-convex glasses yet by the phaenomena of other glasses hereafter mentioned I know it will succeed.[73]

The "phaenomena hereafter mentioned" are those where the light falls obliquely on the mirror, which is equivalent to increasing the

71 *Opticks*, p. $_2$102.
72 Ibid., Bk. II, Pt. IV, Obs. 9, p. $_2$103.
73 Add. 3970, f. 345^r. The phrase "some reciprocating action propagated with the rays" in the first sentence initially read simply "some action propagated." The opening square bracket is Newton's sign for a deletion.

mirror's effective thickness. In revising Observation 9, Newton now added a measurement of the rings' diameters in a thinner mirror, 5/62 inch thick. This experiment indeed confirmed (within 1%) that the diameters vary inversely as the square root of the thickness. He also added a general argument that the rings depend on both surfaces of the mirror and the distance between them. "[T]his dependance is of the same kind" with the colors of thin plates, for the properties of these rings are "such as ought to result from the Propositions in the end of the third Part of this Book, derived from the Phaenomena of the Colours of thin plates set down in the first Part."[74] And in the next paragraph that introduces the final set of observations (10–12) with obliquely incident light, he affirmed that they too "follow from the same Propositions, and therefore confirm both the truth of those Propositions, and the Analogy between these Rings and the Rings of Colours made by very thin plates."[75] Thus the successful investigation of the colors of thick plates, and especially the calculation of the size of the rings based on the theory of fits contained in those propositions, serve to confirm them.

When light falls obliquely on the mirror and not directly through the center, the rings exhibit some unexpected phenomena (Obs. 10–12). As the obliquity of the incident beam increases, the white central spot grows into a white ring that encompasses a new set of colored rings that gradually emerge in the interior, and the central spot alternately becomes dark and colored. Newton was able to explain these new appearances by establishing a correspondence between them and the direct rings by applying the principle of the reversibility of light rays and, once again, symmetry considerations. In this way he was able to predict their size without recalculating them. We need not follow his explanation of the oblique rings, but only recognize that he was able to subsume them within his preceding explanation. He concluded his presentation of oblique rings and thick plates in general by claiming that there are other phenomena, whose calculations are "too intricate to be here prosecuted," that also follow from his theory of fits "and so conspire to confirm the truth of those Propositions." Nonetheless, he is "content" that having discovered the cause of the described phenomena he has been able "to ratify the Propositions in the third Part of this Book."[76]

[74] *Opticks*, Bk. II, Pt. IV, Obs. 9, p. $_2$104; Add. 3970, ff. 344^v–5^v. The reference to "this Book" marks the incorporation of Bk. IV, Pt. I, into Book III during revision.
[75] *Opticks*, p. $_2$104.
[76] Ibid., Obs. 12, pp. $_2$109–10.

The explanation of the colors of thick plates is an unrecognized landmark in seventeenth-century mathematical physics in its combination of physical reasoning, mathematical theory, and quantitative experiment. The physical model and the mathematical theory with its parameters were based on experiment and were able to account for all the observed phenomena. At the age of forty-nine, Newton had more than satisfied his earlier call for a new science of color created by geometrical philosophers and philosophical geometers. In optics only Huygens, with his explanation of double refraction by means of Huygens's principle, had achieved a comparable interplay of physical model, mathematical theory, and experiment. Huygens, to be sure, presented his theory geometrically and in hypothetetico-deductive form, but these differences should not obscure the similarity of their aims and achievements.

4.3. FROM VIBRATIONS TO FITS

Newton developed the theory of fits, as we have seen, during the revision of Book IV, Part I, on the colors of thick plates in the fall and winter of 1691–2. The creation of that theory was a consequence of his investigation of thick plates, but it occurred in two stages. He had worked out his explanation of the colors of thick plates by means of the vibrations – now called "agitations" or "reciprocations" – that he had developed some twenty years earlier to explain the colors of thin films. It was at this point of no consequence for his optical theory that he now considered the corpuscles composing reflecting and refracting bodies to be the vibrating medium rather than the aether, which he had abandoned some years earlier with the completion of the *Principia*. According to his principle of "general induction," he felt that his successful calculation of the diameters and explanation of all other features of the colored rings of thick plates justified, at least temporarily, the transformation of the vibrations from a hypothesis to a theory. He had extended his concept of vibrations to an entirely new phenomenon and confirmed it by experiment and observation. This first stage, from a hypothesis of vibrations to a theory, was only a tentative, transitional one. For the first completed state of the *Opticks*, Newton had drafted a proposition demonstrating the properties of the corpuscular vibrations, but he suppressed not only that but any propositions about the periodic properties of light, no doubt because they were still too hypothetical. Shortly afterward, while revising Book IV, Part I, he eliminated the vibrations altogether and devised the theory of fits, an abstract mathematical set of properties describing the periodic nature of light. He had succeeded in distinguishing the properties of periodicity from their physical cause. This second stage, from corpuscular vibrations to fits, was made principally on methodological grounds, for the corpuscular vibrations and their interaction with light particles were as hypothetical as the existence of the aether.

In his sketch for a planned "Fourth Book" of the *Opticks*, which was apparently composed before his investigation of thick plates, Newton included a number of propositions on the vibrations that are excited in reflecting and refracting bodies when light corpuscles impinge on them (§4.1). The model was much like that of the "Hypothesis" except that the parts of bodies vibrate instead of the aether. Although these were called propositions, we cannot thereby conclude that he then considered (or did not consider) them to be demonstrated propositions and not hypotheses. The "Fourth Book," as I argued earlier, is a programmatic sketch consisting of demonstrated propositions and a wish list of hypotheses that Newton hoped to be able to establish. The first time that he unquestionably, if only temporarily, considered the vibrations of matter as a demonstrated proposition is in the earliest draft (ff. 348–9) of the new propositions on periodicity for the end of Book III, Part III. I have, however, already shown that this draft was composed after his investigation of thick plates in Book IV, Part I, was well underway, if not already completed.[77] Newton first lays down

> Prop. 11 The rays of light in being refracted or reflected cause an agitation in the parts of the refracting or reflecting bodies.

The principal evidence that he adduces in its support is that light heats bodies:

> This is manifest by the heat which light causes in those bodies when collected by burning glasses. For heat argues a vehement motion of ye parts of hot bodies & seems to consist in nothing else y^n^ such a motion. For bodies by vehement motion grow hot, as iron by filing or drilling or hammering.[78]

He then specifies that this agitation is a reciprocating or periodic motion:

[77] See note 23.
[78] Add. 3970, f. 348^v^.

Prop. 12 The motion excited by a ray of light in its passage th[r]ough any refracting surface is reciprocal & by its reciprocations doth alternately increase & decrease the reflecting power of the surface.

. . . [T]he proper argument for ye truth of this Proposition is the alternate reflections & transmissions of light succeeding one another in a thin transparent plate accordingly as the thickness of the plate encreaseth in an arithmetical progression.

A piece of glass 1/240,000 inch thick, for example, alternately reflects and transmits orange light at integral multiples of this thickness.[79]

& the reason I cannot yet conceive to be any other then that every ray of light in passing through ye first surface of ye plate stirs up a reciprocal motion wch being propagated through the plate to ye second surface doth alternately encrease & diminish the reflecting power of y^{t} surface, so y^{t} if ye ray arrive at ye second surface when its reflecting power is encreased by the first impuls or vibration of the motion, it is reflected; but if the plate be a little thicker so that ye reflecting power of ye second surface be diminished before the ray arrives at it, the ray is transmitted . . . & so on.[80]

This is the model developed twenty years earlier, with the vibrations now simply transferred to the parts of the body.

Newton's "proper argument" for the proposition – "the reason I cannot yet conceive to be any other then that" – violates the methodology that he had boldly declared in the "New theory." He then denounced hypotheses "conjectured by barely inferring 'tis thus because not otherwise" in contrast to his own theory, which was "evinced by ye mediation of experiments concluding directly."[81] Newton no doubt must have soon recognized his improper mode of argument here and realized that although he may have experimentally demonstrated periodic properties of light, the mechanism he had proposed to explain it – the interaction of light particles with vibrations – was conjectural and not supported by any empirical evidence.

Having completed his "proof" of the proposition, Newton notes that a number of "things seem to follow" from it. The first consequence is his long-held hypothesis of overtaking waves, "that the vibrations move through the thin plate more swiftly then light." The requirement that the vibrations move faster than the corpuscles is

[79] This value for glass is derived from the "Observations," where the interval for orange light is given as 1/80,000 inch in air. If the index of refraction of glass is 3/2, then the interval in glass is 1/120,000 inch, and the first reflection will occur at half that distance, or 1/240,000 inch.

[80] Ibid., ff. 348^{r}–9^{r}.

[81] Quoted in full at Ch. 1, note 26.

difficult to mathematize and would soon be abandoned, for overtaking waves are unnecessary in the theory of fits. The second consequence ascribes a decrease in frequency or increase in vibration length to the rays as their angle of incidence increases, just as he had in the "Hypothesis": "that ye times between the vibrations excited by the perpendicular rays are shorter then the times between those excited by the oblique ones."[82]

The third consequence begins innocently enough and states that the frequency of violet rays is greater (or the vibration lengths smaller) than that of red rays: "ye times between the vibrations excited by the most refrangible rays are shorter then the times between those excited by the least refrangible ones." He then invoked a crude approximation and asserted that the ratio of these extremes, which he had observed was 14 to 9, is "in round numbers" two to one, or an octave. He could now utilize the same musical division for the vibration lengths of the seven colors as he had for the spectrum, that is, "as y^e numbers $\frac{1}{2}, \frac{9}{16}, \frac{3}{5}, \frac{2}{3}, \frac{3}{4}, \frac{5}{6}, \frac{8}{9}$, 1 that is as the times of the vibrations of y^e air wch cause the eight musical notes or tones in an Octave."[83]

In revising Book IV, Part I, and composing the propositions on fits (namely, Prop. 16), he would later adopt the rule that the intervals of the fits are as the 2/3 power of this scale, which, of course, meant abandoning his cherished octave and a physical foundation for color harmonies. Continuing with his speculations here, he proposed his analogy between the sensation of colors and sounds and suggested that "the agreement & disagreement of colours may thus arise from the same principles wth ye harmony & discord of sounds." He offered some particular examples: "For instance green agrees with neither blew nor yellow for it is distant from them but a note or tone above & below."[84] When Newton completed the *Opticks*, he consigned these conjectures to the queries.[85]

[82] Add. 3970, f. 349^r.
[83] Ibid.
[84] Ibid., ff. 348^r–9^r.
[85] See Queries 12–14, 16. Nonetheless, a number of musical divisions were introduced into the *Opticks*, even if they remained devoid of serious musical or physical significance. Newton's fascination with musical analogies in this period is evident in the so-called "classical scholia" to Props. IV–IX, Bk. III, *Principia*, which are also from the early 1690s. In these commentaries Newton attempted to show that the ancients disguised their knowledge of the inverse-square law of gravitation in such musical analogies as Apollo's lyre, the pipes of Pan, and the harmonies of the spheres. See McGuire and P. M. Rattansi, "Newton and the 'Pipes of Pan' "; and

Newton's draft of Prop. 12 was his first attempt to compose a new series of propositions on the periodic nature of light for Part III. After pondering the problem and no doubt recognizing the hypothetical nature of the corporeal vibrations, he not only suppressed this proposition but any general treatment of the periodic properties of light in the first completed version of Book III, Part III of the *Opticks*. Instead, he concluded Part III of this state with only a new Prop. 10 on the proportion of refractive powers and density. Book IV, Part I, however, still contained references to the vibrations – "some reciprocating action propagated with the rays from one surface to ye other" – that he had used to derive the properties of thick plates.[86] And while he claimed after his calculation of the rings' diameters that "the Theory of deriving these rings . . . agrees with the Observation," he had not set out any general theory.[87] Inasmuch as he devised the theory of fits no more than a few months later, it is reasonable to assume that Newton was searching for a way to formulate the concept of vibrations without introducing hypothetical elements and violating his own methodology.

There is additional evidence that Newton transformed the model of vibrations from a hypothesis into demonstrated principles on periodicity because of confirmation by his investigation of thick plates. This evidence consists of the various drafts of the transitional paragraph between the original portion of Part III on the colors of natural bodies, which derived directly from the "Observations," and the new material added in the preparation of the *Opticks*. In the earliest draft, written before his research on thick plates, he disavowed explaining why light is alternately reflected and transmitted in thin films because it is hypothetical. In later versions, however, written after Book IV, Part I, was revised and he had developed the theory of fits, he announced that he would in fact explain those properties with a new sequence of propositions.

Newton's earliest plan to complete the *Opticks* was to follow Part III with a new "fourth Part. Of the inflexion of light & of the causes of Reflexion Refraction & colours."[88] The title implies that it would be much like the projected "Fourth Book," but it is in fact entirely on diffraction (inflexion). He clearly had trouble writing this transitional

Paolo Casini, "Newton: The classical scholia." It is not clear what causal relation, if any, there is between these two activities; see also note 125.

[86] Quoted in full at note 73.

[87] Quoted in full at note 65.

[88] Add. 3970, ff. 377–8, 328–9; "inflexion of light & of the" were interlined.

paragraph. He deleted his first attempt in which he insisted that it should not "now be expected that I say whence it is . . . that particles of this or that size reflect this or that sort of rays & intromit the rest & why thin transparent plates, as their thickness increases in an arithmetical progression do reflect & transmit one & ye same sort of rays for many successions."[89] He immediately started anew:

> Having shewn the laws of refraction & reflexion & how the transient colours of Prisms & such like transparent substances & the permanent ones of natural bodies are produced by those laws: It may now be expected that I shew why light is refracted & reflected according to those laws & how its several sorts of rays affect ye sense wth severall colours. But this is no part of my designe. I have not undertaken to unfold ye whole nature of light but only to search into its properties so far as I can discover them by experiments & therefore tis not to be expected that I should explain these things by Hypotheses. I shall rather go on to lay a foundation for Hypotheses by making some further discoveries of the properties of this strange subject in ye following Propositions so far as I have attained them.[90]

Newton is not denouncing hypotheses but only renouncing treating them here. It is one of the few later passages where he justifies treating hypotheses experimentally, showing that his views on this point had not changed substantially in twenty years. He returned once again to make another change, but this one need not concern us.

When somewhat later Part III was assuming its published form and Newton was drafting the propositions on fits, he still needed a transitional paragraph from the old to the new propositions. His new version made claims diametrically opposed to the earlier one. Now he reminds us that he has thus far shown that "thin transparent plates, fibres, & particles" reflect different colors depending on their thickness and density, and also explained the colors of natural bodies:

> But whence it is that these plates fibres & particles do according to their several thicknesses & densities reflect several sorts of rays I have not yet explained. To give some light into this matter & make way for understanding the next part of this Book I shall conclude this part wth the following Propositions.[91]

[89] Ibid., f. 377^r^. The concluding paragraph of the "Observations" quite naturally served as Newton's starting point for drafting this transitional paragraph.
[90] Ibid.
[91] Ibid., f. 339^r^. This version is from the draft of the propositions on fits. There is a later draft on f. 370^v^, but for our purposes neither differs significantly from the published *Opticks*, p. $_2$77. The words "part of this" were added between "next" and "Book"; and he originally wrote "I shall conclude this book."

Newton's assessment of explanations of the periodic properties of light changed radically – from hypotheses to demonstrated propositions – during the composition of the final portions of the *Opticks*. My analysis of the sequence and contents of his papers from this period leaves little doubt that this change was caused by his investigations of thick plates, which he cites here as "the next part of this Book." By applying the model of vibrations to thick plates, he confirmed and extended the concepts developed in explaining thin films and showed that periodicity was a general property of light. His line of attack, already clearly stated in the early version (Bk. IV, Pt. I), was to show that the colors of thick plates are analogous to those of thin films and have the same cause. The calculation of the diameters of the rings in thick plates, which yielded such close agreement with observation and was in itself a tour de force of contemporary physical optics, was crucial in confirming that the theory of vibrations applies to both phenomena. The sequence of documents (especially Prop. 12 on ff. 348–9 and the two versions of the transitional paragraph) shows that Newton abandoned the idea that explanations of periodicity were hypothetical only after he had completed his investigation of thick plates. The intermediate stage of corpuscular vibrations in Prop. 12, it should be recalled, was short-lived, for they were considered and dropped within at most six months in 1691 and 1692. The transformation from a theory of vibrations to the theory of fits is particularly well documented. There can be no doubt at all that this occurred between the composition and revision of Book IV, Part I, that is, between August 1691 and February 1692 (§4.1). The initial version was conceived in terms of "some reciprocating action," whereas the revision was couched in terms of fits. Finally, as we shall soon see, in presenting the theory of fits, Newton appeals to the colors of thick plates ("the next part of this Book") as justification, and afterward he concludes that the explanation of those colors "confirms" and "ratifies" that theory.

Newton's attitude toward hypotheses and his use of them in these drafts and outlines for the *Opticks* appear to be inconsistent. In some passages he rejected them and insisted on dealing only with properties discovered by experiment. In many others, however, he not only set forth what he himself called hypotheses, but he proposed propositions that both before and after this period he would consider to be hypotheses. Prop. 12, for example, is just the old hypothesis of aethereal vibrations transferred to the corpuscles of matter, an idea that would again be labeled a hypothesis in the *Opticks*. This process of freely invoking and pursuing his conjectures and speculations in carrying out and interpreting his experiments and then withdrawing them in his published writings is, as we have seen, integral to his methodology, characteristic of his style, and a source of his scientific creativity. There is a recurring tension between his scientific imagination and methodological proscriptions, with the former maintaining the upper hand while he was working out his ideas and the latter coming into play when he formulated them for public presentation. By the time Newton had completed the investigation for the first version of Book IV, Part I, he was sufficiently convinced of the periodicity of light that he proposed the model of corpuscular vibrations as Prop. 12 for the new portion of Part III. He evidently soon recognized that his assumption of corpuscular vibrations and their interaction with light particles was as hypothetical as the earlier aethereal vibrations, and he omitted it from this state of the *Opticks*. As Newton scrutinized his explanation of the colors of thick plates in his effort to complete the *Opticks*, he no doubt recognized that he had demonstrated only the property of periodicity. His more conservative methodological injunctions had come into play. Adopting the technique that he had described to Hooke twenty years earlier (§1.2), Newton "abstracted" the properties of periodicity from his physical models or hypotheses and formulated the theory of fits wholly in terms of the periodic properties of light without any explicit reference to physical causes. We will shortly examine that theory more carefully to see how successful he actually was in eliminating an underlying physical model.

The aether, or rather its elimination, assumes only an incidental role in this interpretation of the development of the theory of fits. Westfall, the only one previously to offer a historical, as opposed to a philosophical, analysis of the theory, has argued that with the aether gone Newton had little alternative to the mathematical and mediumless theory of fits. He dismisses Newton's transitional theory of corpuscular vibrations as "an obviously lame attempt to transfer the vibrations of the late departed aether to a medium obviously unsuited, in the context of Newton's idea of nature, to receive them."[92] I have

[92] Westfall, "Uneasily fitful reflections," p. 94; see also *Never at Rest*, pp. 522–3. He does not elaborate on his two "obvious" claims, but they can be expanded into a valid objection, namely, that the corpuscular vibrations should correspond to acoustic vibrations and propagate too slowly. The "lameness" of invoking corpuscular vibrations to explain optical phenomena was not at all obvious to Newton, for he continued to advocate it for at least fifteen years, or as late as the first edition of the *Opticks*; see note 101. Only in about 1713, when he was preparing an electrical

argued that Newton abandoned the corpuscular vibrations on methodological grounds: They were as hypothetical as aethereal vibrations. But the key to choosing between the two interpretations is, I claim, to ask whether the mere elimination of the aether and a mathematical formulation were sufficient justification for Newton to transform a hypothesis into a theory. Newton, I believe, was too sophisticated to believe that a hypothesis could be converted into a theory solely by a judicious reformulation. I interpret these developments as arising from his investigation of thick plates in conjunction with his methodological rules. The tentative transformation of the hypothesis of aethereal vibrations to the theory of corpuscular vibrations was justified by the new experimental evidence of thick plates, whereby it was "rendered general by induction." The following transformation, to the theory of fits, occurred when Newton abstracted the properties of periodicity from the hypothesis of vibrations so that he could "speake of light in generall terms."

4.4. THE FORMAL THEORY

If Newton's methodological rules compelled him to abandon the theory of corpuscular vibrations, the greater mathematical-physical consistency and simplicity of the theory of fits exerted its own attractions. That theory was perforce formulated in abstract mathematical terms, as his principal aim in developing it was to eliminate references to the hypothetical light particles and seat of vibratory motion. Despite his genuine efforts to eliminate hypotheses from the theory of fits, the emission theory of light remained embedded within it. The theory's terse, abstract formulation, however, also reflected Newton's less than wholehearted commitment to it as a complete physical theory and its unresolved problems. Consequently, the theory of fits is more like the framework for a theory than a fully articulated one and requires a great deal of interpretation.[93] With such an elusive theory,

"Appendix" for the second edition of the *Principia* did he recognize that corpuscular vibrations would propagate too slowly for light. In a draft treating vibrations of the electrical spirit, Newton noted that "the thick particles of bodies can in fact hardly or not at all acquire the velocity of rays of light, so that they cannot communicate that velocity to emitted rays [*Crassae corporum particulae velocitatem radiorum lucis vix aut ne vix quidem acquirunt, ideoque velocitatem illam non communicant radijs emissis.*]" (Add. 3970, f. 597ʳ). For Newton's aborted plan for an "Appendix" to the *Principia* on "the attraction of the small particles of bodies," see Newton to Cotes, 6 January and 2 March 1712/3, *Correspondence*, 5:361, 384.

[93] Biot, for instance, required thirty-four pages for its presentation, whereas Newton laid it down in nine much smaller pages.

one must be careful not to read into it concepts that are not Newton's while properly grasping those that are indeed his. I have attempted to reconstruct his reasoning from a variety of sources: the presentation in the *Opticks* and its drafts, the historical development of the theory, and the optical phenomena themselves.

From the theory's first appearance Newton had settled on the term "fits," although in one early draft he tried out the phrase "fits or passions" before deleting it.[94] Newton's choice of the strange term, drawn from contemporary medical language, helps to illuminate the concept. A fit is one of recurrent attacks of a periodic ailment, in particular, of malaria or, as it was then known, ague or intermittent fever. Malaria was then a common disease in England, especially in the fens of Cambridgeshire, so that Newton and his readers would have been thoroughly familiar with the term. Thomas Willis, in his *De febribus*, as translated by Samuel Pordage in 1684, described the nature of "Intermitting Feaver":

> It hath certain remissions, or times of intermission; that every fit [*paroxysmus*] begins with cold or shaking, for the most part, and ends in sweat; that the accessions or coming of the fits, return at set Periods, and certain intervals of times, that a Clock is not more exact.[95]

The fits of ague have a number of features in common with fits of easy reflection and transmission that make it an apt term: The fits alternate between two opposite phases, cold and hot, which are alternate bodily states and not alternations of some bodily substance such as blood or other humor. The fits are periodic and, depending on the type of malaria, have periods of return of one, two, or three days.[96] Moreover, like Newton's fits, they are only dispositional, that

[94] Add. 3970, f. 339r.

[95] Willis, *Practice of Physick*, "Ch. 3. Of Intermitting Feavers or Agues," p. 56; *De febribus* was originally published in *Diatribae duae medico-philosophicae* (1659). See also Sydenham's contemporary English essay, "Febres intermittentes" (in Kenneth Dewhurst, *Dr. Thomas Sydenham*, pp. 131–9), where the term "fits" is freely used. The cognate for "fit," "paroxysmus," was also in common currency, as is evident from the glossary at the beginning of Pordage's edition: "Paroxisms, *Fits, or the returns of fits, as of an Ague or Feaver*" (*Practice of Physick*, sig. c). In his translation of the *Opticks* Clarke chose not to translate Newton's imaginative "fits" with the medical term "paroxysmus" but used instead "vices" ("alternations" or "reciprocal successions"); *Optice*, Def., p. 240. Pierre Coste more sensitively translated "fits" into French as "accès," as in the contemporary phrase "un accès de fièvre"; *Traité d'optique* (1722), Def., p. 331. Westfall, using the *Oxford English Dictionary* (whose entries "fit" and "paroxysm" one should also see), first called attention to the medical origin of Newton's term, "Uneasily fitful reflections," p. 94.

[96] An entry in Robinson Crusoe's diary reads: "*June* 25. An Ague very violent; the Fit held me seven Hours, cold Fit and hot, with faint Sweats after it" (Daniel Defoe, *Robinson Crusoe*, p. 69); first published in 1719.

is, the chills and fevers may sometimes recur a bit sooner or later or skip a cycle.

Newton set forth the theory of fits at the conclusion of Book II, Part III, in nine terse propositions with minimal commentary.[97] In the first proposition and definition of the sequence, he introduced the fundamental ideas of the theory:

PROP. XII.

Every ray of Light in its passage through any refracting surface is put into a certain transient constitution or state, which in the progress of the ray returns at equal intervals, and disposes the ray at every return to be easily transmitted through the next refracting surface, and between the returns to be easily reflected by it.

DEFINITION

The returns of the disposition of any ray to be reflected I will call its *Fits of easy reflexion*, and those of its disposition to be transmitted its *Fits of easy transmission*, and the space it passes between every return and the next return, the *Interval of its Fits*.[98]

To establish the periodicity of the fits, or that they return at equal intervals, he appeals, just as we expect, to his observations of Newton's rings. These show that "one and the same sort of rays . . . is alternately reflected and transmitted for many successions accordingly, as the thickness of the plate increases in arithmetical progression of the numbers, 0, 1, 2, 3, 4, 5, 6, 7, 8, &c. . . . " Then, to establish that they are permanently impressed in the rays, he invokes his twenty-fourth observation of seeing over a hundred rings through a prism. Moreover, anticipating the results of his investigation of thick films in "the next part of this Book," he can state that these vicissitudes return many thousands of times. "So that this alternation seems to be propagated from every refracting surface to all distances without end or limitation."[99] With this conclusion, Newton is preparing the way for the next proposition, which asserts that light already possesses fits even before it encounters a refracting surface and probably possesses them when they are emitted from a luminous source.

Newton has phrased both the proposition and the definition to

[97] I will follow the presentation of the *Opticks*. The one substantial draft (Add. 3970, ff. 339–40, 363–4) is quite close to the published text, and significant differences and alterations will be noted.

[98] *Opticks*, pp. $_2$78, 81.

[99] Ibid., pp. $_2$78–9.

emphasize that the fits are only *dispositions* to be *easily* reflected or transmitted and not states in which they necessarily occur. As careful as he was to stress the dispositional nature of the fits, he was equally cautious to say virtually nothing about the nature and origin of these dispositions: "What kind of action or disposition this is? Whether it consist in a circulating or a vibrating motion of the ray, or of the medium, or something else? I do not here enquire."[100] After completing a draft of the new propositions on fits, Newton apparently recognized that they were so laconic as to be oracular and that he had to provide some assistance in conceiving of the periodic nature of the fits. He then returned to the conclusion of this proposition and added a paragraph containing a variation upon his old model for "those that are averse from assenting to any discoveries, but such as they can explain by an Hypothesis." He suggested that just as stones thrown into water excite waves in it, so light rays incident upon a body

> excite vibrations in the refracting or reflecting medium or substance . . . that the vibrations thus excited . . . move faster than the rays so as to overtake them; and that when any ray is in that part of the vibration which conspires with its motion, it easily breaks through a refracting surface, but when it is in the contrary part of the vibration which impedes its motion, it is easily reflected; and, by consequence, that every ray is successively disposed to be easily reflected, or easily transmitted, by every vibration which overtakes it.[101]

That Newton explicitly labeled the concept of a vibrating medium *or* substance a hypothesis supports my argument that he abandoned the vibrations for fits because of their hypothetical nature. It is also clear that Newton himself drew on both elements of the dichotomy that he posed in his query as to whether there is "a vibratory motion of the ray, or of the medium." Formally and mathematically he considered the vibrations to be in the ray, as in the proposition itself, but in his physical reasoning he always considered them to be in a medium, as in his "Hypothesis" and the queries. "But," he insisted as usual, "whether this Hypothesis be true or false I do not here consider. I content my self with the bare discovery, that the rays of Light are by some cause or other alternately disposed to be reflected or refracted for many vicissitudes."[102]

[100] Ibid., p. $_{2}$80.
[101] Ibid. The draft for this paragraph was penned on a separate sheet (Add. 3970, f. 472^{v}) from the main draft of the propositions on fits.
[102] *Opticks*, pp. $_{2}$80–1.

Perhaps the best way to understand the nature of the fits is to begin by following Newton's example and compare them with the earlier vibrations (§2.3). The rarefactions correspond to fits of easy transmission and the compressions to fits of easy reflection. In the formation of Newton's rings with an air film, when monochromatic light is incident perpendicularly on the second surface, Newton assumes that at the central dark spot all the rays are transmitted; but at gradually increasing distances from the point of contact and decreasing rarefaction of the vibrations, the central spot passes from "totally dark" to black, as some few rays are reflected back. At a thickness of one quarter of a vibration length, the dark spot passes to a bright one as the vibrations enter a state of compression and reflect more and more rays, until at a thickness of half a vibration length and maximum compression virtually all rays are reflected. At still greater distances fewer rays are reflected, until at a thickness of three quarters of a vibration length the vibrations pass to rarefaction and rays are transmitted, thereby beginning the first dark ring; and so on. As the vibrations successively pass from rarefaction to compression, that is, as their phase continuously varies, their ability to transmit and reflect rays continuously varies. Newton says nothing about the variation of the fits from states of easy reflection to easy transmission. However, from the phenomena, the theory's application and its physical interpretation in terms of vibrations, it is evident that the phase of the fits too must continuously vary. Both Benvenuti and Biot – the two most serious commentators on the theory of fits – represented the fits by sine waves, where the power of the fits varies continuously and at the nodes they are indifferent to reflection and transmission. "One must in the first place note," according to Benvenuti, "that these dispositions do not change instantaneously and by jumps, contrary to the universal law of nature, but gradually."[103]

Now let us see why the fits are only dispositions to be reflected and transmitted and not, like the vibrations, a cause of reflection and transmission. It is important to recall that Newton had established (Prop. I) that the cause of reflection is a difference in refractive density or power between media.[104] The aethereal vibrations caused reflection

[103] Benvenuti, *De lumine*, §46, p. 27; see particularly, Figure 9, Plate I. Benvenuti was a disciple of Bošković, who attributed the fits to oscillations of the light particles. Bošković, in turn, endorsed Benvenuti's treatment of fits in *De lumine*, in particular his geometric representation; see Bošković, *Theory*, §§495–7, pp. 175–6. See also Biot, *Traité*, 4:96–7, esp. Figures 21, 22 on Plates I, II. Bošković first advocated the theory of fits in his *Dissertatio de lumine*, 2 parts (Rome, 1749), esp. Pt. II.

[104] Newton later put this result in terms of fits. By a "fit of refraction," he presumably

or transmission by such a change in density. The fits, however, are not the cause of reflection, but only dispose the rays to be reflected. If we consider a uniform medium, where the difference in refractive density is null and so there is no cause for reflection, the rays will proceed uniformly in a straight line independent of the state of the fits. The fits determine only which rays will be reflected and transmitted at a surface separating media of different refractive power, and not how many rays will be reflected (the intensity) or whether reflection will occur at all; the intensity depends on the refractive power and angle of incidence.

If the refractive power is weak or the light is incident nearly perpendicularly, only those rays very close to a maximum of a fit of easy reflection will be reflected, and all others – including those whose phase is somewhere in a fit of reflection – will be transmitted; and as the refractive power or angle of incidence increases, rays in any phase of a fit of easy reflection will be reflected. If that power is sufficiently strong, even rays in a fit of transmission will be reflected; for instance, beyond the critical angle in total reflection all rays are reflected, even those in fits of easy transmission. The only certainty, as Benvenuti observed, is that it cannot happen that at the same place rays in fits of easy reflection will be transmitted while those in fits of transmission are reflected.[105] The dispositional nature of the fits may in some ways make the theory appear to be like a statistical theory, but the dispositions actually serve to absorb a great deal of ignorance concerning the nature of intensities, partial reflection, and interference. They were used in this way, for example, in his explanation of the central white spot in the rings in thick plates and, as we shall directly see, in his explanation of partial reflection.

Newton attributed fits to light before it falls on bodies because he believed that he had sufficient evidence that periodicity was a general property of light, but his immediate motivation was to explain partial reflection. In 1675, in his "Hypothesis," he had explained it by the vibrations of the rigid aethereal surface separating different media. With the aether eliminated on physical grounds and the vibrating corpuscles on methodological grounds, he now kept just the disembodied vibrations – the fits – as the cause of partial reflection. It was superfluous to have two different vibrations as earlier, one at the

means a fit of easy transmission: "Prop. XIV. Those surfaces of transparent Bodies, which if the ray be in a fit of refraction do refract it most strongly, if the ray be in a fit of reflexion do reflect it most easily" (*Opticks*, Bk. II, Pt. III, p. 282).

105 Benvenuti, *De lumine*, §50, pp. 29–30.

surface of the aether and another within the body. However, because the fits are dispositions, it is not the case that at any surface all rays in fits of reflection will be reflected while all those in fits of transmission will be transmitted. Newton himself made this mistake in what may be the earliest draft of his propositions on fits. Three propositions without proof are written on the verso of the last sheet of his revision of Book IV, Part I, and were then deleted in order to start anew.

Prop. I.

The rays of light both at their first emission from lucid bodies & afterwards at every refraction & reflexion are put into successive fits of easy reflexion & easy transmission: wch fits continue to return at equal distances during the whole progress of ye rays through any uniform Medium.

Prop. II.

The rays of light at their incidence on any refracting surface are either reflected or refracted accordingly as they are then in their fits of easy reflexion or in those of easy transmission. And this is the reason why all refracting surfaces reflect some part of the light incident upon them.[106]

The only reasonable assumption that one can make about the fits is that when light is emitted from a luminous source, it is randomly put into fits of reflection and transmission. Therefore, if all those rays in fits of easy reflection and transmission were correspondingly reflected and transmitted, then every surface would partially reflect and partially transmit half of the incident light, which is contrary to observation.

When Newton rewrote these propositions for the *Opticks*, he was far more cautious. He removed the claim that the fits were in light from emission from the opening Prop. XII, which introduced the nature of fits, and shifted it to Prop. XIII on partial reflection. Moreover, he phrased the new proposition more circumspectly in order to eliminate the objectionable consequence of his earlier version and treated the fits as dispositions:

PROP. XIII.

The reason why the surfaces of all thick transparent Bodies reflect part of the Light incident on them, and refract the rest, is, that some rays at their incidence are in Fits of easy reflexion, and others in Fits of easy transmission.[107]

[106] Add. 3970, f. 307^{v}. Prop. III is an early version of Prop. XIV.
[107] *Opticks*, p. $_{2}$81.

To demonstrate this, Newton again appeals to his twenty-fourth observation, that when he looked through a prism at the white rings formed by thin films, he saw over a hundred alternations of light and dark rings. Although this does not seem to me to establish that the fits are in light before impinging on the body,[108] he concluded:

> And hence Light is in fits of easy reflexion and easy transmission, before its incidence on transparent Bodies. And probably it is put into such fits at its first emission from luminous Bodies, and continues in them during all its progress. For these fits are of a lasting Nature, as will appear by the next part of this Book.[109]

As in Prop. XII, Newton anticipates his investigation of thick plates to show that the fits are lasting, but, as opposed to the draft Prop. I, he considers it only probable that they were in light from its first emission. It may seem strange, as accustomed as we are to the wave theory of light and conceiving of periodicity as an inherent and permanent property of light, that Newton was so cautious in attributing that property to light. It is apparent from its context that he arrived at this conclusion during his investigation of thick plates. Newton's proof that the fits are in light from emission left Benvenuti unconvinced, for he omitted this proposition from his presentation of the theory of fits and did not teach that the fits are in light from their emission. Biot, too, rejected Newton's argument, but not the proposition. He reasoned that because light is put into fits of different phase at the first surface, the only difference in the rays before they arrive at that surface that could cause this is that they are already in fits of different phase.[110]

Despite some significant similarities to a wave theory, the theory of fits was still wedded to the emission theory of light in its use of a single ray that is either reflected or transmitted at the second surface of a thin film. We can immediately recognize this because we know that light possesses no disposition to be easily reflected or transmitted, but rather it is subject to constructive and destructive interference. I raise the wave theory here not because it is "right," but because if

[108] The white light composing these rings is reflected not from the first surface of the plate but from the second surface after the rays have passed through the first surface, where (according to the preceding Prop. XII) they are put into fits.

[109] Ibid., p. 282.

[110] Biot, *Traité*, 4:94. Newton's new definition of "interval" shows how much his views on the nature of periodicity had changed since the 1670s. The term originally referred to the interval between the lenses or thickness of the film, but now it refers to the length of the fits associated with light itself.

Newton's theory is hypothesis-free, as he believed, it should be compatible with any valid physical explanation. Newton proved Prop. XII – that the rays are put into periodically recurring states of easy transmission and reflection by a refracting surface – by appealing to the law of arithmetic progression for the bright and dark rings and justifiably concluding that it "is to be accounted a return of the same disposition." In the course of his proof he argued that the alternate reflection and refraction occurs at the second surface. It "depends" on both surfaces, "because it depends on their distance," and also because the colors grow faint if water is placed on either surface of a thin film. This alternate reflection and refraction, he concluded,

> is therefore performed at the second surface, for if it were performed at the first, *before the rays arrive at the second,* it would not depend on the second.
>
> It is also influenced by some action or disposition, propagated from the first to the second, because otherwise at the second it would not depend on the first.[111]

The flaw in Newton's argument – and we know there must be one, because according to the wave theory it is performed at or above the first surface – is that he does not at all entertain the possibility that it is performed at or above the first surface, *after* the rays return from the second one.[112] Newton violated one of his own methodological canons here, for in a letter to Oldenburg on 6 July 1672 he had laid down the condition for the validity of an argument by elimination: "I cannot think it effectuall for determining truth to examin the severall ways by wch Phaenomena may be explained, unlesse where there can be *a perfect enumeration* of all those ways."[113] It should come as no surprise that after twenty years of using his own model – or hypothesis – of overtaking waves that impede or assist the light particles in breaking through the refracting surface that Newton could no longer conceive of any other explanations of Newton's rings, such as Hooke's. It is an essential requirement of Newton's explanation of the colors of thin films that the alternate reflection and transmission occur at the second surface.

Was Newton aware that his theory of fits incorporated his old hypothesis? There can be no doubt that he recognized that as its source, or much less doubt that he genuinely believed that he had

[111] *Opticks*, pp. $_2$79–80; italics added.

[112] Robert Palter recognized this flaw in "Newton and the inductive method," pp. 169–70.

[113] *Correspondence*, 1:209; italics added.

established the theory of fits in such a way that it transcended its hypothetical origin. He could honestly, albeit mistakenly, hold that he had proved from the phenomena that the alternate reflection and transmission occurs at the second surface. He could also legitimately claim that it does not at all detract from the validity of the demonstration that this idea originated with the emission theory of light. He clearly believed that the theory of fits is independent of the nature of light and is equally valid for the wave and emission theories. In a query added to the Latin translation, in which he launched a vigorous attack on the wave theory of light, Newton did apply it to the wave theory. Among his arguments against the wave theory in Query 20/28 was one based on the theory of fits:

> And it is as difficult to explain by these Hypotheses how rays can be alternately in fits of easy reflexion & easy transmission; unless perhaps one might suppose that there are in all space two ethereal vibrating Mediums & that the vibrations of one of them constitute light & that the vibrations of the other being swifter, as often as they overtake the vibrations of the first, put them into those fits. But to allow two aethers where we have no evidence for so much as one, to suppose that two aethers may be together in all spaces without mixing with one another so as to become one Medium & to suppose also that they may have distinct vibrations without making two sorts of light, are difficulties wch I cannot get over.[114]

That Newton applies to the wave theory the same model of overtaking accompanying vibrations that he applied to his own emission theory shows that he genuinely thought his theory of fits was free of hypotheses on the nature of light and applied equally to both. He simply could not imagine any other way for the successive reflections and transmissions to occur than by some power hindering and aiding the light at the second surface. The charge that his theory contains the hidden hypothesis of the emission theory often carries the implication that he thereby delayed the acceptance of the wave theory and is inconsistent (if not hypocritical) in condemning hypotheses. This sort of criticism is historically irrelevant, as, in the first place, in 1704, no one had a serious alternative to the theory of fits. It should rather have been advantageous that the theory of fits was more compatible with the emission theory – and Newton did not hide that –

[114] Add. 3970, f. 301^{r}; *Optice*, pp. 309–10. After Newton reintroduced an aether in 1716 (§4.5), he had to revise the last sentence for the second English edition: "But how two *Aethers* can be diffused through all Space, one of which acts upon the other, and by consequence is re-acted upon, without retarding, shattering, dispersing and confounding one anothers Motions, is inconceivable" (*Opticks* [Dover], p. 364).

because the latter dominated eighteenth-century optics. Because the theory of fits in fact attracted so few supporters, it is historically more interesting to attempt to explain this anomaly (§4.5). In the second place, Newton's mistaken belief that the theory of fits was hypothesis-free is not due to some surreptitious scheme, but to a methodological error in Prop. XII in the eliminative proof that "interference" occurs at the second surface. To be sure, that assumption in turn derives from the emission theory, and Newton's inability to recognize this represents the failure of his program to construct a hypothesis-free science.

If the theory of fits, like its predecessor, the theory of vibrations, was incompatible with the wave theory, the principles underlying the propagation of fits were more wavelike, mathematically tractable, and consistent than those of overtaking vibrations. These innovations were prompted by the dual challenge of describing the periodic properties of light without a vibrating medium and of explaining the colored circles of thick plates. No doubt, the most notable wavelike characteristic of the fits is that, unlike the vibrations, periodicity is in the rays from their source. The need to follow the rays on their entire path through a thick plate and not simply from the first to the second surface, as in thin films, is largely responsible for the mathematical resemblance of fits to waves.

Two features of the overtaking vibrations made it impossible to follow a particular vibration. First, the vibrations move at some arbitrary but faster speed than the light particles. Newton's vibrational model requires only that at the second surface the particles encounter a compression or a rarefaction, but any number of vibrations may pass them en route to that surface.[115] Second, the vibrations are not continuous across refracting surfaces, as the vibrations in the upper aethereal surface, which govern partial reflection and the ray's entrance into the thin film, are excited by the preceding rays, whereas the vibrations within the film are excited by the ray itself. Moreover, because Newton's account of thin films required that the light path be followed in only one direction, he did not present a rule for how the vibrations are reflected from the second surface; and such a rule was required for thick plates. The theory of fits eliminates both of these features of vibrations (Props. XIX and XX to be considered next). It puts the periodicity in the rays themselves and does away with two separate entities, light particles and aethereal vibrations, moving

[115] See Ch. 2, note 86.

with different speeds. It treats the fits as if they were continuous, so that the path of a particular ray may be followed very much as in a wave theory.

The resemblance of the theory of fits to a wave theory is, however, mostly superficial. Perhaps the most significant difference is that the interval of the fits varies with direction of propagation. The method of determining the location and conditions for "interference" between the light particles and fits (or vibrations) was inconsistent. In thin films the actual path traversed by a single ray determined the appearance of the rings, whereas in thick plates it was the path difference of two rays. Newton drew on his formidable powers as a mathematical-physicist in devising the concept of path difference, but he was motivated by his goal of developing a quantitative model of the rings in thick plates rather than guided by first principles. If this idea turned out to anticipate the wave theory and the principle of interference, it was apparent to hardly anyone, buried as it was in his calculation of the colored rings in thick plates, and it was inapplicable to his earlier analysis of thin films. Physically his concept was totally different from that of the wave theory, for he was not at all considering the interference of two waves. Newton's most fundamental and enduring contribution to the wave theory was in experimentally discovering the laws of periodicity of "interference phenomena," but these were discovered before the theory of fits was formulated, many of them decades before. There is no evidence that the theory of fits played any direct role in Young's or Fresnel's development of a wave theory of light. Their knowledge of wave propagation came directly from eighteenth-century rational fluid mechanics, for which Newton also laid the foundation.[116]

The final two propositions of the theory of fits seem particularly obscure, but their aim is to ensure continuity of the fits within a plate. The two propositions are of a piece, and in the draft Newton initially included XX within XIX.

PROP. XIX.

If any sort of rays falling on the polite surface of any pellucid medium be reflected back, the fits of easy reflexion which they have at the point of reflexion, shall still continue to return, and the returns shall be at distances from the point of reflexion in the arithmetical progression of the numbers 2, 4, 6, 8, 10, 12, &c. and between these fits the rays shall be in fits of easy transmission.

[116] See Naum S. Kipnis, *History of the Principle of Interference.*

This asserts that the fits are continuous at the point of reflection at the second surface of the plate, since "there is no reason why these fits . . . should there cease."[117] Proposition XX states that the obliquity law, which describes the variation of the interval of the fits with the angle of refraction (Prop. XV, to which we will soon turn), is also valid for reflection:

PROP. XX.

The intervals of the fits of easy reflexion and easy transmission, propagated from points of reflexion into any medium, are equal to the intervals of the like fits which the same rays would have, if refracted into the same medium in Angles of refraction equal to their Angles of reflexion.

Newton argues for the truth of the proposition by a plausible symmetry argument and concludes by assuring readers that "these two Propositions will become much more evident by the Observations in the following part of this Book."[118]

These properties are not observable phenomena, nor may they be derived from the phenomena. They are reasonable physical assumptions that are required for his calculation of the diameters of the rings in thick plates and that could only be confirmed by those calculations, which, however, also involved many other assumptions. Newton was unable to present his theory of fits solely in terms of the preceding colors of thin films, which should not surprise us, for we now know that he was prompted to devise it because of his explanation of the colors of thick plates. His statements in the presentation of the theory that its propositions were proposed to aid in the understanding of thick plates suggest as much.[119] At three other points in his presentation, moreover, he appeals to the "next part" on thick plates to justify those propositions, whereas in that part he affirms that the phenomena of thick plates "ratify" and "confirm the truth" of the propositions.[120] There is, of course, nothing wrong with this procedure, but it does not conform to the inductivist approach that Newton otherwise adopts in the *Opticks*, and indeed it violates his methodology. It is the closest that he comes to adopting the hypothetico-deductive method in the presentation of any of his optical theories

[117] *Opticks*, p. $_{2}$85. In fact, in the wave theory a discontinuity, or phase change, must be introduced for reflection from a rare to a dense medium.

[118] Ibid., pp. $_{2}$85–6.

[119] Ibid., p. $_{2}$77, quoted at note 91; Prop. XVIII, p. $_{2}$84, quoted at note 129.

[120] Ibid., Prop. XII, p. $_{2}$79, quoted at note 99; Prop. XIII, p. $_{2}$82, quoted at note 109; Prop. XX, just quoted; and Bk. II, Pt. IV, Obs. 9, p. $_{2}$104, quoted at note 75; and Obs. 12, pp. $_{2}$109–110, quoted at note 76.

and it in part reflects the way in which he actually developed the theory of fits. By choosing to carry out as little revision as possible after he developed the theory – merely inserting the new propositions at the end of Part III and rewriting only portions of Part IV – its nature and relation to the colors of thick plates became evident only to the astute reader. Newton may, of course, have deliberately chosen this form of presentation to avoid the methodological issues involved.

The four remaining propositions that complete the theory of fits are strictly phenomenological and once again in an inductive form. They are all mathematical laws describing the variation of the intervals of the fits and are deduced from the preceding observations of thin films. Although the precise form that the laws assume may be new (except for Prop. XVII), they cover familiar ground. Prop. XVII states that the intervals of the fits vary inversely as the index of refraction, thereby making a law of what Newton earlier considered as only "perhaps" a general rule.[121] In Prop. XVIII he gives the interval at the boundary of yellow and orange for rays perpendicularly incident into air as 1/89,000 inch. Other than for the redetermined value for the interval, which Newton undertook while revising Book IV, Part I, these go back to the early 1670s.

In Prop. XVI, Newton relates the intervals of the fits for the seven principal colors to one another by musical scales. For rays incident at the same angle into the same medium the intervals of the fits "are either accurately, or very nearly, as the Cube-roots of the Squares of the lengths of a Chord, which sound the notes in an Eight, *sol, la, fa, sol, la, mi, fa, sol,* with all their intermediate degrees answering to the Colours of those rays, according to" the musical division of the spectrum described in Book I.[122] Newton came upon this proportion for the intervals, the ⅔ power of a just diatonic scale, to replace the earlier equal-tempered division during his revision of Book IV, Part I, and formulation of the theory of fits.[123] This new scale was of minor observational consequence, but all of them in the *Opticks* were now based on the just diatonic scale.[124] His decision to relate the intervals of the

[121] See Ch. 2, note 51.
[122] Ibid., p. $_2$83.
[123] This proportion first appeared – essentially simultaneously, as best as I can judge – in the draft of this proposition (Add. 3970, f. 363^r; then numbered 15) and in the addition to the conclusion of Bk. IV, Pt. I, Obs. 8 (f. 344), where Newton calculated the diameters of the rings in thick plates. He cites Obs. 13 and 14, Pt. I, in support of this proposition, but this new musical division was not initially in the manuscript of the *Opticks*. He must have added it to Obs. 14 (ff. 144^r, 143^v) at about the time that he was drafting the propositions on fits and revising Bk. IV, Pt. I.
[124] See Ch. 2, note 109.

various colors by musical divisions and the choice of particular scales was motivated primarily by his philosophical fascination with harmonies. Although the new scale is formally independent of the theory of fits, it is probably not entirely coincidental that it appeared while he was developing that theory. Just as the musical division of the spectrum first appeared in the early 1670s, when he was deeply immersed in his investigations of thin films, so the new rash of musical analogies that appeared about 1690 were possibly stimulated by his renewed research into the periodic properties of light.[125]

We have now arrived at the proposition that has been called "the most bizarre fruit plucked from the tree of Pythagoras in the entire 17th century":

PROP. XV.

In any one and the same sort of rays emerging in any Angle out of any refracting surface into one and the same medium, the interval of the following fits of easy reflexion and transmission are either accurately or very nearly, as the Rectangle of the secant of the Angle of refraction, and of the secant of another Angle, whose sine is the first of 106 arithmetical mean proportionals, between the sines of incidence and refraction counted from the sine of refraction.

This is manifest by the 7th Observation.[126]

The proposition sets forth the law for the variation of the interval of the fits with the angle of refraction, namely:

$$I = I_0 \sec u \cdot \sec r = \frac{I_0}{\cos u \cdot \cos r}; \tag{4.4}$$

I_0 and I are, respectively, the intervals when the rays are incident perpendicularly and also at some other angle such that the angle of refraction r results; u is an angle such that

$$\sin u = \sin r - \frac{1}{106}(\sin r - \sin i) = \frac{105 + n}{106}\sin r; \tag{4.5}$$

where n is now the index of refraction into the optically rarer medium.[127]

This law represents the culmination of Newton's twenty-year quest to represent mathematically the variation of the rings with obliquity.

125 See note 85, and also *Optical Papers*, 1:546, n. 28.
126 Westfall, "Newton's coloured circles," p. 190; and *Opticks*, p. $_2$83.
127 The equivalent equation for refraction into an optically denser medium is $\sin u = ([105 + 1/n]/106)\sin i$; see Biot, *Traité*, 4:26–9.

The need for a mathematical law became pressing when he undertook his research on thick films, where he had to be able to calculate oblique paths. It made its first appearance in his calculation of the diameters of the rings in the original version of Book IV, Part I, Observation 8, before he abandoned vibrations for fits.

To understand the law, we should recognize that it originates in his experiments with Newton's rings and films of variable thickness. He justifies the proposition by Observation 7, which contains his measurements of the thickness of an air film producing a ring of a given order at increasing obliquity (§2.2). From these data he derived the law for the oblique thickness $d = d_0 \sec u$, where d_0 is the film's thickness at normal incidence and the angle u obeys Eq. 4.5. The data in Observation 7 are carried over with no alteration from the "Observations" of 1672 and 1675. Newton made no theoretical breakthroughs in the intervening twenty years but simply had devised an empirical formula that satisfied the data. Since 1672, he no doubt knew that the thickness varied very nearly as sec r, for he had proved this earlier in "Of ye coloured circles," though he did not then recognize it. In the *Opticks* he used the sec r approximation for small angles a few times.[128] With the refined measurements of 1672 and 1675, the sec r law was certainly true as far as 60°, and at 75° it still differed by only 4%. Thereafter the deviation from sec r grew immensely; at 90° the observed thickness was only twelve times rather than infinitely greater. To satisfy his observations through their entire range, Newton had to find a law that behaved like sec r to 60° or 70° and then grew much more slowly, but it is not apparent how he came upon his particular law of 106-mean-proportionals.

The problem confronting Newton in Prop. XV (as we saw in §2.3) was to explain the appearance of a ring of the same order at varying obliquity. In his model a bright ring of order $(m + 1)/2$ appears by reflection when m compressions overtake the light particle after it has traversed the film, where m is an odd number. Because the path of the particle through the film continually increases as the obliquity increases, Newton had no alternative but to assume that the vibration lengths similarly increase in order to be able to fit the same number of intervals into the increasing path. Let us now see how Prop. XV can be readily derived, with the vibrations, or compressions, being replaced by fits of easy reflection. If a bright ring of order $(m + 1)/2$ appears at perpendicular incidence at thickness d_0, then m fits of

[128] See note 60.

reflection of length $I_0/2$ must equal that thickness, or $mI_0/2 = d_0$. When the same ring is produced by oblique rays that are refracted within the air film at angle r, it will appear at a thickness $d = d_0 \sec u$. Because the rays now follow an oblique path $s = d \sec r$, then

$$mI/2 = s = d_0 \sec u \cdot \sec r$$

or

$$I = I_0 \sec u \cdot \sec r.$$

Newton had at last succeeded in mathematizing the increase of the intervals with angle of refraction, the feature that so strikingly distinguishes his explanation of interference phenomena from a wave theory. If this did not trouble him in 1675, when he was still working with aethereal vibrations, then its successful application to the colors of thick plates must have convinced him that he had uncovered a fundamental property of the periodic behavior of light and not created a hypothetical one.

These four propositions complete my account of the formal theory of fits. As Newton noted in summarizing them, they provide the calculational basis for his theory:

> From these Propositions it is easy to collect the intervals of the fits of easy reflexion and easy transmission of any sort of rays refracted in any Angle into any medium, and thence to know, whether the rays shall be reflected or transmitted at their subsequent incidence upon any other pellucid medium. Which thing being useful for understanding, the next part of this Book was here to be set down.[129]

This was in fact Newton's only commentary on these four propositions other than for the identical statement following each one: "This is manifest by the *n*th Observation." Though his commentary on the other more physical propositions was more expansive, it was still unequal to the task of explicating such an elusive theory.

Why was Newton so reluctant to elaborate on the physical content of his theory? In the first place, his methodology would not allow him to mix hypotheses with demonstrated principles. When he conceived of the theory of fits as a way to present the periodic properties of light, this meant that he had to formulate those properties "abstractedly," "in generall terms," and purge it of all hypothetical elements, in particular, vibrating media and the emission theory of light. As Newton himself had recognized when he composed his "Hy-

[129] *Opticks*, p. $_2$84.

pothesis," this practice leads to a loss of intelligibility, and his theory of fits certainly bears out his insight.

In the second place, the vibrational models were no longer adequate for explicating the theory of fits. After 1692 and the formulation of the theory of fits, I believe that it was no longer possible for Newton to have written an interpretive account along the lines of the "Hypothesis," for the theory of fits is not just a mathematical formulation of the earlier vibration models, though it obviously evolved from them.[130] The two most irreconcilable differences are, first, that the fits are in the rays and proceed at the same velocity rather than propagating as a separate entity, whereas the vibrations move faster than the rays and overtake them; and, second, that periodicity or the fits probably begins with the emission of light and certainly before its incidence on any bodies, whereas the vibrations, whether of the aether or corpuscles of bodies, are created by their encounter with the surfaces of bodies. Although it is a more subtle difference, there is no way that the *same* vibration could cause both partial reflection and transmission at the first surface and also alternate reflection and transmission at the second, as is the case with the fits. Because the two were not fully equivalent, it was simply not possible for Newton to give an adequate interpretation of the theory of fits in terms of vibrations. On a number of occasions in the next twenty-five years, he set forth variants of the hypothesis of vibrations, yet he made no attempt to motivate or justify, let alone derive, any of the mathematical properties of fits by means of vibrations.

Newton's investigation of fits came to a standstill with the conclusion of his work on thick plates. After he had formulated the theory, he seems never to have been especially concerned with developing it further or pursuing such problems as the dispositional nature of the fits. His initial aim appears to have been to devise a series of propositions on periodicity that conformed with his methodology and were adequate to justify his calculation of the colored rings of thick plates. He seemed sufficiently satisfied with his accomplishment in this respect that he did not afterward attempt to modify it. He was not, however, content with the strictly mathematical theory of fits. During the course of the next twenty-five years he proposed one hypothetical cause of the fits after another, just as he did with gravity, which he

[130] Koyré, *Newtonian Studies*, p. 49, for example, makes the common assertion that the fits are equivalent to the aethereal vibrations.

proved "does really exist." The three models Newton set forth during this period were fundamentally variants on the overtaking aethereal vibrations that he had been using at least since the early 1670s. The only essential differences between the proposals were in the nature of the vibrating medium and in the seat of short-range forces. Consequently, they are more informative about the vicissitudes of his natural philosophy than his optics.[131]

From the completion of the *Principia* and abandonment of the aether through the Latin translation of the *Opticks* in 1706, Newton attributed the vibrations to the corpuscles of matter. In the new queries added to the translation he vigorously argued that optical phenomena arise from forces acting at a distance between particles of light and matter. To explain the fits he proposed in Query 21/29 that

> [n]othing more is requisite for putting the rays of light into fits of easy reflexion & easy transmission then that they be small bodies wch by their attractive powers or some other force stir up vibrations in what they act upon, wch vibrations being swifter then the rays, overtake them successively & agitate them so as by turns to increase & decrease their velocity & thereby putt them into those fits.[132]

This suggestion differs from those of the early 1690s only in explicitly mentioning forces. Newton seems to be unaware that his model of corpuscular vibrations and the theory of fits are inconsistent concerning the existence of fits in light before its incidence on bodies (Prop. XIII). Without an aether, vibrations cannot commence until light encounters a body.

Shortly after Newton completed these new queries, Francis Hauksbee, acting under Newton's direction, began a series of electrical experiments at the Royal Society that would lead Newton to adopt a new medium – the electrical spirit – as the seat of short-range forces and the activity of matter. Between about 1709 and 1716, Newton composed various essays, drafts of queries, and parts of an "Appendix" to the *Principia* that elaborated on the nature and effects of the electrical spirit. The final paragraph of the General Scholium to the *Principia* contains his only public proposal of the widespread role of this "most subtle spirit" in all bodies. This spirit, he suggested, may

[131] For overviews of Newton's speculative natural philosophy and its evolution from the composition of the *Principia* through the last edition of the *Opticks*, see McGuire, "Force, active principles, and Newton's invisible realm"; McMullin, *Newton on Matter and Activity*, esp. Ch. 4; and Westfall, *Force in Newton's Physics*, Ch. 7; and *Never at Rest*.

[132] Add. 3970, f. 257^{r}; *Optice*, pp. 317–18; unchanged in *Opticks* (Dover), pp. 372–3.

cause the attractions of the corpuscles of bodies, electrical attraction and repulsion, and animal sensation and motion and be the medium by which "light is emitted, reflected, refracted, inflected and heats bodies." "But," he concluded, "these are things that cannot be expounded in few words, nor are we furnished with that sufficiency of experiments which is required to an accurate determination and demonstration of the laws by which this electric and elastic spirit operates."[133] In contrast to the early aether, this spirit is present only in bodies and their immediate vicinity and so at great distances from bodies it could neither cause gravitational attraction nor act on light and vibrate.

Although the vibrations of the electrical spirit were conceptually similar to the aethereal vibrations and overtaking waves, the concept of force required a fundamental change in the way that the spirit caused the rays to be reflected or transmitted. The traditional dense aether that Newton used earlier acted directly by changes in density and pressure gradients. The aether was assumed to be rarer in substances with greater refractive power – for instance, rarer in glass than air – so that light particles were propelled toward the rarer region of lesser pressure. Thus, in corpuscular vibrations, the rays are impelled toward rarefactions and transmitted. The contrary occurs with the electrical spirit, which is a seat of attractive force. The electrical spirit is denser in substances with greater refractive power, so that light particles are attracted to it. In vibrations of the spirit, particles are attracted to the region of compression and transmitted. In this way Newton was able to explain the origin of fits:

For, where the electric spirit is denser it attracts light rays to itself. Thus where the denser part of the vibrations follows the rays it retards them, where it precedes the rays it accelerates them, so that the rays are alternately [*per vices*] accelerated and retarded by the electric spirit; and according to the thickness of the transparent skin, they fall upon its farther surface where they are slower or faster, and in the former case they will be reflected and in the latter they will pass through.[134]

[133] *Mathematical Principles*, p. 547. The Latin text has only "this spirit [*hujus spiritus*]" and lacks the phrase "electric and elastic." Motte's interpolation is, however, justified by Newton's insertion "electrici & elastici" in his annotated copy of the *Principia;* see Koyré and Cohen, "Newton's 'electric & elastic spirit.' " More generally on the electric spirit, see Guerlac, "Newton's optical aether"; R. W. Home, "Newton on electricity and the aether"; and J. L. Heilbron, *Physics at the Royal Society during Newton's Presidency*, pp. 58–66.

[134] "Nam spiritus electricus ubi densior est radios lucis ad se attrahit, ideoque densior pars vibrationis ubi radios sequitur ipsos retardat, ubi praecedit ipsos accelerat, et

Because the vibrations are periodic, and condensations always both precede and follow a particle, Newton evidently means that the rays are attracted to the nearer of the two.

Newton made his last major statement on natural philosophy at the age of seventy-four in new and revised queries for the second English edition of the *Opticks* in 1717. He was no longer enamored of the electrical spirit to explain all short-range forces and now limited it to producing electrical phenomena alone. In a somewhat surprising move he reintroduced a new, sleek aether to explain optical phenomena, gravity, heat radiation, and animal sensation and motion. Unlike the dense Cartesian aether of the "Hypothesis," this new aether was exceedingly rare and elastic, but, like the traditional aether, it operated directly by density gradients. Newton suggested, though, that the aether's elasticity and rarity arose from a repulsive force in its particles (Query 21). His motivation for reintroducing an aether, albeit of a different variety, has puzzled historians.[135] Insofar as he used it to explain the fits in Query 17, it offered nothing new.

Perhaps Newton was most comfortable with the model he had developed in his early years of discovery and used for the next two decades, for he has essentially returned to it. His papers show no evidence of his attempting to apply this, his last model, to resolving any outstanding optical problems. The explanation of alternate reflection and transmission by overtaking vibrations remained remarkably unchanged in the course of half a century, despite the introduction of the concept of force and the theory of fits. The only significant difference was in the seat of the vibrations. Like the wave theorists, Newton always demanded some medium for his vibrations. A physical model of fits appropriate to an emission theory of light was first proposed by Bošković, who located the vibrations in the light particles without requiring any accompanying medium.

4.5. CONCLUSION AND HISTORICAL POSTSCRIPT

Newton was perfectly comfortable with hypotheses and publicly expounded them throughout his career from the "Hypothesis" in 1675

sic radij a vibrationibus spiritus electrici accelerantur & retardantur per vices et pro crassitudine cuticulae pellucidae incident in ulteriorem ejus superficiem ubi vel tardiores sunt vel velociores et priore casu reflectentur posteriore pertransibunt" (MS in private possession).

[135] See, for example, McMullin, *Newton on Matter and Activity*, p. 96; and also P. M. Heimann and McGuire, "Newtonian forces and Lockean powers," pp. 237–46.

to the queries added in the last edition of the *Opticks* in 1717. These works satisfied his requirements that hypotheses be empirically founded and kept distinct from demonstrated principles. By following Newton's research on periodic colors through the course of twenty-five years from the mid–1660s through the early 1690s, we have seen how fruitfully and creatively he utilized his hypothesis of vibrations in his investigations and finally transformed it into a theory of fits, which consisted of demonstrated propositions on the periodic properties of light. He had informed Pardies that hypotheses could legitimately be used to explain the properties of things and to suggest further experiments, but we have seen him use them in a much richer way than implied by those generalities. In one of his drafts for a preface to the *Opticks,* Newton succeeded in capturing more of his actual practice. Expounding the method of resolution and composition (later called analysis and synthesis), he explained:

> The method of Resolution consists in trying experiments & considering all the Phaenomena of nature relating to the subject in hand & drawing conclusions from them & examining the truth of those conclusions by new experiments & drawing new conclusions (if it may be) from those experiments & so proceeding alternately from experiments to conclusions & from conclusions to experiments untill you come to the general properties of things, [& by experiments & phaenomena have established the truth of those properties].[136]

The continual interplay of experiment and theory (interpreting "conclusions" broadly) and the broad scope of his inquiry, encompassing all relevant phenomena (e.g., heat and the structure of matter), characterized his investigation of periodic colors.

Newton used his hypotheses very much as in the hypothetico-deductive method. He started from laws and properties (both observed and assumed), formulated a model suggested by the phenomena, made deductions from it, put them to experimental tests, eliminated those features that did not agree with the phenomena, and further refined the model until it agreed with all observations. In his earliest account of the colors of thin films in his essay "Of Colours," he utilized both light particles and vibrations. With the pulses, he was able to establish the periodic nature of the colors of Newton's rings and calculate the pulse length. His attempt to use light particles to deduce how the size of the rings varies with a change

[136] Add. 3970, f. 480ᵛ; McGuire, "Newton's 'principles' of philosophy," p. 185; the brackets are Newton's. Newton eventually expounded his method of analysis and synthesis in the *Opticks*, Query 23/31; quoted at Ch. 1, note 73.

in the obliquity of the incident light led nowhere, and he crossed out his derivation. Five years later, in his essay "Of ye coloured circles," he had refined his hypothesis of vibrations and deduced a series of propositions on the properties of Newton's rings, which he then attempted to prove by experiment. He had also refined his speculations on the motions of the light particles and their interaction with the vibrations. This hypothesis, as we saw, now led him totally astray. He had derived an erroneous obliquity law that he declared to have been proved by his experiments, even though, as he later discovered, they demonstrated an entirely different (and correct) law. After this misadventure, Newton no longer attempted to use the motion of light particles to deduce the mathematical properties of periodic colors. Henceforth the particles served only as the entities that are reflected, transmitted, or absorbed in encounters with the aether or corpuscles of bodies. The first stage in Newton's investigations of periodic colors and development of his vibrational hypothesis culminated with his "Observations" and "Hypothesis" in 1675.

The second stage, his work on thick plates in the early 1690s, vindicated his continued pursuit of the hypothesis of vibrations. By developing his model of vibrations still further, he was able to calculate the diameters of the rings and explain all the other features of thick plates. Newton had confirmed his hypothesis by successfully extending it to an entirely new phenomenon, and he felt justified in considering it a demonstrated theory. This brief phase is represented by the draft on corpuscular vibrations (Prop. 12), composed and then rejected for the first complete state of the *Opticks*. At this point, in preparing his formal account of the theory, Newton's methodological strictures came into play, and he attempted to eliminate all traces of the hypothetico-deductive method and the hypotheses that he had utilized in discovering the theory. He now formulated the theory of fits as a set of "abstracted" properties, so that it appeared that they were all derived from experiment. Newton recognized that this practice would result in a loss of intelligibility, and so many commentators have since bemoaned it.

In attempting to eliminate all hypotheses from his theory of fits, Newton encountered two problems. He remained unaware of the first one, that his theory of fits was still permeated by the emission theory of light in a latent form. The location of the physical activity of the fits at the second surface, the variation of the interval with direction of propagation, and the consideration of rays rather than wave fronts are features indelibly wedded to the emission theory.

The second problem was that he was unable to present the theory of fits solely as a set of properties discovered by experiment, the method adopted in the rest of the *Opticks*. At a number of points he had to appeal to the forthcoming part on thick plates, which afterward confirmed those propositions. Mathematical properties such as the laws of continuity in Props. XIX and XX simply could not be directly derived from or confirmed by phenomena. Rather, they were justified by their successful predictions from detailed calculations. Newton's actual method was complex and varied according to the scientific problem confronting him.

Concern for Newton's use of hypotheses has in the main been imposed on subsequent generations by Newton's own frequent claims that they have no place in a properly formulated scientific theory. The charge of hypocrisy improperly lingers over his work. That hypotheses could be used to develop a theory Newton freely allowed, and we have seen him use hypotheses in this way. Where he ran into problems was in the next stage of his methodological program, in his attempt to create an "abstracted" theory without any hypotheses. We could ask here whether it is actually possible to formulate a theory without hypotheses. Certainly Newton did not succeed despite a sustained effort. Rather, we can ask what were the consequences of his attempt to achieve a science free of hypotheses. The subsequent history of his theory suggests that they were deleterious. Although Henry Brougham's description of the theory of fits in 1796 as "the smoke of unintelligible theory" is excessive,[137] it accurately reflects the laments about it over the centuries. Newton's methodological requirement that a theory not contain any hypotheses and thus be set apart from them is, I claim, in no small part responsible for the theory's terse formulation and physically unsatisfying nature. Would the theory of fits have fared better in the eighteenth century had Newton presented it in the form in which he arrived at it, or if he had otherwise reformulated it? Perhaps, but a number of other factors appear to have contributed to its poor reception. Because I have not made a thorough study of the reception of the theory, the following comments should be taken as suggestions to be further examined.

Although one problem with the theory of fits was its abstract presentation, in none of the hypotheses that Newton did propose in order to make it more intelligible was he able to interpret it in terms of forces and particles. All of his models were based on some vibrating

[137] For the full quotation see Ch. 7, note 31.

medium, yet the dominant school of natural philosophy in the eighteenth century, especially on the Continent, followed Newton's program for a new physics based on short-range forces. The approach of Bošković and Benvenuti, which was based on particles and forces, was too mathematical for most until the end of the century.

Relative to the state of eighteenth-century optics the theory of fits was intrinsically difficult and mathematically sophisticated. Few grasped that it was a mathematical theory and that its real test was its quantitative, predictive power. When the theory was presented at all in eighteenth-century texts, it was only as a qualitative disposition of the rays to be easily reflected or transmitted.[138] Very few took any notice of the colors of thick plates, which is the theory's true measure. The parallel history of Huygens's wave theory of light leads me to believe that the mathematical nature of these theories significantly impeded their comprehension and adoption. Huygens's theory was the only other seventeenth-century optical theory similar to Newton's in its mathematical structure, in its mathematical-physical assumptions about properties of light that were only indirectly confirmable, and in its consequent dependence on making quantitative predictions about a relatively obscure new class of optical phenomena (periodic colors and double refraction). It, too, was all but ignored in the eighteenth century.[139]

The theory of fits, like Huygens's wave theory, began to be seriously studied and garner supporters only at the end of the eighteenth century just as the mathematical and quantitative style of physics, whose seeds were planted by Newton and Huygens, began to flourish. By this time the abstract mathematical nature of the theory of fits was more of a virtue than a liability as the French physicists led by Biot

[138] See, for example, Thomas Rutherforth, *A System of Natural Philosophy* (1748), 1:475–6; and John Rowning, *A Compendious System of Natural Philosophy* (1744), Pt. III continued, 1:163–7 (see Ch. 5, note 30, for the complex publication history of this work). Even as mathematically sophisticated a work as Robert Smith's *Compleat System of Opticks* (1738) omitted the theory of fits. Juval Le Roy, the French translator of Smith's *System*, thus found it necessary to add copious notes on the colors of thin films, but he explained the theory of fits only very briefly and qualitatively; *Traité de optique* (Brest, 1767), p. 244. In contrast, Henry Pemberton showed a real grasp of the theory of fits by applying it to a new phenomenon. In 1723, he attempted to explain the formation of supernumerary rainbows by means of fits, though he did not derive any quantitative results; "A letter to Dr. Jurin," pp. 245–8. See Carl B. Boyer, *The Rainbow*, pp. 277–9; and on Pemberton's *View of Sir Isaac Newton's Philosophy*, which presents a popular account of Newton's theory of fits (pp. 371–7), see Ch. 5, note 28.

[139] See Shapiro, "Kinematic optics," pp. 252–8; and Jed Z. Buchwald, "Experimental investigations of double refraction," pp. 334–5.

provided the theory a new physical interpretation (or "hypothesis") utilizing particles and forces. That the fates of two similar sorts of quantitative physical theories paralleled the development of quantitative physics itself in the eighteenth century suggests that the mathematical nature of the theory of fits significantly affected its reception.

At mid-century, when little enthusiasm for the theory of fits had yet manifested itself,[140] two experimental studies of the colors of thin films and thick plates appeared that cast doubt on the validity of Newton's research in Book II of the *Opticks*. In 1752, the Abbé Guillaume Mazéas reported that when he was rubbing two flat pieces of glass against one another he found colored rings that appeared like Newton's rings. He concluded that these colors do not depend on either the thickness or the nature of the substance between the plates, which is utterly contrary to Newton's (and Hooke's) conclusions on the origin of the colors of thin films. After carrying out numerous experiments, such as heating the plates, placing suet and sediment of urine between them, and putting them in a vacuum chamber, he rejected Newton's theory of fits and called for more experiments.[141] Mazéas's work was solidly in the tradition of qualitative experimental natural philosophy. He was even skeptical of Newton's measurements of the thickness of thin films because he thought it "very difficult" to make exact measurements of such small distances as millionths of an inch.[142] In fact, Mazéas was not observing the colors of thin films but diffraction from the scratches caused by rubbing the two glasses together.[143]

The work of the duc de Chaulnes, in a paper read in 1758, was ultimately of greater significance. While repeating Newton's experiments on the colors of thick plates, he happened to breathe on the mirror and noticed that the rings became much more distinct and vivid. He found that he could make the effect permanent by coating

[140] For example, after presenting the theory of fits, Rowning declared: "It is too much clogged with Suppositions; neither is it consonant to that Simplicity, Uniformity, and Regularity, with which Nature is every where observed to act" (*Compendious System*, 1:167). D'Alembert in 1754 said of the theory that "however ingenious it is, it is not nearly sufficient to entirely convince and satisfy the mind" ("Couleur," p. 330).

[141] Mazéas, "Observations sur des couleurs engendrées par le frottement des surfaces planes et transparentes" (1755), pp. 35–6. This paper, published by the French Academy of Sciences, is a revised and expanded version of a paper with the same title published by the Berlin Academy in 1752, p. 259.

[142] Mazéas, "Observations" (1755), p. 37; not in the earlier version.

[143] In his *Course of Lectures*, 2:625, Young described Mazéas's investigation as being on "striated surfaces."

the upper surface of the mirror with a mixture of milk and water and then allowing it to dry into a haze. From his experiments he concluded, correctly and contrary to Newton, that the rings are formed by the first surface, and that the second one only serves to reflect a sufficient number of rays to make the rings visible.[144] He attributed the cause of the colored rings to diffraction in the pores of the first surface. Substances placed on or before the first surface served, according to him, to increase the quantity of pores, and thus the brightness of the rings.[145] Thus, rather than simply shifting the seat of the scattering from the second to the first surface, the duc altogether abandoned Newton's explanation by scattering.[146] Although his experiments were more disciplined than Mazéas's, they too remained entirely qualitative. He gives no evidence that he appreciated Newton's quantitative explanation of the colors of thick plates or felt it necessary to develop an alternative method of calculating the rings. This is a bit surprising. The duc later devoted much effort to improving precision instruments, especially to designing machines for graduating measuring scales.[147]

These papers attracted much attention, particularly Mazéas's, coming as it did with an endorsement by Euler, whose paper on the colors of thin films immediately followed Mazéas's in the *Histoire* of the Berlin Academy of Sciences.[148] In his influential history of optics in 1772, Priestley devoted half of his chapter on eighteenth-century research on periodic colors to a paraphrase of it. He began the chapter

[144] Michel-Ferdinand d'Albert d'Ailly, duc de Chaulnes, "Observations sur quelques expériences de la quatrième partie de deuxième livre de l'Optique de M. Newton" (1755), p. 140. Although it is in the 1755 volume (which was published in 1761), the paper is dated 1 March 1758, p. 136.

[145] Ibid., pp. 143–4.

[146] The duc's discovery that the rings become more intense when the first surface acquires a haze was known to the early wave theorists, Young, Herschel, and Stokes, who invoked it in their explanations. Biot, *Traité*, 4:195–229, repeated the duc's experiments and defended Newton's theory.

[147] See Maurice Daumas, *Scientific Instruments of the Seventeenth and Eighteenth Centuries*, pp. 197–200; and also the *éloge* by Fouchy (1769).

[148] Euler's paper is discussed in §5.2. Pieter van Musschenbroek in his influential textbook, *Cours de physique expérimentale et mathématique* (1769), §§1837, 1842, 2:499–500, 501–2, cited and extended Mazéas's experiments and declared the cause of the colors of thin films to be unknown. This is a translation of his *Introductio ad philosophiam naturalem* (1762). Etienne François Dutour criticized Mazéas's experiments and carried out new ones without coming to any general conclusion; "Recherches sur le phénoméne des anneaux colorés" (1763). The French translation of Smith's *System*, pp. 241–4, adds a full summary of Mazéas's paper; Johann Samuel Traugott Gehler, in his *Physikalisches Wörterbuch*, "Farben," 2(1789):147–8, rejects the theory of fits and cites the work of Mazéas, Musschenbroek, and Dutour.

with a harsh judgment of Newton's research: "In no subject to which he gave his attention does he seem to have overlooked more important circumstances in the appearances he observed, or to have been more mistaken with respect to their causes."[149] Priestley also presented a paraphrase of the duc de Chaulnes' memoir in this chapter. To a community already little committed to the theory of fits, the work of Mazéas and the duc de Chaulnes could serve as a justification for ignoring it.

Young's assessment of Priestley's *History* in 1807 nicely captures the change that physics had undergone in the intervening thirty-five years:

> Dr. Priestley rendered an essential service to the science of optics, considered as a subject for the amusement of the general reader, by an elegant and well written account of the principal experiments and theories. . . . But this work is very deficient in mathematical accuracy, and the author was not sufficiently master of the science to distinguish the good from the indifferent.[150]

The emphasis on mathematics is one of the hallmarks of the new physics and was applied as well to experimental research. Although Young's optical research laid the groundwork for the rejection of Newton's theory of fits two decades later, his understanding of that theory and carrying out of controlled experiments in search of mathematical laws mark the return to a quantitative physics pioneered by Newton.

It is perhaps ironic that for most of the eighteenth century the theory of fits was considered implausible and unconvincing, while the equally implausible – or, if you prefer, equally plausible – theory of colored bodies was widely accepted. Then, as the theory of colored bodies came under attack, the theory of fits was being more respectfully received.[151] The theory of fits was largely unchanged throughout

[149] Priestley, *History*, p. 498. These experiments became still more widely known after Charles Hutton included extensive extracts from Priestley in his article "Chromatics," pp. 724–8, in the *Encyclopaedia Britannica* (1797); for the identification of Hutton as the author, see Ch. 7, note 23. In Germany this critical tradition was propagated by Georg Simon Klügel's translation of Priestley's *History* in 1775–6.

[150] Young, *Course of Lectures*, 1:480–1. His judgment of Mazéas and Dutour was not much more favorable: "The experiments of Mazéas on the colours of thin plates are mere repetitions of those of Newton under disadvantageous circumstances; Mr. Dutour has, however, considerably diversified and extended these experiments . . . yet without obtaining any general results of importance" (p. 480).

[151] In France the theory of fits was widely adopted from about 1790 to 1820 as part of the Laplacian program of restoring Newtonian physics. In England Herschel and

the period. What did change, I believe, was the context in which it was received, for a community of physicists committed to mathematical theories and quantitative experiments had developed in the interim. This will become evident from the history of the theory of colored bodies in the eighteenth and early nineteenth centuries, presented in the next part of this book.

Brewster utilized it, while Young appreciated it (see note 3). It also garnered support in Switzerland and Italy; see Pierre Prevost, "Quelques remarques d'optique" (1798), p. 330; and Giovanni Battista Venturi, "Indagine fisica sui colori" (1799), p. 702.

[illegible] happened. What did change, I believe, was the context in which it was received, for a community or public [illegible] committed to [illegible] political [illegible] and [illegible] arguments had [illegible] in the interim. This [illegible] from the history of the theory [illegible] explored [illegible] in the eighteenth and early nineteenth [illegible] examined in the first part of the book.

PART II

PHYSICS AND CHEMISTRY: THE THEORY OF COLORED BODIES, THE CHEMISTS' REVOLT, AND ABSORPTION SPECTROSCOPY

5

THE GLORY YEARS: 1704–1777

Newton's theory of colored bodies had a long and tumultuous history before it was finally refuted in the early 1830s. For nearly three quarters of a century it was widely accepted by natural philosophers and served as a basis for research. Just as the theory was at its high point and apparently being confirmed and extended by Delaval, chemists subjected it to an aggressive attack. The revolt of the chemists can be considered to have broken out in 1776 with a paper by Christophe Opoix that was read to the Academy of Sciences by the doyen of French chemists, Pierre-Joseph Macquer. To chemists, it was, on the face of it, misguided to think, as the Newtonians did, that the color of a body depends solely on the size and density of its corpuscles and is independent of its chemical composition. Chemists, and an increasing number of physicists, considered light to be a chemically active substance that possessed elective affinities for the matter that composed bodies and entered into chemical composition with it, just as any other material substance – a view that Dorian Brooks Kottler has aptly called "the chemistry of light."[1] Because sunlight, as Newton showed, is a heterogeneous mixture of different colors, advocates of this view explained the colors of bodies by assuming that the constituents of bodies had different affinities for the variously colored particles of light, some of which were selectively absorbed. At this time affinities were conceived of as Newtonian forces acting between corpuscles of matter, in this case between those of bodies and of light, so that the way was now set for a conflict between a genuine Newtonian explanation of the cause of coloration and Newton's own theory. The situation is fraught with irony. Perhaps no other physicist

[1] Kottler, "Jean Senebier and the emergence of plant physiology," p. 67.

of Newton's stature was so deeply immersed in (al)chemical experiment and study.

The principal evidence supporting the chemistry of light and leading to the chemists' restiveness was drawn from phosphorescence (which will not concern us) and the increasing number of photochemical reactions that were being discovered and studied: the darkening of silver chlorides by light; the discoloration of dyes, wood, and other substances in sunlight; and, most notably, photosynthesis. Jean Senebier's research on photosynthesis and photochemistry, which began just as Opoix launched his attack and culminated in his *Mémoires physico-chymiques* in 1782, was the most important source for diffusing knowledge of the chemistry of light. Also contributing to the chemists' revolt against Newton's theory was the rise of dye chemistry, one of the great success stories of eighteenth-century industrial chemistry. Dye chemists were able to alter the colors of bodies with a range of dyes and chemical processes and quite naturally interpreted the cause of color and color change as chemical and not mechanical. The leading critics of Newton's theory at the height of the conflict at the turn of the century were both dye chemists, Claude-Louis Berthollet in France and Edward Bancroft in England.

Beginning about a decade after the chemists launched their attack on Newton's theory, optical experiments were brought to bear against it. In 1785, Delaval, hitherto the theory's most prominent advocate, published a careful experimental refutation of the central tenet of the phenomenological theory, that colored bodies reflect colored light. A still more important experimental technique, whose significance extended beyond coloration, was the observation of the absorption spectra of colored substances. When a narrow beam of sunlight is transmitted through a transparent colored body and observed through a prism, its spectrum is found to consist of colored regions separated by dark bands. To generations of scientists who were conditioned to see a classical Newtonian spectrum consisting of a continuous gradation of all the colors of the rainbow, it was difficult to conceive of a discontinuous spectrum and to grasp its physical significance. To many it was also surprising to find that the color transmitted by, say, a blue glass or green solution was not in general a simple blue or green but rather a complex mixture of widely separated colors. John Herschel's description in 1822 of the absorption spectra of colored glasses vividly captures

> the singular phenomena . . . of an almost total obliteration of some of the colours by certain glasses, while others intermediate between, or sharply

bordering on those obliterated, appeared to be transmitted in all their brilliancy, thus producing an image . . . not consisting of a broad band of gradually varying colour, but an assemblage of more or less sharply defined coloured streaks of different breadths and colours, separated by intervals, in some cases absolutely black, – in others only feebly illuminated.[2]

The appearance of absorption spectra depends on the nature and intensity of the illumination and, most crucially, the nature and thickness of the colored substance. As the thickness or density of a substance increases, the intensity of each transmitted color decreases at a rate independent of the others; and the colors successively vanish, until only one, and finally none, remains. By observing absorption spectra in different substances at various thicknesses, the processes of selective absorption and coloration could be experimentally studied. As it turned out, Newton's theory predicts similar compound spectra with missing bands of colors (§2.4). Consequently, once absorption spectra were observed, the theory initially appeared to some, such as Thomas Young, to be even more prescient. To those who supported a chemical view, however, the missing colors in absorption spectra directly represented those colors absorbed by elective affinities, and a theory of coloration by elective affinities and absorption was opposed to Newton's theory of coloration based on thin films. Absorption spectra appeared to allow a decisive test of Newton's theory to be made by comparing its definite predictions of the composition of such spectra with observations of absorption in various colored bodies. When such tests were made about the turn of the century by Young, Giovanni Battista Venturi, Jean-Henri Hassenfratz, and Claude-Antoine Prieur-Duvernois (Prieur de la Côte-d'Or), all of them declared it to have partly or totally failed. Nonetheless, many physicists – at least those in a mathematical and quantitative tradition – did not abandon the theory. In France, René Just Haüy and Biot put up a spirited defense against the chemists Berthollet and Hassenfratz. In Britain, in this same period, there was no such dramatic confrontation; in fact, there was no confrontation at all. No one, physicist or otherwise, stepped forward to defend Newton's theory.

Not only did many physicists not abandon Newton's theory, neither they nor chemists carried out further experimental investigations of selective absorption after the outburst at the turn of the century. This first birth of absorption spectroscopy has totally eluded historians of spectroscopy. And it must be considered as genuine spectroscopy, for by means of prisms spectra of absorbed light were being examined.

[2] Herschel, "On the absorption of light," p. 446.

The study of absorption spectra was independently taken up again in Britain in 1822 and entered the mainstream of physics through the efforts of David Brewster and John Herschel to develop monochromatic light sources using colored filters. Their method of observation was virtually identical to that of Venturi, Young, and Hassenfratz. Between 1827 and 1831, Brewster and Herschel at last refuted Newton's theory of coloration and established selective absorption – interpreted phenomenologically without affinities – as the principal cause of the colors of bodies. Thus our history ends with the elimination of both Newton's theory of colored bodies and the rival affinity theory, though selective absorption, which was in part suggested by the chemists' affinities, remained.

To understand the abortive development of absorption spectroscopy at the turn of the century and the refusal of many physicists to abandon Newton's theory, it will not suffice simply to invoke the stubborn sway of Newton's authority. Rather, a number of independent factors deflected further pursuit of these areas. Although the chemists' concept of elective affinities promoted the investigation of selective absorption, their approach to optics differed fundamentally from that of physicists. Chemists rarely used prisms and investigated spectra, for that was the realm of the physicist. Instead, they devoted themselves to determining and isolating "coloring principles" and to studying photochemical reactions such as the darkening of silver salts and the action of chlorine. Moreover, in the first decade of the nineteenth century, the prevailing concept of affinities as developed in the preceding century ceased to be a viable research program as a result of Berthollet's wide-ranging critique of that concept and Dalton's atomic theory. Quantitative physicists, in turn, had difficulty accepting selective absorption as an independent physical phenomenon because of the unlawlike and arbitrary behavior of affinities, and so they would not study it. A number of optical discoveries in the first fifteen years of the nineteenth century – infrared and ultraviolet radiation, polarization and crystal optics, and diffraction and interference – no doubt appeared more promising avenues of research to physicists than colored bodies. Finally, because all early observations of absorption spectra were carried out within the context of determining the cause of the colors of bodies, when absorption spectra were rediscovered in a different, physical context, the earlier literature was an entirely alien one. These factors contributed to the nearly complete discontinuity in research traditions at the turn of the century and in the 1820s.

The chemistry of light has been uniformly ignored by historians of chemistry and physics alike. When William Whewell wrote his *History of the Inductive Sciences* (1837), the struggle to establish the new wave theory in Britain had just concluded, and he was very much aware of the historical significance of the chemistry of light. In describing the earlier struggle between the wave and emission theories in the era before Young and Fresnel, he explained that Newton's emission theory

> was still more firmly established, in consequence of *the turn generally taken* by the scientific activity of the latter half of the eighteenth century; for while nothing was added to our knowledge of optical laws, the chemical effects of light were *studied to a considerable extent* by various inquirers; and the opinions at which these persons arrived, they found that they could express most readily, in consistency with *the reigning chemical views*, by assuming the materiality of light.[3]

Like other partisans of the wave theory, Whewell went on to dismiss the chemistry of light as "inevitably vague and doubtful" reasoning with no place in optical theory. Nonetheless, his assessment of its earlier dominance is valid, and it demands an independent study.

Virtually all the participants in this history adhered to the emission theory of light. That is not because of some intrinsic incompatibility between Newton's theory of colored bodies and the wave theory but rather because the emission theory predominated until the last few years of our story, and the chemistry of light itself reinforced confidence in its truth. In 1789, Gehler's *Physikalisches Wörterbuch* clearly expressed the symbiosis between the chemistry of light and the emission theory:

> It seems surely that a closer acquaintance with chemistry must incline everyone toward the system of emanation, because most chemists not only admit a *light matter*, but employ it as an essential ingredient in their best theories. To be sure, this is far from being a proof for its true existence, because both theories are only hypothetical, and some things can perhaps also be reconciled with Euler's system. Yet in fact there are phenomena where light displays an affinity for other substances and produces changes in the combination

[3] Whewell, *History of the Inductive Sciences*, 2:400. In a footnote Whewell enumerated some of the "various inquirers": "As Scheele, Selle, Lavoisier, De Luc, Richter, Leonhardi, Gren, Girtanner, Link, Hagen, Voigt, De la Metherie, Scherer, Dizé, Brugnatelli. See Fischer, vii. p. 20." The reference is to Johann Carl Fischer, *Geschichte der Physik*, 8 vols. (Göttingen, 1801–8). Fischer's omission of Senebier and Berthollet is striking even with a German bias.

and decomposition of matter that are difficult to ascribe to mere vibrations of the aether.[4]

Although the principal wave theorists considered here – Leonhard Euler, Young, Fresnel, and John Herschel – all opposed Newton's theory of colored bodies, Young and Herschel initially supported it. The wave theorists, however, could not adopt selective absorption by means of affinities, because light waves could not be imagined to be attracted to the corpuscles of bodies as light particles were. With the acceptance of the wave theory, the chemistry of light (though certainly not photochemistry) was no longer a viable research program, and alternative physical models based on the transfer of motion or energy, rather than matter, had to be developed.[5]

5.1. BACKGROUND: CHEMISTRY AND PHYSICS IN THE EIGHTEENTH CENTURY

One of the features that makes the history of Newton's theory of colored bodies particularly fascinating is the way in which it illuminates the development and relation of physics and chemistry in this period and their fundamentally different status in France, where physicists defended Newton's theory, and in Britain, where the chemical view was widely adopted with little resistance. It is therefore useful to sketch briefly the growth of these sciences in the two countries, though my account will not be confined solely to France and Britain. The various trends, traditions, and schools that I will describe do not represent a neat set of parallel developments, for as my story unfolds we shall see that there were tensions and conflicts among them.

Physics as a coherent mathematical, experimental, and quantitative discipline barely existed in Newton's day and was largely a creation of the late eighteenth and early nineteenth centuries.[6] By the end of the Scientific Revolution, the physical sciences had separated into two distinct groups. One, consisting of the mathematical sciences of me-

[4] Gehler, *Physikalische Wörterbuch*, "Licht," 2:902–3.

[5] Johann Wolfgang von Goethe, who also rejected Newton's theory of colored bodies in his *Zur Farbenlehre* (1810), will not be treated here, for he simply was not a participant in the debate that I describe. Goethe's rejection of Newton's theory of colored bodies was part of a much broader attack on the entirety of Newton's optical work and the very nature of Newtonian science. He thus remained largely outside the communities described in this book.

[6] Thomas S. Kuhn, "Mathematical versus experimental traditions in the development of physical science," in his *Essential Tension*, pp. 31–65; and Heilbron, *Electricity in the 17th and 18th Centuries*, Pt. I, which presents an overview of physics in this period.

chanics, astronomy, and geometrical optics, developed out of the intermediate sciences or mixed mathematics (§1.2). These mathematical sciences underwent profound conceptual transformations during the Scientific Revolution and formed one focus of that revolution. Galileo, Newton, Huygens, Leibniz, and the Bernoullis, among others, were progenitors of the new mathematical physics, but Newton's *Principia* was above all the exemplar for the "mathematical way." The second group, consisting of a cluster of experimental areas, evolved out of the old *physica*, the qualitative study of the motions and qualities of natural bodies. It was transformed by the rise of the new "experimental philosophy," which formed a second focus of the Scientific Revolution.

Experiment had played only a minimal role in the conceptual revolution of the mathematical sciences. Some fields, like physical optics, acoustics, and fluid mechanics (including hydrostatics and pneumatics), partook of the developments of each group. During the eighteenth century, electricity, magnetism, heat, and physical optics grew into distinct experimental sciences. By the first decades of the nineteenth century, all of them had been mathematized, thereby combining what had been two distinct streams of the Scientific Revolution and producing quantitative experimental physics. This should not be conceived of as a sudden development at the turn of the nineteenth century but rather as the culmination of a century-long process. Newton and Huygens had succeeded in combining mathematics with precise measurements in the newly emerging experimental fields. But if we consider the example of Newton's research on periodic colors or "interference" and Huygens's on double refraction, we see that they remained isolated (and largely ignored) exemplars of quantitative physics (§4.5).

The endeavor to broaden the experimental scope of the mathematical sciences and to extend mathematics to the experimental branches of physics and make them quantitative demanded the ability to make precise, reliable, and meaningful measurements. During the last decades of the eighteenth century new standards for carrying out experimental physics evolved: utilization of specially designed and constructed instruments, adoption of carefully conceived and controlled experimental procedures, and expression of data by mathematical laws. This process of creating a quantitative experimental physics was centered in France and was essentially complete there by about 1815. In the battle between physicists and chemists over Newton's theory of colored bodies, it turns out that it was the representatives of the new, quantitative physics who defended Newton's theory. After 1775, natural philosophers (or "physicists" in the original, older meaning of the term), who were primarily concerned with a qualitative, causal account of the physical world, increasingly abandoned Newton's theory. This group remained an important one in Britain into the 1820s.

The development of physics in France illustrates many of these trends. Science was centralized in Paris at the Academy of Sciences and other key institutions. Because of this centralization there tended to be a uniformity in outlook that was punctuated by clashes between opposing groups. British science throughout our period of concern was more diffused – in London, Cambridge, Edinburgh, and Manchester. At Newton's death, in 1727, French science was dominated by Cartesianism, but by the end of the next decade the Newtonian synthesis of mechanics and astronomy was successfully introduced by Pierre-Louis Moreau de Maupertuis, Pierre Bouguer, Alexis-Claude Clairaut, and Jean Le Rond d'Alembert.[7] Britain simply did not possess mathematical physicists of this caliber, and the disparity would only increase through the course of the century as Pierre-Simon de Laplace, Joseph-Louis Lagrange, and Charles-Augustin Coulomb, for instance, entered the Academy of Sciences. Whether they were mathematicians, astronomers, or engineers, all had a common commitment to mathematics and mechanics, and they gradually extended this approach to optics, acoustics, and electricity. Qualitative experimental physics, or natural philosophy, as exemplified by the electrician Jean-Antoine Nollet, who was a member of the academy in the middle third of the century, was at the same time gradually suffering a retreat until it all but vanished in the first decades of the nineteenth century. Coulomb's demonstration in 1785 by means of his torsion balance that the electrostatic and magnetic forces obeyed an inverse square law represents the rise of the new quantitative physics and served as a model for its further refinement. Its hallmarks are evident in his specially designed and carefully constructed instrument, measurements, search for sources of error, and determination of mathematical laws.

The nature and reality of Newtonian forces were at first troublesome issues for physicists and philosophers, although chemists more readily accepted them in the guise of chemical affinities. However, in the

[7] See Pierre Brunet, *L'introduction des théories de Newton en France;* and John L. Greenberg, "Mathematical physics in eighteenth-century France."

early 1780s, Laplace declared his commitment to the program advocated by Newton in the preface to the *Principia* and queries of the *Opticks* to create a physics based on short-range forces between corpuscles. Short-range forces, or affinities when applied to chemical processes, were a unifying concept of French physical science for a generation and would serve as one of the principal bonds between physics and chemistry. During the first two decades of the nineteenth century, Laplace and his followers dominated both the conceptual and institutional structure of French physics. A veritable torrent of fundamental contributions to optics, electricity, and heat poured forth from André-Marie Ampere, Dominique-François-Jean Arago, Biot, Jean-Joseph Fourier, Fresnel, Etienne-Louis Malus, and Siméon-Denis Poisson in the first quarter of the century. Although this activity was not all Laplacian or Newtonian, it was mathematical and quantitative.[8] The rapid development of high standards for quantitative experimental physics between the 1780s and 1800s is evident in optics in the investigations of double refraction and polarization by Haüy, Malus, Laplace, and Biot.[9] In brief, between 1704 and about 1825, mathematics had extended its reach from mechanics, astronomy, and geometrical optics to the experimental sciences of physical optics, heat, and electricity and magnetism, and in the process the discipline of physics had been transformed.

Chemistry in France, on the other hand, in the first half of the eighteenth century, was undeveloped and still closely tied to pharmacy, medicine, and metallurgy. Only in midcentury did French chemistry develop into an independent physical science and broaden its scope.[10] The prime agency for this transformation were the inspiring and seminal lectures delivered by Guillaume-François Rouelle at the Jardin du Roi from 1742 to 1768.[11] In his popular lectures (which were attended by Denis Diderot, Antoine-Laurent Lavoisier, Macquer, and Gabriel-François Venel, among others) Rouelle introduced

[8] On French physics, particularly optics, and Laplacianism around the turn of the century I have found the following most useful: Robert Fox, "The rise and fall of Laplacian physics"; Robert H. Silliman, "Fresnel and the emergence of physics"; and especially Eugene Frankel, "J. B. Biot and the mathematization of experimental physics."

[9] Buchwald, *Rise of the Wave Theory*, Ch. 1.

[10] For surveys of eighteenth-century chemistry, particularly in France, see Guerlac, "Some French antecedents of the chemical revolution"; Maurice Crosland, "The development of chemistry"; and "Chemistry and the chemical revolution." On chemistry as an independent science, see Meinel, "Theory or practice?"

[11] Rhoda Rappaport, "G. -F. Rouelle"; and "Rouelle and Stahl"; and Douglas McKie, "Guillaume-François Rouelle."

and developed the ideas of Georg Ernst Stahl and other German chemists. The chemists' revolt against Newton's theory of color can be related to the spread of Stahlian chemistry – in Britain as well as France – in the quarter-century preceding its outbreak. Although Stahl is now closely associated with phlogiston, his principal goal was to reform chemistry and establish it as an autonomous science with its own methods, concepts, and subject matter. The Stahlian independence movement was accompanied by a polemic against the contemporary practice of physics. Rouelle's students, Venel and Macquer, disseminated the new chemistry in France.

The Chemical Revolution, instituted by Lavoisier in the 1770s, marked the maturity of French chemistry as well as the resolution of the Stahlian conflict with physics. With its reliance on instruments and measurements, chemistry was becoming increasingly allied with experimental physics. In an anonymous review of his own book, *Opuscules physiques et chimiques* (1774), Lavoisier declared that he had "applied to chemistry not only the apparatus and method of experimental physics, but the spirit of exactitude and calculation that characterizes this science."[12] The new status of chemistry is perhaps best symbolized by the long collaboration of Lavoisier and Laplace on heat that began in 1777.[13] In the next decade, as Lavoisier drew physicists, such as Gaspard Monge and Alexandre-Théophile Vandermonde, and chemists of the stature of Berthollet, Jean-Antoine Chaptal, Antoine-François de Fourcroy, and Louis-Bernard Guyton de Morveau into the antiphlogiston ranks, both the stature of chemistry and its alliance with physics continued to grow. At the beginning of the nineteenth century, the chemists' ranks were further enhanced by the younger generation of Pierre-Louis Dulong, Louis-Joseph Gay-Lussac, and Louis-Jacques Thenard.

After Lavoisier's execution in the Reign of Terror during the French Revolution, Berthollet assumed the leadership of French chemistry, and he joined forces with Laplace to continue to bring the two disciplines together. In his attempt to found chemistry on affinities, a program that had been underway since midcentury, Berthollet was as avid a proponent of short-range forces as Laplace. In the informal

[12] Lavoiser, *Oeuvres*, 2:96; originally published in *Histoire de l'académie des sciences* (1774) [1778]. Lavoisier is identified as the author of the review in *Oeuvres*, 5:320.

[13] Guerlac, "Chemistry as a branch of physics." The relation of physics and chemistry in the chemical revolution has recently been much discussed; see Evan M. Melhado, "Chemistry, physics, and the chemical revolution"; and it is touched upon in many of the papers in Arthur L. Donovan, ed., *The Chemical Revolution*, especially the editor's own "Lavoisier and the origins of modern chemistry."

Society of Arcueil, which met in Berthollet's home, he and Laplace encouraged collaborative work among the bright young generation of physicists and chemists, which included Arago, Biot, Chaptal, Gay-Lussac, and Malus.[14] The chemists' attack on Newton's theory of colored bodies can in part be interpreted as reflecting the growing maturity and autonomy of chemistry. During the first half of the eighteenth century, chemistry was, for the most part, so far removed from physics that there was insufficient common ground for controversy. The aggressive attack launched by Opoix with Macquer's encouragement marked both an assertion of chemistry's independence as a science and an effort to capture the common intellectual territory of color. By the turn of the century, when physics and chemistry were still more closely allied, the dispute may have become less vituperative, but it continued, as the two fought over common ground with fundamentally different approaches.

In Britain the development and relation of physics and chemistry was altogether different.[15] From Newton's death in 1727 until the beginning of the nineteenth century, there were scarcely any mathematical physicists – Brook Taylor, Colin Maclaurin, Henry Cavendish, and Thomas Young – let alone a physics community committed to mathematics and quantitative experimentation. Young, whose career straddles the turn of the century, marks the revival in Britain of physics as an exact science, as he was in the succeeding decades followed by Brewster, John Herschel, Michael Faraday, and George Biddell Airy. Chemistry, on the contrary, continually flourished from Boyle's day: Newton himself, in the queries of the *Opticks*, and Stephen Hales, in the first half of the century, and Cavendish, Priestley, Joseph Black, John Dalton, and Humphry Davy, reaching into the beginning of the nineteenth century, represent a notable succession of chemists. Chemistry also possessed a different intellectual status in Britain than in France, for throughout this period it was an integral part of natural philosophy.

In the first part of the eighteenth century John Keill and John Freind attempted to develop a Newtonian chemistry based on short-range forces between corpuscles, much as Newton himself had suggested

[14] Crosland, *The Society of Arcueil*.

[15] For British natural philosophy, physics, and chemistry in this period, see Robert E. Schofield, *Mechanism and Materialism;* Thackray, *Atoms and Powers;* Heimann and McGuire, "Newtonian forces"; Heimann, " 'Nature is a perpetual worker' "; and "Ether and imponderables"; Richard L. Ziemacki, "Humphry Davy and the conflict of traditions," Ch. 4; and Donovan, *Philosophical Chemistry in the Scottish Enlightenment*.

in his queries. In the second half of the century an aether with repulsive forces, which Newton proposed in his last queries, largely replaced the earlier corpuscular model. This shift occurred partly under the pressure of the growth of chemistry into an independent branch of natural philosophy. The active aether that they adopted was variously conflated with the electrical fluid, light, and especially the chemists' fire and phlogiston. These imponderable fluids were quality-bearing substances and intrinsically chemical in their conception. Indeed, as Heimann has shown, British natural philosophy in the late eighteenth century had become almost "alchemical," as these substances were capable of transforming one into the other and were a self-sustaining source of nature's active powers. Much of the concern of this school was with developing a comprehensive, causal account of the physical world, an approach that was rapidly vanishing in France at the same time. Thus, by the end of the century, British natural philosophy had assumed a chemical outlook with no counterbalance from a physics community.

Even from this very brief sketch at least one reason should be evident why at the turn of the century physicists did not defend Newton's theory of colored bodies and the chemical alternative was so widely accepted: There were virtually no physicists, and British physical science had adopted a chemical outlook. It should also be apparent that in both countries, around 1800, physics and chemistry were more closely allied than they were before and would be afterward.

The chemists' attack on Newton's theory of colored bodies represents not only disciplinary development and rivalry but also deeper, epistemological and methodological differences, introduced by the Stahlians, whereby chemistry was attempting to distinguish itself from physics.[16] By about 1730, it was evident to chemists that mechanical explanations of chemical phenomena – or what Newton called "vulgar," or "common," chemistry – based on primordial corpuscles of a single sort of qualityless matter that differed only in size, shape, and motion were unfruitful, and a reaction against them set in.[17] Chemists proposed as an alternative that their science be based on specifically chemical criteria and reintroduced quality-bearing sub-

[16] I am endebted to Arthur Donovan for indicating the epistemological issue underlying the chemists' attack on Newton's theory.

[17] Martin Fichman, "French Stahlism and chemical studies of air"; and J. B. Gough, "Lavoisier and the fulfillment of the Stahlian revolution." For Newton's terms, see Ch. 2, note 67.

stances whose properties were defined in terms of chemical operations. Chemical explanations were to be in terms of particles of matter that possessed enduring chemical qualities that could be isolated and transmitted by chemical analysis and synthesis.

The progenitor of this approach was the German physician and chemist Georg Stahl, who vigorously rejected mechanical constructs in favor of experimentally derived chemical principles. To speak of primordial, qualityless corpuscles in the manner of Descartes, Boyle, and Nicolas Lemery, Stahl held, was to introduce the "occult," for we possess no empirical knowledge of them.[18] The physical principles are assumed "a priori," whereas the chemical principles are determined "a posteriori" as those "into which all Bodies are found reducible by the chemical operations hitherto known."[19] The new chemistry thus was avowedly (and self-righteously) more inductive than the mechanical chemistry in admitting only directly observable properties, such as the acidity, alkalinity, and combustibility of the particles. Like mechanical chemistry, it assumed that the elements were particulate, but the mechanical properties of the particles (the size, shape, and motion) did not enter into its explanations, only the experimentally determined qualities.

Stahl taught that the "Mechanical Philosophy, though it vaunts itself as capable of explaining everything most clearly, has applied itself rather presumptuously to the consideration of chemico-physical matters . . . it scratches the shell and surface of things and leaves the kernel untouched."[20] He introduces here a related idea in distinguishing chemistry from physics that would also be widely adopted: that physics deals with only the gross, aggregate properties of bodies, such as their weight, elasticity, and hardness, and not with their inner nature. Thus, to many chemists, the colors of bodies belonged to chemistry, not physics, and Newton's theory of colored bodies was certainly not a chemical but a physical theory. Indeed, it was a mechanical theory characteristic of the 1660s and pre-*Principia* physics, insofar as it depended solely on the size, density, and arrangement of homogeneous corpuscles. The dispute over colored bodies highlights not only the development of chemistry but also the transfor-

[18] J. R. Partington, *A History of Chemistry*, 2:665; translated from Stahl, *Specimen Beccherianum* (Leipzig, 1738), p. 18.

[19] Stahl, *Philosophical Principles of Universal Chemistry*, p. 4; which is Shaw's translation of *Fundamenta chymiae dogmaticae & experimentalis* (Nuremberg, 1723).

[20] Partington, *History*, 2:665; translated from Stahl, *Fundamenta chymiae*, Preface, sig. 2^v.

mation from pre- to post-*Principia* physics and the integration of the concept of force.

A consequence of the efforts of the French Stahlians to distinguish physics from chemistry is that their writings assumed a decidedly anti-physics tone, as is evident in the preceding quote from Stahl himself.[21] By claiming to possess a sounder methodology than physics and by appropriating the study of the inner nature of things, while relegating physics to external properties, chemists were effectively claiming theirs to be the superior science. Venel's *Encyclopédie* article "chymie," in 1753, is notorious for its antiphysical and antimathematical attitude. His purpose, and that of the other Stahlians, was not in fact to condemn physics as a science but rather to restrict it to its proper realm and establish chemistry as an autonomous science. Venel called for a "revolution that would raise chemistry into the position that it deserves and place it at least alongside of quantitative physics [*la Physique calculée*]."[22]

Despite great differences in precisely what they accepted as a chemical element or principle, most chemists in the second half of the eighteenth century accepted the Stahlian concept of an experimental, quality-based chemistry and rejected a mechanical approach. On methodological grounds they held that one could not talk about lower level entities than the principles. Peter Shaw, England's principal exponent of Stahlian chemistry, argued that the "genuine chemistry" should leave to "other Philosophers the sublimer Disquisition of primary Corpuscles, or Atoms," for these "remain undiscernible to the Sense," and are just "metaphysical Speculations" that lead to "a corrupt Fountain of Hypothesis and Illusion."[23] In 1763, William Lewis, a physician who pioneered the development of applied chemistry in England, clearly drew the distinction between the mechanical and the chemical in the preface to his *Commercium Philosophico-Technium:*

> Natural or mechanical philosophy seems to consider bodies chiefly as being entire aggregates or masses; as being divisible into parts, *each of the same general properties with the whole;* as being of certain magnitudes or figures, known or investigable; gravitating, moving, resisting, etc. with determinate forces, subject to mechanic laws, and reducible to mathematical calculation.
> *Chemistry considers bodies as being composed of such a particular species of matter;*

[21] Crosland, "Development of chemistry," pp. 409–10; Fichman, "French Stahlism," p. 101; and Gough, "Lavoisier," pp. 26, 28–9.
[22] Venel, "Chymie," p. 409.
[23] Shaw, *Chemical Lectures* (1755), p. 146; the first edition of 1734 is substantially the same.

dissoluable, liquefiable, vitrescible, combustible, fermentable, etc. impregnated with colour, smell, taste, etc. or consisting of dissimilar parts, which may be separated from one another, or transferred into other bodies. The properties of this kind are not subject to any known mechanism, and seem to be governed by laws of another order.[24]

He continually stressed that in chemical reactions "the laws of the mechanical philosophy have no place," and that the mechanical and chemical must "be kept distinct, as many errors have arisen from applying to one such laws as obtain only in the other."[25]

In France, we see that Macquer at midcentury likewise insisted that it would simply "tire our minds with vain conjectures" to describe the corpuscles that compose the elements, for "our senses cannot possibly discover the principles of which they are themselves composed."[26] Thus, whereas the chemists' attack on Newton's theory of colored bodies and their contention that color is a chemical property represent an assertion of the independence of their discipline, on a deeper level they represent the rejection of reductive, mechanical explanations that was initiated by the Stahlians. At the same time that chemists were attempting to distinguish their science from physics and launching their attack on Newton's theory, British natural philosophy generally was moving toward a chemical view, so that the two movements supported one another and worked against Newton's theory of colored bodies.

5.2. NEWTON'S THEORY AND NATURAL PHILOSOPHY

Most Newtonian natural philosophers accepted the theory of colored bodies as an integral element of Newton's optical theory. Only his attempt to assign the size of a body's corpuscles by its color encountered any consistent opposition. This phase of widespread acceptance culminated in two publications by Delaval, who carried out a series

[24] Lewis, *Commercium Philosophico-Technicum* p. iv; italics added.

[25] Ibid., p. ix. See also F. W. Gibbs, "William Lewis," and Nathan Sivin, "William Lewis (1708–1781)."

[26] Macquer, *Elements of the Theory and Practice of Chemistry* (1758), 1:2; a translation of his *Elémens de chymie théorique* (1749) and *Elémens de chymie-pratique* (1751). Lavoisier accepted and refined the Stahlian conception of an element. In his *Traité élémentaire de chimie* (1789), he wrote: "If, by the term *elements*, we mean to express those simple and indivisible atoms of which matter is composed, it is extremely probable we know nothing at all about them; but, if we apply the term *elements*, or *principles of bodies*, to express our idea of the last point which analysis is capable of reaching, we must admit as elements, all the substances into which we are capable, by any means, to reduce bodies by decomposition" (*Elements of Chemistry*, p. xxiv).

of experiments to demonstrate that the colors of opaque bodies varied when the size and density of their corpuscles were altered, just as Newton's theory required. Delaval's work was widely known and highly regarded throughout Europe. One year after his second work was published and just as a French translation appeared, the chemists' revolt broke out in France.

The initial support for Newton's theory of colored bodies came, not unexpectedly, from the British Newtonians. George Cheyne, a member of the Scottish Newtonian medical circle centered about David Gregory and Archibald Pitcairne, offered what is probably the earliest endorsement in a book that appeared just one year after the *Opticks*. Cheyne, like many to follow, simply paraphrased and summarized Newton's account in the *Opticks* without providing any insights.[27] Henry Pemberton's *View of Sir Isaac Newton's Philosophy* (1728) was a more influential work that was translated into French, German, and Italian and helped to make the theory a better known part of Newton's optical work. Pemberton, a physician who was close to Newton in the later years of his life and who edited the third edition of the *Principia*, attempted to disarm the readers' resistance to the theory by warning that it was "no small paradox," "almost incredible," and "almost past belief; notwithstanding that the arguments, by which they are established are unanswerable."[28] In a chapter devoted to the theory of colored bodies, he presented an intelligent popularization. Richard Helsham, in lectures delivered probably in 1729 at Trinity College, Dublin, and published posthumously in 1739, presented Newton's theory,[29] as did John Rowning in his *Compendious System of Natural Philosophy*. Rowning's account is unnotable except that in expounding Newton's hierarchical matter theory, he introduced the "nut-shell" phrase that Thackray had made so well known, though he mistakenly attributed it to Priestley. The hierarchical theory, according to Rowning, "shews that the whole Globe of Earth, nay, all the known Bodies of the Universe, *for any thing that appears to us to the contrary*, may be composed of no greater Quantity of Matter than

[27] Cheyne, *Philosophical Principles of Natural Religion* (1705), Ch. 1, §42, pp. 88–90; which is also contained in the second edition of 1715, Ch. 2, §40.
[28] Pemberton, *A View of Sir Isaac Newton's Philosophy*, Bk. III, Ch. 2, pp. 338–9. See also Cohen's preface to the reprint.
[29] Helsham, *A Course of Lectures*, pp. 300–3. See Schofield, *Mechanism*, p. 31, on Helsham and the date of his lectures.

what might be reduced into a Globe of an Inch Diameter, or into a *Nut-shell.*"[30]

Robert Smith, Plumian Professor of astronomy at Cambridge, represented the more mathematical wing of Newtonians. His *Compleat System of Opticks* (1738) was the most influential optical treatise of the eighteenth century. Although he was primarily interested in geometrical and applied optics, he devoted the opening book of his work to "a popular treatise" on light and vision. In one chapter he presented Newton's theory of transparency, opacity, and colored bodies by means of verbatim extracts from the *Opticks*.[31] However, he passed silently over Newton's attempt to conjecture the size of the corpuscles of bodies from their color (Prop. 7). Thomas Rutherforth, in his lectures on natural philosophy delivered at St. John's College, Cambridge, likewise supported Newton's theory but went even farther than Smith and criticized Newton's attempt to establish the corpuscles' size. He argued in 1748 that it is most likely that the corpuscles are even denser than glass, so that Newton's table of colors (Figure

[30] Rowning, *A Compendious System* (1744). This work, which was first published in 1735–7, is a bibliographical nightmare, since the four parts were published separately and different "editions" include parts from earlier and later (!) "editions." In Vol. 1 of this 1744 edition (at Cambridge University Library) *Part III Continued. Containing Catoptrics, and the Doctrine of Light and Colours* bears London 1738 on its title page and continues the pagination of *Part III* (London, 1737); the "nut-shell" quotation appears in Ch. 13, p. 157; the first set of italics are mine. The account of Newton's theory of colored bodies is in Ch. 12, pp. 147–53. In a London 1753 edition at the Bodleian Library *Part III Continued* is dated London 1755; Rowning eliminated the "nut-shell" phrase here (p. 157) and altered the final few words to "reduced into any determinate Space how small soever." The "nut-shell" phrase remained in a London 1744–5 edition at the British Library, where *Part III Continued* is dated 1743. Although I was unable to locate a *Part III Continued* dated between 1743 and 1755, the only safe conclusion is that Rowning made this change between those years. Priestley is quite clear that he drew the phrase from elsewhere: "It has been asserted . . . *that for any thing we know to the contrary*, all the solid matter in the solar system might be contained within a *nut-shell*" (*Disquisitions Relating to Matter and Spirit* [1777], p. 17; italics added); and the similarity of the words in italics to Rowning's make him almost certainly the source. The bibliography in Priestley's *History* includes the 1744 edition of Rowning's *System*. See Thackray, "Matter in a nut-shell," p. 29. No doubt the ultimate source for the phrase, as P. M. Harman pointed out to me, is *Hamlet*, who proclaims, "I could be bounded in a nutshell and count myself a king of infinite space" (Act II, Sc. II, ll. 260–1).

[31] Smith, *Compleat System*, Bk. I, Ch. 8, 1:95–9. Translations were published in Dutch (1753), German (1755), and twice in French (1767). Francesco Algaroti, who was actually knowledgeable in optics, gave a charming account of the theory of colored bodies in *Sir Isaac Newton's Philosophy Explain'd for the Use of the Ladies* (1739), Dialogue V, 2:108–18. The original Italian, *Il Newtonianismo per le dame* (1737), was a European best-seller and was also translated into French (1738), German (1745), Dutch (1767), and Swedish (1782).

2.8) is useless.[32] At this point alone did the supporters of Newton's theory frequently balk. A few years later those ardent and sophisticated proponents of Newton's optics, Bošković and Benvenuti, also drew the line here. Benvenuti judged Newton's position to be "dubious," inasmuch as the particles of a body can have any density greater than that of the body itself, depending on the ratio of the size of the particles to the size of the pores. Therefore, he concluded, "any color at all can be connected to any density of the total mass or of the individual particles."[33] Joseph Priestley, however, in his history of optics, endorsed Newton's theory and enthusiastically supported his estimate of the size of the corpuscles as "one of the most astonishing, and yet one of the clearest deductions from the phenomena of thin plates."[34]

The works of the Dutch Newtonian Willem Jacob 's Gravesande, were a crucial vehicle for the diffusion of Newtonian science on the continent and were popular in England as well. In his *Mathematical Elements of Natural Philosophy . . . or, an Introduction to Sir Isaac Newton's Philosophy* he set forth Newton's theory of colored bodies, but he too omitted Newton's estimate of the corpuscles' size. 's Gravesande showed a sophisticated grasp of Newton's doctrine by correcting his statement (Prop. 2) that the "least parts" of bodies are transparent.[35] "The Least Parts of all Bodies" 's Gravesande explained, "that is, those which separate the ultimate or smallest Pores, are perfectly solid; of these we do not here speak."[36] He recognized the dependence

[32] Rutherforth devotes a section to arguing that "we cannot from the colour of a body make any conjecture about the size of the particles, upon which its colours depend" (*System of Natural Philosophy*, 1:480).

[33] Benvenuti, *De lumine*, §172, p. 87; see also Bošković, *Theory*, §472, p. 167; and Ch. 4, note 103, on the relation of the two scholars.

[34] Priestley, *History*, p. 296. Bryan Robinson likewise endorsed Newton's determination of the corpuscles' size in *A Dissertation on the Aether* (1743), pp. 86–9.

[35] See Ch. 3, note 76.

[36] 's Gravesande, *Mathematical Elements* (1747), §3444, 2:236; the colors of bodies is in Bk. V, Ch. 26, pp. 273–6. 's Gravesande's modification of Newton's theory was made in the final, posthumous edition of his *Physices elementa mathematica* (1742); a French translation appeared in 1746. The other influential Dutch Newtonian, Pieter van Musschenbroek, was more noncommittal. He explained that the colors of bodies are caused by corpuscles shaped like thin plates that produce colors just as with Newton's rings, but he did not set out Newton's theory; *Cours de physique* (1769), §1843, 2:502–3. This loose version of Newton's theory was often adopted. Charles Bonnet, for example, in discussing color changes caused by sunlight, appealed to it: "The infinitely small plates which compose the surfaces of bodies variously break up and reflect the sun's rays. Hence, the different colors of bodies: when these plates come to change position, or when their thickness increases or decreases, they reflect other colors" ("Lettre sur les moyens de conserver diverses especes d'insectes" [1774],

of Newton's theory on the hierarchical matter theory and explained that it is rather the small, "exceeding little" *compound* parts that must be transparent.

In France, at midcentury, Newtonian science had just been introduced, and the theory of colored bodies came as part of the ensemble. Writing in the *Encyclopédie* in 1754, d'Alembert gave a succinct account of Newton's theory and explained that according to it "the color of a body depends on the density and thickness of the particles" – an epitome that would later be much bandied about.[37] He rejected Newton's attempt to deduce the size of the corpuscles of colored bodies as "conjectural." The Abbé Jean-Antoine Nollet, an experimental natural philosopher who retained a residual commitment to the Cartesian system, sketched Newton's theory without endorsing it in his classic textbook.[38] By the turn of the century, though, Newton's theory had clearly taken the field among French physicists, as we will soon see from our examination of the vigorous attack of the chemists and the defense by Haüy and Biot.

Many of the treatises that I have cited were available in various European languages, but undoubtedly the principal source for knowledge of Newton's theory was his own *Opticks*. It was available in English, Latin, and French and was perhaps the most widely read scientific book of the eighteenth century. The criticism of the theory on various points shows that it was not merely being parroted but genuinely adopted. An even stronger indication of its acceptance is that by midcentury it had become the foundation for further research.

Thomas Melvill exhibited an unusual comprehension of Newton's optics in the two papers that he wrote before his death at age twenty-seven in 1753. His "Observations on light and colours" is largely an analysis of various problems in the Newtonian system and, like the *Opticks* itself, ends with a series of queries. In the seventh query he asks:

> Is it not possible to prove by experiment what Sir Isaac Newton takes for granted as a reasonable supposition, that thin transparent plates, of any uniform colour, divided into smaller fragments, would compose a powder of like colour? And would not this tend to strengthen the analogy between the colours of such plates and those of natural bodies?

p. 300). See Ch. 6, note 4, for Peter Shaw's use of a similar version of Newton's theory.

[37] D'Alembert, "Couleur," p. 330.

[38] Nollet, *Leçons de physique* (1777), 5:425–6; the first edition appeared from 1743 to 1748.

To prove this central assumption of Newton's theory, Melvill undertook a truly Swiftean experiment: "I have tried to freeze soap-bubbles; but could never make any stand till they were turned to ice . . . however, I doubt not, but, with due care, the thing might be done."[39] He then proceeded to describe his various attempts to carry out the proposed test. Although Melvill recognized that Newton's untested experiment remained a supposition, we should note that he did not seriously question Newton's methodology. Even had the experiment succeeded as Newton predicted, the supposition would have remained an "analogy," for what is true at the macroscopic level of thin plates of glass or ice can be extended to the corpuscular level only if the method of transduction is accepted.

The philosopher and physician David Hartley urged further research on Newton's theory of colored bodies in his *Observations on Man* (1749), in which he attempted to use Newton's theory of a vibrating aether to explain human psychology and physiology. He exhorted

> Philosophers to inquire into the Orders of the Colours of natural Bodies, in the Manner proposed and begun by Sir Isaac Newton; and *particularly to compare the Changes of Colour, which turn up in chemical Operations,* with the other Changes, which happen to the Subjects of the Operations at the same time. Nothing seems more likely than this to be a Key to the Philosophy of the small Parts of natural Bodies, and of their mutual Influences.[40]

Delaval, who would become the most prominent spokesman for Newton's theory, appears to have carried out the research program advocated by Hartley. He began his research on the theory of colored bodies by fully accepting Newton's theory and then attempted to confirm some of its consequences. Delaval was a Cambridge-educated, experimental natural philosopher who also carried out research on electricity and phosphorescence in collaboration with Benjamin Wilson. His first two works on coloration were of a piece. In a paper in 1765, "Experiments and observations on the agreement between the specific gravities of the several metals, and their colours when united to glass," he showed that the colors of opaque bodies vary in proportion to the density of their corpuscles in accordance

[39] Melvill, "Observations," p. 66; he has a footnote here to Newton's *Opticks*, Bk. II, Pt. III, Prop. V.

[40] Hartley, *Observations on Man*, Pt. I, Ch. 2, Sect. IV, Prop. 60, 1:208–9; italics added.

with Newton's theory; and in a book in 1777 he demonstrated that their colors vary, as Newton's theory predicts, according to the size of the corpuscles.

In order to test Newton's principle that a body's color varies with a change in the density of its corpuscles, Delaval extracted two principles from Newton's experiments and observations that served as the foundation for his own experiments: "that bodies have their *refractive and reflective powers* nearly proportional to their *densities;* and that the *least refrangible* rays require the *greatest* power to reflect them." "In like circumstances," Delaval reasoned, denser bodies, which have a greater reflective power, ought to reflect the less refrangible rays and appear red, whereas less dense bodies "should reflect rays proportionably *more refrangible,* and thereby appear of several colours in the order of their density," that is, orange, yellow, green, and so on.[41]

Delaval used metals for his investigation because their specific gravities were accurately known. Metals, he explained, consist of "inflammable or sulphureous matter" (phlogiston) and a fixed matter or calx, whose specific gravity and other properties differ in each metal. Because phlogiston acts strongly on light, as Newton showed in his measurements of refractive powers, it was necessary to remove the phlogiston in order to study the action of the calx on the light rays.[42] To examine all the metals *"in like circumstances"* and eliminate the phlogiston, he exposed the metals with pure glass to the greatest heat they could endure without losing their color. It indeed turned out from these experiments and other observations on the metals:

> that they actually do, without any exception, exhibit colours in the order of their densities, as follows,
>
> Gold —— Red.
> Lead —— Orange.
> Silver —— Yellow.

[41] Delaval, "Experiments and observations," pp. 11–12; which is reprinted in his *Experimental Inquiry into the Cause of the Changes of Colours* (1777). See also Charles Taylor, "Biographical memoranda," though he fails to recognize that Delaval ultimately abandoned Newton's theory. Because of Newton's inconsistent terminology, Delaval has confused optical density or index of refraction n and mass density or specific gravity ρ. In Bk. II, Pt. III, Prop. X, Newton established that $n^2 - 1 \propto \rho$ which does not yield the simple proportionality between color and "density" ρ claimed by Delaval. Delaval's critics did not note this misinterpretation, and in any case it did not affect the thrust of his qualitative investigation.

[42] For Newton's experiment, see Ch. 6, note 13.

Copper —— Green.
Iron —— Blue.[43]

In the 1790s, Berthollet and especially Bancroft vigorously criticized Delaval's "gratuitous and fallacious" chemical interpretations of his experiments as part of their attack on Newton's theory.[44] Berthollet undermined Delaval's experiments by noting: "We have no means to determine either the thinness or density of the molecules of bodies; the specific weight cannot indicate them, since small molecules with very numerous pores can give the same specific weight as a more considerable extent of pores and a smaller number of larger molecules."[45] Delaval had arbitrarily assumed that the density of the corpuscles of each metal was proportional to the density of the metal itself. Bancroft observed that pure metals refute these experiments, as platinum, which is the densest metal, instead of being the most red is white like tin, the lightest metal. But if Delaval would object that this arises from the phlogiston in the metals, then, Bancroft argued, he is conceding at the start that their colors depend on their chemical composition.[46] Both Berthollet and Bancroft pointed to the colors of metallic oxides as contradicting Delaval, for metals with only the smallest change of density (which is smaller than the differences of the densities of different metals) assume all colors as they are oxidized.[47] Nonetheless, in the very different physico-chemical world of 1765, which was then in the midst of a transition from a mechanical to a dynamical and chemical worldview, Delaval's paper was greeted with great acclaim and was awarded the Copley Medal by the Royal Society. Although Delaval had successfully tested a part of Newton's theory for opaque bodies, he had not dealt with the heart of the theory, the analogy to the colors of thin films. This he did in his second work a decade later.

In his book, *Experimental Inquiry into the Cause of the Changes of Colours in Opake and Coloured Bodies*, Delaval published a vast number of experiments to demonstrate Newton's theory with respect to the size of the corpuscles. He opens by observing that no experiments have been carried out on permanently colored bodies to see if their colors do in fact vary with the corpuscles' size in the same way as they do

[43] Delaval, "Experiments and observations," p. 13.
[44] Bancroft, *Experimental Researches concerning the Philosophy of Permanent Colours* (1813), 1:22, the first edition appeared in 1794.
[45] Berthollet, *Eléments de l'art de la teinture* (1791), 1:3.
[46] Bancroft, *Experimental Researches* (1813), 1:13–14.
[47] Ibid., pp. 14–17; Berthollet, *Eléments* (1791), 1:3–4.

for Newton's rings. All of Newton's experiments were conducted on colorless transparent substances, so that his theory remained "merely speculative and unestablished by any Proofs."[48] This is directed only against Newton's argument by analogy and not against transduction, which Delaval himself invokes. If Newton's theory of colored bodies is valid, as the corpuscles' size increases or decreases the colors of the bodies should change in exactly the same sequence as those of Newton's rings and descend or ascend the table of colors in Figure 2.8. To alter the size of the corpuscles, he used such chemical methods as dissolution in acids and alkalis, heating, and precipitation. From his experiments and observations he found "that the *Changes* of Colour in *Permanently Coloured* bodies are made according to the same Law, which is shewn by Sir Isaac Newton's experiments to have taken place in *Pellucid Colourless* substances."[49]

Let us follow Delaval in his repetition of Boyle's experiment on color indicators, which Newton also recounted – that is, on the change of color of a solution of purple flowers to red when an acid is added and then to green with the addition of an alkali. After careful observation, Delaval found that the addition of an alkali causes the solution not to pass directly to green, which "would be an irregularity inconsistent with the Laws of Opticks," but rather through all the gradations of purple, violet, and blue before reaching green. When a stronger acid, oil of vitriol (sulfuric acid), was added, the red ascended to yellow. "Thus," he concluded, "all the primary Colours are exhibited in their regular Order."[50] These observations are consistent with Newton's claim that the purple of flowers is of the third order.[51] Like Newton, Delaval interpreted the cause of these chemical changes mechanically, namely, that the corpuscles are "united into larger Masses by the Alcali, and Divided into Smaller, by the Acid."[52] All the sequences of color changes that he investigated followed Newton's table of colors and were hailed as impressive evidence for his theory of colored bodies. Over half a century later, Herschel and Brewster would still invoke color changes in chemical reactions in behalf of Newton's theory.

Delaval's book, with a reprint of his earlier paper, was soon translated into French (1778), Italian (1779), and German (1781).[53] An ear-

[48] Delaval, *Experimental Inquiry into the Cause of the Changes of Colours*, p. 6.
[49] Ibid.
[50] Ibid., p. 15.
[51] See Ch. 3, note 62.
[52] Ibid., p. 16.
[53] See the very favorable reviews by William Bewley, "[Review of] 'Experiments and

lier, abbreviated version was read to the Berlin Academy of Sciences and published in its *Mémoires* for 1774 with an enthusiastic introduction by Johann Castillon, the editor of the fine three-volume edition of Newton's *Opuscula* (1744). He successfully repeated several of Delaval's experiments and declared him to be following the method of Bacon, Galileo, and Newton.[54] Delaval's 1777 book was the culmination of Newton's theory of colored bodies.

Just as Delaval's second study appeared, the purely mechanical conception of Newton's theory was being openly challenged by chemists, and in the next decade Delaval himself would adopt a chemical theory of color. The criticisms of Delaval by Berthollet and Bancroft some fifteen years later show the nature of this changing conception of the cause of coloration. Bancroft's objections in particular were part of a broader attack on mechanically conceived chemistry. Acids and alkalies, he insisted, are "*chymical agents*" that form "new compounds, which are totally foreign to those mechanical effects by which, Sir Isaac Newton intended to explain the changes of colour in question."[55] Delaval's very conception of changing the size of the particles is misguided and confuses the mechanical and the chemical:

> for when, in operating upon, or with different matters, he professes either to increase or diminish the sizes of their particles, and *to do nothing more,* (to shew that the changes of colour produced in them, accord with the thicknesses stated in the table of Sir Isaac Newton,) instead of choosing and employing *mechanical* means, which alone are suited to produce these, and *only these effects,* he has recourse to mere *chymical* agents, whose action in the ways which he supposes, must have been always doubtful at least, though their powers of producing other, and very different effects from any supposed by him, is most certain.[56]

Before we turn to the chemists, let us briefly look at two physical theories offered by Euler and Dutour as alternatives to Newton's.

Euler's explanation of colored bodies was based on his wave theory of light and invoked entirely different physical models than Newton's. Nonetheless, it shows how strong the hold of Newton's views were

observations,' " and Thomas Bentley, "[Review of] *An Experimental Inquiry.*" The reviewers are identified by Benjamin Christie Nangle, *The Monthly Review, First Series,* pp. 3, 4, 88, 173.

[54] Delaval, "Recherches expérimentales," pp. 156–7. The translation ends at p. 55 of the book, where Delaval explains in a note that "a few copies" of that part of the book had been privately printed.

[55] Bancroft, *Experimental Researches* (1813), 1:10. See also Ch. 7, note 10.

[56] Ibid., p. 22.

in this era, for Euler unquestioningly accepted the idea that the cause of the colors of natural bodies and that of thin films are the same, although he reversed the direction of Newton's analogy. Euler's perceptive critique of Newton's theory, as well as his misapprehensions, also serves to illuminate the state of knowledge in midcentury. Euler first set forth his explanation in his classic account of his wave theory, "A new theory of light and color" (1746),[57] but I will turn to his more thorough treatment in his 1752 essay on the colors of thin films. This essay was published immediately after Mazeas's paper, which attacked Newton's experimental investigation of thin films, and began by endorsing Mazeas's work.[58] It must therefore be seen as part of Euler's broader attack on Newton's emission theory.

Newton's explanation of the colors of opaque bodies by those of thin films, where some rays are reflected and others are transmitted, was rejected by Euler because "the reasons given to support this supposition seem too feeble to be admitted in such important research."[59] In particular, he argues that opaque bodies are not perceived by (specularly) reflected rays. If rays were in fact reflected from them, we would see an image of the luminous source rather than the surface of the body itself; the body should change its appearance as it is viewed from different positions, just as images in a mirror; and the image of the body would not appear at its surface but elsewhere (§§14–16). Further evidence that opaque bodies are not colored by reflected rays is provided by viewing a highly polished opaque body, for we see objects reflected from it just as in a mirror and also the colored body itself: "The double appearance of such a polished body is so distinct – the one variable, the other permanent – that the cause of one must be entirely different from the cause of the other; therefore if one is the effect of reflection, the other will have an entirely different origin."[60] In his *Letters to a German Princess,* Euler explicitly notes that rays reflected from a mirror always maintain the color of the body from which they initially proceeded – "the mirror which reflects makes no change in this respect" – whereas rays illuminating a colored body assume the color of the body.[61] Although Euler did not grasp the

[57] Euler, "Nova theoria lucis et colorum," *Opera omnia,* ser. 3, 5:1–45. See Casper Hakfoort, *Optica in de eeuw van Euler.*
[58] See Ch. 4, note 141.
[59] Euler, "Essai d'une explication physique des couleurs engendrées sur des surface extrêmement minces," *Opera omnia,* ser. 3, 5:159, §12.
[60] Ibid., §20, p. 161.
[61] Euler, *Letters* (1833), Letter XXIV, 1:100. The *Lettres à une princesse d'Allemagne* were published in 1768 and 1772.

distinction between specular and diffuse reflection, his analysis is nonetheless original and trenchant, for Newton's theory demands that the rays responsible for the color of a body be specularly reflected from its transparent corpuscles. Later in the century, the nature of reflection from colored bodies, which had been repeatedly demonstrated in the seventeenth century, would become one of the foci for experimental refutations of Newton's theory.

Euler's own solution to coloration starts from his analysis of the failure of Newton's explanation. It is evident that an opaque body must reradiate light of its proper color in all directions from every point of its surface (§22). To explain the nature of this reradiated light Euler applies his wave theory, which he closely modeled on a theory of sound. The particles of a luminous source such as the sun vibrate with a diversity of frequencies corresponding to all the spectral colors, and these vibrations are propagated in all directions throught the surrounding aether. From studies of resonance it is known that when a sound wave falls on a body capable of vibrating, such as a string, the body will start to vibrate if the frequency of the wave is equal to the natural frequency of the body. Similarly, when sunlight falls upon a colored body it sets its molecules in vibration at their proper frequency. As ingenious as Euler's model is, it suffers from two (corrigible) problems: It assumes that the colors of bodies are simple, or of a single frequency, rather than compound; and it ignores the existence of absorption.

When Euler extended his reasonance model to the colors of thin films, it turned out to be incapable of accounting for many of the principal phenomena. Euler fell into error from the start by holding that, just as with colored bodies, light is radiated and not specularly reflected from a thin film. Indeed, he argued that because there is no reflection, Newton's hypothesis of fits of easy reflection must be rejected as imaginary (§13). Euler evidently never carried out any experiments with the colors of thin films and appears to have been confused by rings in films of varying thickness (like Newton's rings) where the same order can be seen at any angle of reflection. By adopting Newton's argument that all opaque bodies become transparent when sufficiently thin, so that there is no fundamental difference between transparent and opaque bodies, Euler was able to apply his resonance model to the molecules composing thin transparent films. To explain how the color varies with the film's thickness, he again appeals to an acoustic analogy: Just as the natural frequency of a string varies with its length, so the frequency of vibration of the film – and hence the color – will continually decrease as the film grows thicker (§§32–4).[62] The higher order colors are harmonics of the first order or one half, one quarter, and so on, of the fundamental frequency (§§43–5). There is no need to analyze the details of this model, other than to note that it requires that the transmitted and reflected rings be the identical color.

Despite his many differences with Newton, Euler was still operating within a Newtonian framework, for he never doubted that the periodic colors of thin films and those of bodies had the same cause. He simply inverted the direction of the analogy by explaining the former in terms of the latter. Euler's theory of coloration, which was as unabashedly a physical theory as Newton's, was restricted to followers of the wave theory; and as the evidence for the chemical nature of light increased in the coming decades, the wave theory itself was driven into temporary eclipse.

Although he reported a number of original experimental results, Dutour's research on the colors of natural bodies is more interesting as an indication of the direction of future research than in its own right. His paper on the coloration of bodies was the penultimate one in a series of twelve memoirs on physical optics published in the *Journal de physique* between 1773 and 1776. Dutour (or Du Tour) requested that his papers be published anonymously because, the editor noted, they contained new views that were "opposed . . . to the ideas of the celebrated Newton."[63] It was not difficult to guess the name of the author, as Young did, because in a number of the papers he was identified as "D. T. Correspondant de l'Académie des Sciences."[64] Dutour was the correspondent of the Abbé Nollet (until

[62] Euler's model cannot distinguish between vibrations of the individual molecules composing the film and the whole thickness of the film.

[63] Dutour, "Considérations optiques" (1773), p. 339. The *Observations sur la physique, sur l'histoire naturelle et sur les arts,* edited and published by François Rozier, was commonly known as Rozier's *Journal* or the *Journal de physique,* even before it formally adopted that title in 1793; I will henceforth cite it by its later name. "Physique" in the title was used in the older and broader sense of physical science, which included, for example, chemistry, electricity, and optics, but not mathematical physics. The *Journal de physique* was one of the earliest scientific journals not associated with a scientific society or academy and represents the forerunner of the modern specialized journal, such as *Annales de chimie* and *Annalen der Physik.* It will play an important role in the French part of my history through the late 1780s, when the *Annales de chimie,* which was more sympathetic to Lavoisian chemistry, assumed the role of rapidly publishing articles of current interest. See Douglas McKie, "The 'Observations' of the Abbé François Rozier"; David A. Kronick, *A History of Scientific Periodicals,* pp. 106–12; and James E. McClellan III, "The scientific press in transition."

[64] Young, *Course of Lectures,* 2:316.

Nollet's death in 1770) and his closest ally in the defense of his neo-Cartesian electrical theory against Benjamin Franklin's. Dutour's optics was similarly neo-Cartesian and invoked refracting fluids and optical atmospheres rather than attractive or repulsive forces to explain refraction and diffraction; in a memoir directed against Newton he argued that reflection occurred from the parts of bodies.[65] This approach, which drew upon the work of Dortous de Mairan and Nollet from the 1730s and 1740s, was by 1776 essentially outmoded among French physicists.[66]

Dutour was probably the first person to describe absorption spectra, though he was not particularly concerned with the process of absorption or the absorbed colors, but rather with those not absorbed. His aim was to demonstrate and analyze the compound nature of the colors of natural bodies and to locate where the process of coloration occurs, as is evident from the title of his paper, "Experiments on the decomposition and combinations of the rays to which both transparent and opaque bodies owe their colors." To study the composition of the colors of bodies he viewed through a prism the light transmitted through and reflected from colored glasses, as well as that reflected from pigments painted on very narrow strips of white paper (1½ lines or an ⅛-inch wide) from small bits of flower petals, and from leaves.

Although to the naked eye all of these objects appeared to be uniformly colored, Dutour found that their spectra were "divided into more or fewer bands of different colors." The pigment ultramarine, for example, was divided into violet, blue, and green bands, with a little dull red at the ends, and cinnabar (which is red) into green and red bands.[67] The spectra (except for yellow) were observed to be the same for the same color, whether it was reflected from a pigment or a flower, or reflected or transmitted through glass; that is, the composition of a given species of perceived color was the same whatever sort of body it came from. He also found that the missing colors are

[65] Dutour, "Considérations optiques. VIe mémoire" (1774); and "Septieme mémoire" (1775).

[66] Near the end of his career Nollet advised Dutour in a letter on 13 March 1769: "It seems to me that you often call on the configurations of the ultimate parts of bodies, on the arrangement of their pores . . . , on an unknown matter to which you assign a large role, etc. I ought not hide from you that the Academy is getting more and more difficult about this way of philosophizing" (quoted by Heilbron, *Electricity in the 17th and 18th Centuries*, pp. 76–7). See Home, "The notion of experimental physics."

[67] Dutour, "Considérations optiques. XIe mémoire" (1776), §2, p. 231.

always from the ends of the spectrum and not the middle (§9). Dutour did not pursue the study of absorption itself – for example, how the spectra varied with the thickness of coloring matter traversed – for his aim was rather to show that the colors of most natural bodies are compound. Newton, as Dutour noted, had shown this, but, as we shall see, as late as the first decade of the nineteenth century few understood this, especially chemists.

Dutour concluded from his observations that (1) white light alone is specularly reflected from the surface of bodies (§§23–30); (2) some white light is also transmitted by transparent colored bodies (§17); (3) the light "primarily" responsible for the color of a body enters it (§34); and (4) the transmitted and reflected colors are the same, as their spectra demonstrate, and not complementary (§19).[68] He arrived at these results by rather simple means. The closely related first and third points were deduced from his observations that light specularly reflected from colored bodies is uncolored and that when transparent colored glass is viewed in a dark room, it appears almost black with a dark object placed behind it, but of its proper color with a white object behind it (§16).

Dutour's experimental technique is characteristic of much contemporary natural philosophy: His experiments were casually designed and uncontrolled and thus unreliable and inconclusive. By not thoroughly excluding scattered light, he arrived at his erroneous second conclusion and the qualification ("primarily") in the third. Less than a decade later Delaval would arrive unambiguously at the same (valid) conclusions with a much larger number of more carefully executed experiments. Likewise, because of his lax method of observation and failure to vary the experimental conditions, both of his generalizations about absorption spectra are mistaken, namely, that spectra of the same perceived color are always the same, and that absorption occurs only at the two ends. Dutour was unaware of the most fundamental property of selective absorption, that as the thickness increases more and more colors vanish, and so the composition of the spectra continually changes. Of course, because none of his contemporaries had

[68] Dutour concluded that "although transparent colored bodies are disposed both to transmit and return more rays of certain species than of other species, one cannot maintain that they transmit only the species of rays that they do not reflect: Observation contradicts such an assertion, because it almost always happens that the species of rays that are reflected by a transparent body more appreciably than the others are precisely those that are also transmitted more appreciably than others" (ibid., §19, p. 238). Newton's archetypal phenomena, such as lignum nephriticum, now became exceptions to the rule, rather than the rule (§20); see Ch. 3, note 21.

carried out these or similar experiments, the unreliability of his results would not have been apparent. It does show, however, why when related research was later undertaken, Dutour's work, when it was known, would have been of little consequence.

No doubt another reason why Dutour's work had little influence, at least among physicists – as we shall see was the case with Venturi – was his own explanation of coloration. He held that light could be decomposed into colors only by refraction, exactly as in a prism, and he tried to reduce all appearances of color to this one cause. He had earlier argued that the colors of thin films are produced by refraction in the wedge-shaped film of air between two glasses and thereby rejected the insight of Hooke and Newton that the colors of thin films are a fundamentally different phenomenon from refractive dispersion.[69] To reduce the colors of natural bodies to the same cause, Dutour assumed that the pores of bodies were shaped like prisms and filled with a fluid whose refractive power differed from the surrounding medium (§36). The diversity of colors of natural bodies could then be explained by differences in the shape and density of the pores. Dutour was drawing on an idea of Nollet, his mentor, who had also attributed coloration to the pores (§38). Some chemists later adopted this concept, but the reaction of Venturi, who devoted the beginning of his memoir on coloration to refuting it, was probably more typical of physicists.[70]

The principal point of Dutour's memoir, that the colors of bodies are compound, was made by Newton, and even if many people had lost sight of it, it was not especially novel. Dutour's most important innovation was his observation of absorption spectra, but for him they were actually a means to study the composition of the transmitted colors, and not absorption. It was difficult, as our subsequent story will show, for scientists to reorient themselves conceptually to focus on the missing colors and recognize that absorption is an independent physical process deserving of study. When true studies of absorption were first introduced around 1800, such a reorientation was provided by the chemistry of light, according to which particular colors, depending on their elective affinities, were imagined to combine with bodies and vanish. Dutour had no concept of selective absorption, and in his physical explanation of coloration, he could only vaguely propose that

[69] Dutour, "Considérations optiques" (1773).
[70] See §8.1.

the lost or extinguished rays are those which, although borne in the same direction as the others [i.e., reflected and transmitted rays], are then scattered, because the small planes where they land are found to be inclined in an infinity of different directions; or they are extinguished and intercepted, because after having pierced the interstices of the proper parts, they are carried on to their walls and tossed about there.[71]

Because Dutour himself had no concept of selective absorption and was scarcely concerned with it, it is not surprising that others did not recognize the significance of his absorption spectra. Moreover, he apparently did not realize – and certainly did not mention – that his observations contradicted some of the fundamental tenets of Newton's theory; in particular, that no colored light is specularly reflected from colored bodies, and that the colors (or spectra) of the reflected and transmitted light in transparent bodies are identical. To contemporary physicists or natural philosophers, Dutour's memoir would not have seemed notable: His experimental technique was casual, his physical explanations outmoded, and his conclusions unrelated to current issues, namely, to Newton's theory of colored bodies. To chemists, who would launch their attack on Newton's theory in the next volume of the *Journal de physique,* Dutour's physical theory would have been as objectionable as Newton's.

[71] Dutour, "Considérations optiques. XIe mémoire," §21, p. 239.

6

THE CHEMISTRY OF LIGHT IN FRANCE: 1776–1790

6.1. THE CHEMISTS' REVOLT

Christophe Opoix began his "Physico-chemical observations on colors" by charging that the colors of bodies have not yet been treated from a "true point of view," but only by physicists (*physiciens*) who are concerned with the common properties of bodies such as weight, motion, and elasticity rather than with the principles (*principes*) that compose each body in particular and distinguish it from others.

It should not therefore be surprising that they would be mistaken when they wanted to explain natural effects that depend on the constitutive principles of bodies. This is what appears to us to have happened to them, particularly on the cause of the colors of bodies. Newton's opinion on the colors of light is without doubt the most interesting, best conceived, and best developed system.... Time, in respecting the work of this superior genius, appears to have affixed the seal of truth to it, but Newton's system is not always exempt from the reproach that we can bring against other physicists. When it is a question of explaining the cause of the colors of natural bodies, it appears that he has known these only very imperfectly, in the manner of other physicists, so that there must result explanations which are more clever than solid and consequences which are, at the least, rash.

After quickly sketching Newton's and Nollet's explanations, he emphasized that "these two physicists agree in regarding colors as totally indifferent to the nature of bodies." And he resumed his attack:

This assertion supposes a profound knowledge of chemistry and could only be the result of a long series of observations on the nature of colored bodies, but not one experiment has been done. Physicists, who have always regarded chemistry as a foreign science, have not at all attended to it; if they had consulted it, they would have seen that it does not confirm their opinion. On the contrary, it would appear to demonstrate by a multitude of facts that bodies are colored only inasmuch as they contain an inflammable principle, to which the name phlogiston has been given. The discovery of this truth seems to us to open a new course and throw great light on the nature of colors – perhaps it could even lead to a revolution in this part of physics.[1]

Although Opoix's revolution would ultimately be frustrated, his paper marks the beginning of the extended battle in France between physicists and chemists over Newton's theory of colored bodies. Opoix, an apothecary who had earlier published analyses of mineral water, should not be viewed as some lone crank engaged in a diatribe against physics. His paper was read to the Academy of Sciences by France's leading chemist, Macquer, who would later propound Opoix's ideas in his influential *Dictionnaire de chymie*. Rather, Opoix was characteristic of the early generation of Stahlians, like Venel, who were trying to establish chemistry as an independent physical science and restrict the scope of physics. If this group's polemics against physics are set aside, most chemists subscribed to the Stahlian goal of establishing chemistry as a science with its own concepts, methods, and subject matter, which most certainly included color. This program demanded that color be explained chemically and not mechanically. Thus we can understand why even such a stalwart of a closer union of chemistry and physics as Berthollet attacked Newton's theory nearly as ardently as Opoix did.

Before turning to Opoix's theory of color, we should recognize that the idea that the colors of bodies are caused by the fire, or phlogiston, in them is an ancient one. It goes back at least to Aristotle, but its eighteenth-century source is Stahl, who introduced the term phlogiston for the matter of fire or inflammable principle that he held was incorporated or fixed in bodies and responsible for its color (and other properties).[2] By means of phlogiston, calcination, which was Stahl's primary concern, combustion, and respiration (all oxidation processes) could be explained: Calcination involved the escape of phlogiston from the metal; and combustion involved the release of phlogiston, which was the source of the attendant heat and light. The concept of phlogiston and its role in chemical processes grew ever more complex through the

[1] Opoix, "Observations physico-chymiques sur les couleurs" (1776), pp. 100–2.
[2] Rappaport, "Rouelle and Stahl." The classic, four-part study on phlogiston by Partington and McKie, "Historical studies on the phlogiston theory," is still useful, especially for its copious references.

course of the eighteenth century, and there was a great diversity of views on its chemical nature – whether it be, for example, sulfur (as Stahl held), solar matter, fire, electricity, or light. Fortunately, to understand the chemical theories of colored bodies most of these subtleties need not concern us. The caricature that phlogiston is negative oxygen will largely suffice, for in the first flush of conversion to Lavoisian chemistry, his followers essentially just replaced phlogiston with oxygen and made as extravagant claims for its role in coloration as Stahl and Opoix had done for phlogiston (as we shall see in §6.2).

Stahl's phlogiston chemistry was introduced (and transformed) into France in midcentury through the popular lectures of Rouelle. In his lectures Rouelle adopted Stahl's teaching that the colors of bodies depend on the amount of phlogiston they contain.[3] Nearly as important as phlogiston was the idea of the Dutch physician and chemist Hermann Boerhaave, that fire is the active principle of nature and pervades the universe. Its presence is known through heat and light. Boerhaave's ideas were particularly influential in the second half of the eighteenth century in Britain, where fire was closely identified with light and played a role in natural philosophy much like Newton's aether.[4] Although Newton's views were compatible with, and indeed stimulated, much of eighteenth-century chemical theory, he did not believe that heat was a material substance such as phlogiston or caloric. Like most seventeenth-century mechanical philosophers, he held that heat was a vibrating motion of the corpuscles of bodies.[5]

[3] Rappaport, "Rouelle and Stahl," p. 78. For Stahl on phlogiston as the cause of color, see Hélène Metzger, *Newton, Stahl, Boerhaave et la doctrine chimique*, pp. 164–6, which includes quotations from Stahl's *Traité du soufre*, trans. Baron d'Holbach (Paris, 1766). See also L.J.M. Coleby, *The Chemical Studies of P. J. Macquer*, pp. 52–4.

[4] Boerhaave set forth his concept of fire in his frequently reprinted and translated *Elementa chemiae* (Leyden, 1732), which was translated into English and annotated by Peter Shaw, *A New Method of Chemistry* (1741). The first edition of Shaw's translation in 1727 was from an unauthorized version, *Institutiones et experimenta chemiae* (Paris, 1724), published by Boerhaave's students from their notes; and the ideas on light and fire in the two editions differ dramatically; see Cohen, *Franklin and Newton*, pp. 215–16. In a note Shaw presented a watered-down version of Newton's theory of colored bodies according to which "in the surface of every colour'd body, we may conceive innumerable, small thin lamellae, or plates"; *New Method* (1727), 1:229; (1741), 1:212. Muschenbroek and Bonnet invoked a similar weak version; see Ch. 5, note 36. Boerhaave's ideas were widely disseminated by his countrymen 's Gravesande and Muschenbroek. On Boerhaave's concept of fire, and especially its role in theories of light, see Cantor, *Optics after Newton*, pp. 95–7; and also Metzger, *Newton, Stahl, Boerhaave*, Pt. III; and Cohen, *Franklin and Newton*, pp. 214–34.

[5] See, for example, *Opticks*, Query 5. In a footnote Shaw clearly contrasts Boerhaave's belief in the materiality of heat with the classical English view of Bacon, Boyle, and Newton; Boerhaave, *New Method* (1727), 1:220; (1741), 1:207.

However, his suggestion in the queries that absorbed light was converted into gross matter and contributes to the activity of bodies was similar to the role of phlogiston and was often cited by advocates of the new chemistry of light.[6]

The last essential concept required to understand the chemical theories of colored bodies is that of the affinity, elective attraction, or *rapport*, between the corpuscles of different chemical substances. In the queries Newton attempted to extend the concept of force to the invisible corpuscles of matter to explain such phenomena as chemical reactions, capillary attraction, and reflection and refraction. In Query 23/31 he asked:

> Have not the small Particles of Bodies certain Powers, Virtues, or Forces, by which they act at a distance, not only upon the Rays of Light for reflecting, refracting, and inflecting them, but also upon one another for producing a great Part of the Phaenomena of Nature? For it's well known, that Bodies act one upon another by the Attractions of Gravity, Magnetism, and Electricity; and these Instances shew the Tenor and Course of Nature, and make it not improbable but that there may be more attractive Powers than these. For Nature is very consonant and conformable to her self.[7]

Newton then marshaled a broad range of chemical evidence in this longest of queries to argue for the existence of such attractive forces. Unlike gravitational attraction, which depends only on the distance and mass of two bodies, chemical reactions depend on the nature of the bodies. In addition, chemical reactions saturate or terminate, whereas gravitational forces act continually. To explain this, Newton introduced the concept of elective attractions between the particles of different substances. Invoking a series of displacement reactions with acid and metal solutions, he asked, "Does not this argue that the acid Particles of the *Aqua fortis* [nitric acid] are attracted more strongly by the *Lapis Calaminaris* [zinc carbonate] than by Iron, and more strongly by Iron than by Copper, and more strongly by Copper than by Silver," and so on?[8]

In England the idea of a Newtonian chemistry based initially on attractive, and later, also, on repulsive, forces ("as in Algebra, where affirmative Quantities vanish and cease, there negative ones begin; so in Mechanicks, where attraction ceases, there a repulsive Virtue ought to succeed") was immediately picked up and developed by the

[6] Query 22/30 quoted at Ch. 3, note 72.
[7] *Opticks* (Dover), pp. 375–6.
[8] Ibid., p. 381.

British Newtonians – Keill, Freind, Hales, and others.[9] In France, Etienne-François Geoffroy, in 1718, prepared the first table of affinities summarizing a large number of reactions, but he avoided the suspect concept of attraction by using the neutral term relation (*rapport*). After Macquer introduced the "mutual conformity, relation, affinity or attraction" of bodies as the effect "that will enable us to explain *all the phenomena* furnished by chemistry and connect them together," in 1749, affinities played an ever-increasing role in chemistry.[10]

Because affinities were directly measurable and treated independently of the size, shape, and arrangement of the component particles, they were consistent with the new chemical methodology. They also promised the creation of a quantitative chemistry and thus fostered closer relations between chemistry and physics, especially in France. In 1782, Lavoisier expressed the hope that

> perhaps one day the precision of the data will lead to the point that the geometer will be able to calculate in his study the phenomena of any chemical combination in the same way, as it were, that he calculates the motions of celestial bodies. The views of M. Laplace on this subject and the experiments that we have planned according to his ideas in order to express numerically the force of the affinities of different bodies, already allow one not to consider this hope to be a total chimera.[11]

By the first decade of the nineteenth century, things appeared quite differently. The quantitative foundation for chemistry turned out to be weight rather than affinity, and affinity theory itself confronted fundamental problems as a consequence of Berthollet's research.

Because chemists considered light to be simply another chemical substance – and most physicists came to think in this way too – the concept of affinities applied equally to light particles. This was a radical break with the mechanical theory of optical phenomena that Newton had presented and utilized in the *Opticks* and *Principia*, for mechanical forces are independent of the chemical nature of bodies and depend only on the distribution of mass in space. Nonetheless, Newton himself, as we have seen, had seriously pondered a chemical approach to optics in his private papers and set forth some of his

[9] Ibid., p. 395. For the development of affinities and their relation to Newtonian forces, see Thackray, *Atoms and Powers;* A. M. Duncan, "Some theoretical aspects of eighteenth-century tables of affinity"; and Martin Carrier, "Die begriffliche Entwicklung der Affinitätstheorie."

[10] Macquer, *Elémens de chymie théorique*, Ch. II, p. 20; italics added.

[11] Lavoisier, "Mémoire sur l'affinité du principe oxygine avec les différentes substances auxquelles il est susceptible de s'unir" (read 1783), *Oeuvres*, 2:550.

speculations in the queries.[12] Moreover, new research he had carried out on the refractive force when he was composing the *Opticks* provided a more promising, empirical foundation for the chemistry of light. He had discovered that the refractive power of bodies $n^2 - 1$ (where n is the index of refraction) depends not simply on the quantity of matter but also on its chemical nature. His measurements showed that

> all Bodies seem to have their refractive powers proportional to their densities, (or very nearly;) excepting so far as they partake more or less of sulphurous oyly particles, and thereby have their refractive power made greater or less. Whence it seems rational to attribute the refractive power of all Bodies chiefly, if not wholly, to the sulphurous parts with which they abound. For it's probable that all Bodies abound more or less with Sulphurs. And as Light congregated by a Burning-glass acts most upon sulphurous Bodies, to turn them into fire and flame; so, since all action is mutual, Sulphurs ought to act most upon Light.[13]

Sulfurous, oily (and also fatty and unctuous) bodies, such as olive oil, spirit of turpentine, amber, and diamond ("which probably is an unctuous substance coagulated"), in the terminology of eighteenth-century chemistry contain phlogiston.[14] Newton's discovery that refraction depends on the phlogiston in bodies was fundamental to the new chemistry of light, though it was often misinterpreted and directly applied to the colors of bodies. As we shall see, Opoix fell into this trap.

Although the difference between the behavior of the force of affinity and that of gravitational attraction was recognized, it was frequently held – for instance, by Buffon, Laplace, and Berthollet, though with varying degrees of conviction – that the two forces were the same but that because of the exceedingly small distance between corpuscles, their shapes intervene to disturb the simple inverse square law.[15] Few were dogmatic on this point, but in practice chemists treated affinities as characteristic properties of chemical substances without attempting a further reduction. This difference, though, would ultimately prove to be a fundamental one, for affinities that were characteristic of each substance and not further reducible were considered by physicists to be arbitrarily assigned without "law and order."

[12] See, for example, Ch. 2, note 68; Ch. 3, note 73; Ch. 4, note 16.
[13] *Opticks*, Bk. II, Pt. III, Prop. X, p. $_2$76.
[14] Ibid., p. $_2$75.
[15] Fox, "The rise and fall of Laplacian physics," pp. 95–9; and Thackray, *Atoms and Powers*, Ch. 5, §5, and Ch. 7.

The aim of Part I of Opoix's memoir, "On colors considered in bodies," was to show that the colors of bodies depend directly on the amount of phlogiston they contain. According to his theory, white bodies contain little or no phlogiston, black ones the most, and bodies of intermediate, chromatic colors contain a decreasing quantity according as their color approaches red; that is, darker colors contain more and lighter ones less phlogiston. "Chemistry," Opoix assures us, "is full of facts that prove this truth, and it suffices to rarefy the phlogiston of a body to make it assume the varied colors of the rainbow."[16] To demonstrate this he invokes a large number of chemical observations, all of which are qualitative, as he has no measure for the quantity of phlogiston. For example, when lead is heated for an extended period, it gradually loses its phlogiston, and the resulting calx passes from yellow, to orange, and finally red, just as his theory predicts.[17] His arguments, however, are in part circular; he uses the predicted sequence of color changes to infer the change of phlogiston, which is not independently demonstrated other than by its agreement with phlogiston chemistry. Let us consider another example, which was treated by Newton and then Delaval: Opoix argues that syrup of violets turns green when an alkali is added because alkalies have a great affinity for phlogiston and remove some from the syrup, whereas Newton and Delaval argue that the alkali thickens the corpuscles.[18]

Having demonstrated to his satisfaction that the colors of bodies depend on the quantity of phlogiston they contain, in the second part of his memoir, "On colors considered in light," Opoix explains how bodies reflect light of their proper color. He assumes that rays of different spectral colors contain the same proportion of phlogiston ("the principle of their colors") as bodies of that color and are attracted to bodies containing a like proportion of phlogiston. For example, red bodies, whose phlogiston is very rare, "*attract*" red rays, which proportionally have the rarest phlogiston, because those bodies have a greatest "relation (*rapport*)" with the like rays. "This body," Opoix explained, "will become the center where all red rays of light converge from all directions and from which they are incessantly reflected," and it will appear red.[19] He believed that sunlight is a pure substance,

[16] Opoix, "Observations," p. 104.
[17] Ibid., p. 106.
[18] Ibid., p. 111. For Newton's and Delaval's interpretations, see Ch. 3, note 62, and Ch. 5, note 52.
[19] Ibid., p. 190.

but when it passes through the atmosphere it acquires a subtle earthy matter which is its coloring principle. When light combines with bodies, "it is this fixed light, . . . light perfectly saturated with the earth of bodies, that I call the *phlogiston* of the chemists."[20] Because phlogiston is attracted to phlogiston, light gradually dissolves the phlogiston in bodies, which readily explains why colored bodies fade in sunlight. In plants, however, fixed light causes their green color, for plants grown in the dark are white, or etiolated.[21] Opoix introduced one sophisticated feature into his theory by proposing that the density of phlogiston need not be uniform, for the colors of most bodies are compound and not simple.[22] This was not then generally understood, especially by chemists, and it is most likely that he drew the idea from Dutour's paper in the preceding volume of the *Journal de physique*.

Opoix's paper certainly does not derive its significance from his theory, which is a speculative account of a broad range of phenomena with no experimental tests, but rather as the opening salvo of a chemical attack on Newton's theory that would grow increasingly sophisticated. The paper immediately stimulated debate on phlogiston, light, and color. A few months after it appeared, Etienne Claude de Marivetz, writing under the pseudonym "Madame T.E.S.A.V.L.M.O.R.," welcomed Opoix's theory because it provided an alternative to the unacceptable view of Newton.[23] Marivetz related that he had been trying to develop a theory of light based on fire when Opoix published his theory, which he at once embraced. His only serious difference with Opoix was on the precise nature of the substance with which pure light combined to form phlogiston. Perhaps Marievetz chose to publish his letter to the *Journal de physique* pseudonymously because as a neo-Cartesian and anti-Newtonian he did not believe that his views would otherwise receive a fair hearing, at least from the physics community.

One year after Opoix's paper appeared, the *Journal de physique* published a brief commentary on the paper by "a simple amateur of physics." He is not entirely critical of Opoix; in fact, he accepts his demonstration that the colors of bodies depend on phlogiston and attempts to reconcile Opoix's theory with Newton's (which he does

[20] Ibid., p. 193.
[21] Ibid., pp. 192–3.
[22] Ibid., p. 208.
[23] [Marivetz], "Lettre adressée a l'auteur de ce recueil" (1777). Opoix identifies Marivetz as the author in his *Théorie des couleurs et des corps inflammables* (1808), p. iii. See also Kottler, "Jean Senebier," pp. 69–70, who identifies his other pseudonym, Madame V***.

not fully understand). The "simple amateur" suggests that adding phlogiston to a body alters the nature of the corpuscles (*lamella*) of bodies by changing their size, number, and refracting power. He then turns to more fundamental points and criticizes an apparently natural inference made by Opoix from Newton's observation that refraction is greater in inflammable, or phlogisticated, bodies. Opoix argued that because violet rays are refracted more than red, it follows that there must be more phlogiston and a greater refraction in violet than red bodies. The "simple amateur" observes that this need not at all follow, and it must be experimentally demonstrated that indigo and violet bodies have a proportionally greater refraction, other things being equal, than red ones.[24]

Although Newton's observation on the greater refraction of inflammable bodies was widely invoked to establish the chemistry of light, it has no relation to the coloration of bodies. Refraction acts on all rays by continually varying amounts, but what is called for to explain coloration is an effect that acts selectively on rays of different color. Such a concept of a distinct absorptive affinity gradually emerged only in the 1790s. More seriously, when a refraction model is utilized to explain the appearance of, say, a yellow body, one necessarily considers the force attracting the yellow rays, but it is precisely those rays that are either *not* attracted or are least attracted in a body of that color. The "simple amateur" made this fundamental point when he asked:

> One clearly sees that bodies where the phlogiston is most abundant must attract the light more strongly and retain it. Do they therefore not reflect it? Do they therefore appear black? But why when the phlogiston is ever so less abundant in a body, does it reflect the color violet? And why when it is rarest do bodies reflect the color red?[25]

In an "Addition," written after the second part of Opoix's paper appeared, he still more pointedly noted the principal contradiction of Opoix's theory: "But if wax *attracts* the yellow ray, it does not repel it nor therefore *reflect* it; and if it does not reflect it, how can it appear *yellow?*"[26]

Neither Marivetz nor the "simple amateur" was in the mainstream of the French scientific community, which was not at all receptive to Opoix's ideas. In 1808, in the preface to a book on the same subject,

[24] "Propositions et demandes sur les couleurs des corps," pp. 67–8.
[25] Ibid.
[26] Ibid., p. 71.

Theory of Colors and Inflammable Bodies, Opoix explained the problems that he had encountered with the physicists nearly thirty years earlier:

I had exhibited a lack of tact in these two memoirs in objecting to some errors of Newton and Nollet. This was enough to antagonize the physicists of the Academy, who moreover could not sufficiently appreciate my proofs based on chemical experiments, because then physicists were little concerned with chemistry. Physics and chemistry were then considered . . . as two sciences that possessed many great connections, but which could be cultivated separately.[27]

One cannot, in fact, imagine the senior d'Alembert or the young Laplace being favorably impressed with such a speculative theory or being sympathetic to Opoix's own lack of understanding of physics. Opoix had concluded his paper by going full circle in not only rejecting Newton's application of the colors of thin films to colored bodies but even in rejecting his explanation of thin films. He adopted Dutour's idea that Newton's rings were caused solely by refraction in the prismatic wedge of air between the two plates.[28] Physics and chemistry were certainly more closely related in the beginning of the nineteenth century, as Opoix recognized, but they were not brought together by his approach.

In his preface Opoix explained that his work fared better among those who possessed the two sciences and he cited Marivetz and Macquer, who had read his paper to the academy.[29] Moreover, in the new edition of his widely read chemistry dictionary in 1778, Macquer adopted Opoix's ideas on phlogiston and color.[30] Jean-Baptiste Lamarck was also in his camp: "It appears now beyond doubt that the principal cause of the colors of bodies must be sought in the nature of their constutive elements; this is the opinion of the most celebrated chemists, and M. Opoix has presented it in his excellent dissertation

[27] Opoix, *Théorie des couleurs,* p. iii.

[28] Opoix, "Observations," pp. 209–10; the "learned memoir" in the May 1773 issue of the *Journal de physique* that he here refers to is the first of Dutour's "Considérations optiques."

[29] Opoix, *Théorie des couleurs,* pp. iii–iv.

[30] Macquer declared in the article "phlogiston" that "I am very much inclined to believe *with the majority of chemists* that it [the fixed matter of light] . . . becomes the cause of all colors. The opinion that M. Opoix has expounded in two good memoirs published in Abbé Rozier's collection appears to have considerable probability. This skilled chemist has there assembled and compared in a satisfactory manner a great number of phenomena, the entirety of which is most appropriate to prove not only that light is the material principle of all colors; but also that in becoming the phlogiston of bodies by its fixation, it produces each species of color according to the way in which it is combined" (*Dictionnaire de chymie* [1778], 3:143–4; italics added).

on colors."[31] If Opoix had quoted the remainder of Lamarck's sentence, it would have become evident that Lamarck supported his theory only insofar as it was chemical in opposition to Newton's physical theory.

In Lamarck's own theory, color depends on the quantity of free fire in a body, and not simply on the total quantity of fire, as with Opoix's, and the differences increase from there. Lamarck composed his *Recherches physiques*, which treats a broad range of subjects, in 1776, though the part on colored bodies, he tells us, was written after Opoix's paper. When he submitted his work to the academy in 1780 or 1781, it suffered the same fate as Opoix's: rejection.[32] This is not surprising, for judging by the section on colored bodies, it was even more speculative and less experimental than Opoix's memoir. What is even worse, he supported a modification theory of color. He did not accept chemical affinities and held that the free fire in bodies modified the incident light to produce their colors.[33] By 1794, Lamarck's work was so far out of the mainstream that it no longer contributed to the debate on colored bodies. In 1781, however, he was not the outcast that he would become in the 1790s, but a rising star of French natural history. Through Buffon's patronage he was elected to the academy in 1779.[34] Lamarck's support in the early 1780s was no doubt welcomed by Opoix and other chemical critics of Newton's theory. When the *Recherches physiques* was published in 1794, it was hopelessly out of date, especially as Lamarck did not (and never did) adopt Lavoisier's oxygen theory. By this time, though, Berthollet's *Eléments de l'art de la teinture* was published, and the chemical attack on Newton's theory was emanating from the opposite wing, the physical one, of French chemistry.

Opoix also claimed that "I possess proofs of the value that M. Lavoisier had of my work," but I think he was bluffing, or at least reading too much into what were perhaps some polite comments.[35]

[31] Opoix, *Théorie des couleurs*, p. iv. Opoix's quote is from Lamarck's *Recherches sur les causes des principaux faits physiques* (1794), 2:133–4.

[32] Lamarck, *Recherches*, 1:vii, 2:160. Leslie J. Burlingame, "Lamarck's chemistry," p. 68, shows that the date 22 April 1780 given by Lamarck for submission of the manuscript to the academy is mistaken, and that it was submitted some time between that date and 3 May 1781. On Lamarck's chemistry, see also Richard W. Burkhardt, Jr., *The Spirit of System*, pp. 38–44, 96–100.

[33] Lamarck, *Recherches*, 2:132–83, especially the concluding "Remarque," pp. 180–83. In his *Mémoires de physique et d'histoire naturelle* (1797), pp. 60–2, he abandoned his modification theory of color.

[34] Burkhardt, *Spirit of System*, pp. 23–7.

[35] Opoix, *Théorie des couleurs*, p. iii. Opoix also cites endorsements by Jean Saury (or

By 1776, Lavoisier was already well on his way to his oxygen theory, and his experimental and physical style of chemistry was the antithesis of Opoix's. Opoix was nothing if not persistent, and when half a century later he submitted yet another memoir in the same vein to the Academy of Sciences, the scientific milieu had changed so much that Fresnel was bewildered by it (§9.2).

6.2. THE CHEMISTRY OF COLOR

The issue of the *Journal de physique* preceding the one with Opoix's paper carried the first installment of a series of six memoirs by Senebier on phlogiston and light that would appear over the course of three years. Senebier, who graduated from the University of Geneva with a degree in theology, was at first drawn to philosophy and natural history, but in the mid–1770s his interests turned to chemistry and experimental plant physiology, a field that he in large part created. He was the chief librarian of the Republic of Geneva for most of his professional career and was an active member of the vital Genevan scientific community. His mentor was the eminent natural historian Bonnet, who initially planned to become a physicist, and he studied chemistry under Pierre-François Tingry, who had attended Rouelle's lectures. Because Senebier wrote in French, published in French journals, and closely followed French scientific developments, it is appropriate that I treat him with the French.

Senebier never set forth a general explanation of the colors of bodies, although he carried out all his work with the assumption that its cause was chemical. Nor did he directly attack Newton's theory, although he attempted to modify it to include the chemical nature of bodies. Nonetheless, his work is of extreme importance for our story. By systematically investigating photochemical reactions, especially photosynthesis and other processes involving color changes, he made the chemistry of light a central concern of contemporary science. Once it was judged to be conclusively demonstrated that light entered into chemical composition with matter and indeed could alter its colors, and that the various colors of light had different affinities for different substances, Newton's theory of colored bodies became increasingly improbable. So successful would this aspect of the chemists' revolution be that, by the turn of the century, it would be difficult to find

Sauri), professor of philosophy at the University of Montpellier, in his *Précis de physique* (Paris, 1780), and *Dictionnaire de chimie,* neither of which have I seen.

even a physicist who did not subscribe to some extent to the chemistry of light.[36]

Senebier's first three memoirs, identically titled "On phlogiston considered as the cause of the development, life, and destruction of all living beings in the three realms," were in the speculative natural philosophical tradition and only minimally dealt with light.[37] However, after the third memoir appeared, Marivetz (now as Madame V*** and converted to Opoix's views) criticized Senebier's conception of light and its relation to phlogiston. As a consequence of Marivetz's criticism, Senebier's interests in the remaining three memoirs, all addressed now to Marivetz/Madame V***, shifted from phlogiston to light. He devoted the fifth memoir to attacking the wave theory of light (which Marivetz supported) and defending Newton's emission theory, with much of his evidence being drawn from the chemical effects of light.[38]

In the final, and by far the longest, memoir of the series Senebier left the realm of speculative science and undertook his own experiments on the discoloring effects of light and etiolation. He now adopted Opoix's and Marivetz's positions that light contained phlogiston (as did flame, fire, and electricity), that "light and phlogiston are the cause of colors," and that they combined with bodies with which they have "reciprocal affinities."[39] Much of the paper is devoted to color changes brought about by light. That light caused discoloring or fading was long known. Opoix, for instance, had given an explanation of it. New photochemical reactions were continually being discovered, the most important of which (leading to photography) was the discovery of the sensitivity of silver salts to light. Johann Heinrich Schulze in 1727 discovered that silver nitrate darkened in

[36] The chemistry of light has not yet been studied as an independent historical movement. Such a study promises new insights on the relation of physics and chemistry, especially the influence of chemistry on physics, as most research has been concerned with the flow in the opposite direction. Works treating aspects of the chemistry of light are Joseph Maria Eder, *History of Photography;* Metzger, *Newton, Stahl, Boerhaave;* Partington and McKie, "Studies on phlogiston," III, IV; Kottler, "Jean Senebier"; Cantor, *Optics after Newton;* and Hakfoort, *Optica in de eeuw van Euler,* Ch. 5.

[37] Senebier, "Sur le phlogistique" (1776–7). Kottler's fine dissertation, "Jean Senebier," is an indispensable guide to his oeuvres.

[38] [Marivetz], "Lettre de Madame de V***, à M. Senebier" (1777); for Marivetz's identification, see note 23. Senebier, "Quatrième mémoire sur le phlogistique, ou réponse à la lettre de Madame de V***" (1778); and "Réponse à la lettre de Madame de V***" (1779).

[39] Senebier, "Seconde lettre à Madame de V***" (1779), pp. 368, 369.

light; and in 1757, Giacomo Battista Beccaria found that light turned silver chloride (horned silver) violet.[40] In his memoir Senebier studied the color changes of ribbons, papers, wood, and silver chloride. He investigated, for example, the effect of the different spectral colors on silver chloride and found the violet rays to act most strongly and swiftly, which suggested that they contained more phlogiston than the other colors.[41]

Senebier concluded his memoir with a few pages discussing some experiments on etiolation that soon blossomed into an epoch-making investigation of photosynthesis and plant physiology. The title of his three-volume work, *Physico-chemical Memoirs on the Influence of Sunlight in Modifying the Beings of the Three Realms of Nature, Especially those of the Vegetable Realm* (1782), indicates its broad scope and connection to his earlier papers. In the first volume, drawing on the work of Bonnet, Priestley, and especially Jan Ingenhousz, Senebier established that in sunlight plants took up fixed air (carbon dioxide) from the atmosphere and emitted pure air (oxygen). Here was indeed a prodigious chemical effect of light essential for the economy of nature.

Senebier then turned to etiolation and established that light is the cause of the green color of plants. Light, especially violet, which contains the most phlogiston, colors plants green by combining with their resinous, or phlogisticated, part: "I can therefore still more vigorously conclude that light colors plants green only by combining with this green part, with which it has a definite affinity. . . ." To explain the chemical reaction involved in producing the green color, Senebier applied Macquer's analysis of the new dye Prussian blue: Plants contain fixed alkali (sodium carbonate), iron, and acid of fixed air (carbonic acid). The acid dissolves the iron, and the phlogiston in

[40] On discoloration by sunlight, see Eder, *History of Photography*, pp. 84–6, 88–9; and on silver salts, pp. 60–83, 86–8, 95–9. These experiments on fading originated in Dufay's test for the fastness of dyes described at note 51.

[41] In these experiments Senebier used a prism as the source of his spectral colors, which is one of the few instances that I have come across in the eighteenth century of chemists using prisms; Senebier, "Seconde lettre," pp. 379–80. Scheele (whom Senebier cites on p. 378) had earlier carried out a similar experiment on silver chloride with a prism; "Chemical treatise on air and fire," in *The Collected Papers*, pp. 85–178, on p. 131; which is a translation of his *Chemische Abhandlung von der Luft und dem Feuer* (Upsala/Leipzig, 1777). Senebier included this experiment in his *Mémoires physico-chymiques sur l'influence de la lumière solaire* (1782), Mémoire XIV, 3:192–205. Here he used colored filters to obtain the "prismatic colors" (3:201–2), a solution of carmine for red, a solution of tournesol or litmus for violet, and a solution of turmeric for yellow (1:13).

light combines with the alkali and causes a precipitate, which contains the coloring part.[42] As he had earlier, Senebier turned to other instances of color changes caused by light, but he now marshaled far more experiments and evidence from other sources and produced a virtual encyclopedia of photochemistry. In reflecting on all this evidence, he concluded:

> All the experiments that I have reported seem to concur to show light to be a body similar to those which we know: It has, like them, the same properties, especially the affinities which are proper to them. It would undoubtedly be a very interesting work, very important to physics and chemistry, that would reveal to us all the affinities of light. . . .
>
> Nevertheless, a step has been taken in this science by having proved that light is not merely the vibration of an aethereal fluid, but that it is truly a compound of small bodies, which certainly combine with larger bodies.[43]

To support his case for the chemical activity of light, Senebier invoked Newton's support. He appealed to Query 22/30, where Newton proposed that light and matter are mutually transformed, for the light absorbed by bodies enters into their composition. "My experiments," Senebier affirmed, "show this combination, since they demonstrate a sensible alteration, produced by the action of light alone, and an inner alteration which can only be an effect of its combination."[44] He then turned to Book II, Part III, Prop. V – the heart of Newton's theory of colored bodies – but he avoided directly attacking it, no doubt out of respect for the great man.[45] Rather, he simply suggested that the change of color with density that Newton observed could be explained by the altered attractions caused by the combined light, which in turn disposes incident light to combine more or less strongly with the body. "Leaves, fruits, and wood," Senebier proposed, "successively change their color when they are exposed to light, *because* the sun's rays in combining with these bodies supply to their parts constituents which determine their properties and thus change their refractive powers

[42] Senebier, *Mémoires physico-chymiques*, Mémoire IV, 2:200, 253–72. See Kottler, "Jean Senebier," pp. 197–8.

[43] Senebier, *Mémoires physico-chymiques*, Mémoire XVII, 3:245–6.

[44] Ibid., pp. 303, 304. The query is quoted at Ch. 3, note 72. Senebier (p. 308) also appeals to Bk. II, Pt. III, Prop. X, on the greater refraction of sulfurous bodies, quoted at note 13.

[45] Senebier treated his mentor, Bonnet, equally kindly when they differed; Kottler, "Jean Senebier," pp. 115, 153. We can also note that Senebier was familiar with Delaval's work in support of Newton's theory and likewise chose not to criticize it; "Seconde lettre," p. 383.

while changing their fabric."[46] While avoiding a confrontation with Newton's theory, Senebier has offered a chemical alternative to it, a wholly legitimate Newtonian position, if not Newton's own.

Senebier's achievement in establishing the chemistry of light was impressive. Because his views not only typify those of the period but helped to form them, we should pause to consider them more carefully. Most of his evidence in behalf of a chemical theory of color in his *Mémoires physico-chymiques* came from color changes caused by light. Underlying that approach is the implicit assumption that if light causes a change of color, it must be a cause of color. This assumption follows directly from the prevailing emission theory of light, specifically, that light is a corpuscular substance that may therefore enter into chemical combination with ordinary matter. If light through its varying affinities behaves like any other chemical substance, then after it enters into new combinations the resulting compound will have new properties, in particular, a new color. In what was still essentially a Newtonian world of corpuscles and forces or affinities, the overwhelming evidence for the chemical activity of light provided by this program served in turn as proof for the Newtonian corpuscular theory of light. Casper Hakfoort has shown that the wave theory of light, which then flourished in Germany alone through Euler's advocacy, was eclipsed by the emission theory in the 1790s principally because of the demonstrated chemical influence of light.[47] This situation would

[46] Senebier, *Mémoires physico-chymiques*, 3:305; italics added.

[47] Hakfoort, *Optica in de eeuw van Euler*, Ch. 5. Although I have not thoroughly studied the German literature, I can find no evidence of a battle between physicists supporting Newton's theory of colored bodies and chemists advocating a theory of affinities. Both supported a chemical theory of coloration. This pattern is what one would expect, because Germany had no strong, continuous tradition of Newtonian optics. By midcentury, according to Hakfoort, advocates of Euler's wave theory came to outnumber supporters of Newton's emission theory. After about 1785, however, the tables were turned, and Newton's emission theory was now widely adopted in place of Euler's wave theory. Because this later turnabout was largely a consequence of the chemistry of light, the chemical theory of coloration was quite naturally adopted at the same time. For example, Friedrich Albrecht Carl Gren, founder of the *Journal der Physik* (later the *Annalen der Physik*), explained the color of bodies by affinities: "The different substances in nature have a force to bind and fix certain species of homogeneous lights and cancel their expansive force more than other species" (*Grundriss der Naturlehre* [1793], Pt. II, Sect. 2, §596, p. 456). A similar explanation of absorption by a selective attractive force is found in Johann Tobias Mayer, *Anfangsgründe der Naturlehre* (1805), §637, p. 533.

Placidus Heinrich's answer to the prize question put by the Bavarian Academy of Sciences – "Does the Newtonian or Eulerian system of light agree better with the latest experiments and knowledge of physics?" – was a resounding yes for Newton, and it garnered first prize. His essay was a cornucopia of the chemistry of light and, according to Hakfoort, appears to have contributed to the precipitous decline of

change in the next century when more compelling evidence for the wave theory of light became available and concepts of the transfer of energy rather than matter became more widespread. By focusing his attention on change of color rather than on permanent color, Senebier's account remained essentially chemical and did not treat the optical concept of absorption. He explained, for example, that phlogisticated violet rays combine with (or were absorbed by) the coloring parts to create the green of plants, for without light they are white. This is a marked advance over an Opoix, who would have focused attention precisely on the green rays, but Senebier does not explain why the new green parts are seen as green or what happens to all the other light rays other than for a general statement that they combine with the body.

The year after his *Mémoires physico-chymiques* appeared, Senebier felt compelled to defend chemistry as an independent physical science in his *Investigations on the Influence of Sunlight* (1783). He had introduced modern chemistry to plant physiology and contrasted his approach with that of his predecessors – "the Grews, Malpighis, Duhammels, and Bonnets" – which was based on "the laws of movement alone," that is, physics. To understand bodies, Senebier explained, one must know their components, for the qualities of bodies necessarily result from the qualities of the elements composing them. Only the experimental methods of the "enlightened chemist" can provide this knowledge. Because all effects in the physical world are due either to mechanical or chemical means, and "almost always" both together, both physics and chemistry are required to comprehend nature.[48] Senebier represented the new chemistry that drew concepts, knowledge of phenomena, and methods from physics and applied them to the chemical properties of bodies. He declared chem-

Euler's theory in Germany. He rejected both Euler's and Newton's theory of colored bodies in favor of a chemical explanation; "Ueber die Preisfrage" (1789), pp. 249–50. Heinrich entered a similar contest announced at the St. Petersburg Academy of Sciences in 1804 but this time took only second place to Heinrich Friedrich Link; see Link and Heinrich, *Ueber die Natur des Lichts* (1808); Eder, *History of Photography*, pp. 145–7; and also Ch. 5, note 4. Johann Gottfried Voigt's explanation of the colors of bodies, that they are caused by the relative quantities of light and heat matter (*Wärmestoff*) they contain, appears to be typical of German theories in this period; "Beobachtungen und Versuche über farbigtes Licht" (1796); for similar theories see Partington and McKie, "Historical studies," IV, esp. pp. 132, 137.

[48] Senebier, *Recherches sur l'influence de la lumière solaire*, pp. iv, vi, x. Kottler quite plausibly suggests that Senebier's preface defending chemistry was a reply to Bonnet's accusation in a letter in September 1782 that his *Mémoires physico-chymiques* devotes "too much to the chemistry of your subject and not enough to the physics"; "Jean Senebier," p. 226.

istry to be "nothing other than a branch of experimental physics, and one of its most useful branches" and lamented physicists' lack of knowledge of its resources and means; "they see in chemistry only the metallurgist or the vulgarity of pharmacists without considering that the true chemist . . . is above all constantly a good physicist, *the only one truly in a position to study the operations of nature*" and to discover the "principles" of bodies, the laws of their combination, and the resultant effects.[49] In attacking physicists' prejudice rather than physics, Senebier was asserting chemistry's status – indeed, superior status – as an independent branch of physical science with its own intellectual territory, the internal composition of bodies. Although this territory would continually shift, the colors and composition of bodies are common to both sciences and consequently constantly subject to outbreaks of territorial disputes.

Most of Senebier's evidence for the chemical origin of color came from color changes caused by light, and this remained one of the major themes in the case against Newton's theory. He had modeled his explanation of the green color of plants on Macquer's analysis of the dye Prussian blue and had cited analyses of indigo by Torbern Bergman and others. During the course of the eighteenth century the textile and dyeing industries assumed a significant role in the European economy, especially in France and Britain, with large increases in output and major technological innovations.[50] Indeed, they were foci of the Industrial Revolution and not without consequence for the fate of Newton's theory. Although much of the progress was based on practical knowledge, this was one of the earliest instances in which applied science contributed to industrial practice. Throughout the course of the eighteenth century, leading natural philosophers and chemists successively became director of the state-organized French dyeing industry: Charles-François de Cisternay Dufay, Jean Hellot, Macquer, and Berthollet, who together laid the foundations of the

[49] Senebier, *Recherches*, pp. vii, xii; italics added.
[50] For Britain see Archibald Clow and Nan L. Clow, *The Chemical Revolution*, Ch. 10; Albert Edward Musson and Eric Robinson, *Science and Technology in the Scientific Revolution*, Ch. 9; and Susan Farlie, "Dyestuffs in the eighteenth century." For France see Guerlac, "Some French antecedents," pp. 77–81; H. Wescher, "Great masters of dyeing"; and John J. Beer, "Eighteenth-century theories of the process of dyeing." My account also draws upon Franco Brunello, *The Art of Dyeing*. A useful bibliography is the Jewish National and University Library, *Catalog of the Sidney M. Edelstein Collection*. L. G. Lawrie, *A Bibliography of Dyeing and Textile Printing*, is amateurish and unreliable, but of some value because of its chronological arrangement.

modern chemistry of dyeing. New dyes, tests, and processes were introduced, traditional methods improved, dyes chemically analyzed, and general theories of dyeing proposed.

Dufay, the noted electrical investigator, was asked by the controller general of France to assist in revising the earlier regulations for dyeing that his predecessor, Jean-Baptiste Colbert, had instituted in 1669 and 1671. The old regulations strictly enforced the distinction between dyers of *grand et bon teint* from *petit teint*, or colorfast (and expensive) from fugitive (and cheap) dyes. The old tests to distinguish the two were inadequate and Dufay was charged with developing a new one. After much experimentation, he developed a test for fastness that exposed dyed pieces of wool to the sun together with a rapidly fading standard for a fixed time (twelve days in summer and eighteen in winter).[51] This is the origin of the discoloration experiments carried out later in the century by Bonnet, Senebier, and others. Dufay also succeeded in changing government regulations so that indigo could be freely used in place of the less effective blue dye then used, woad or pastel. Indigo would become the most important vegetable dye of the eighteenth century. Accordingly, it became the center of great attention, together with the other principal blue dye, Prussian blue: Hellot, Bergman, and Jean-Baptiste Le Blond carried out analyses of indigo; and the French Academy offered prizes for its analysis and use and issued reports on it.[52]

In 1740 and 1741, Hellot, Dufay's successor, presented a mechanical theory of the process of dyeing in which the atoms of the dye entered the pores of the fiber and became trapped there "like a diamond set in a ring." In a fast dye the atoms were small enough to enter the pores, whereas in a poor one they were too large and remained on the surface of the cloth. He also recounted this theory in 1750 in his important book on the dyeing of wool, the first systematic treatise on

[51] See Wescher, "The French dyeing industry and its reorganization by Colbert"; Brunello, *Art of Dyeing*, pp. 223–5; and Stanley D. Forrester, "The history of the development of the light fastness testing of dyed fabrics."

[52] On the prize contest for indigo see the preface to *Mémoires de mathématique et de physique, présentés à l'académie royale des sciences, par divers savans, & lûs dans ses assemblées* 9(1780): iii–v; this volume also contains one of the prizewinning papers by Hecquet d'Orval and de Ribaucour (the prize was split with Quatremère-Disjonval, who translated Delaval's 1777 book into French) and Bergman's paper, which won honorable mention. Bergman accurately captured the approach of dye chemists in this era with his remark: "It is nearly universally acknowledged that phlogiston, as a principle of bodies, is the principal cause of their color in modifying the passage and reflection of light" ("Analyse et examen chimique de l'indigo," p. 132). See also Le Blond, "Essais sur l'art de l'indigotier" (1791).

the subject.[53] Hellot's successor as director, Macquer, devoted himself especially to silk and in 1763 wrote a pioneering book on it. He developed a method to dye silk with Prussian blue (till then only a painter's pigment) that opened up a new market, as hitherto only red silk possessed the two desirable qualities of both rustling and being colorfast.[54] His analysis of Prussian blue, which Senebier utilized, was further pursued by Scheele, Berthollet, and others. "There are few subjects," Berthollet observed, "which have so exercised chemists as the coloring principle of Prussian blue."[55] Undoubtedly, the scientific discovery with the greatest economic significance was Berthollet's development of a process (based on a discovery by Scheele) to bleach cloth – anti-dyeing, if you will – with dephlogisticated marine acid (chlorine). Later we will turn to his treatise on dyeing, which is considered to mark the beginning of the modern scientific study of dyeing.

Though not as intensively as in France, the scientific study of dyeing was pursued throughout Europe and in America. In Britain, the textile industry assumed major importance in the second half of the century, especially cotton textiles whose manufacture was centered in the north. The dyeing and calico printing industries also grew. The British government, unlike the centralized French state, did not provide resources to organize and develop these industries, but scientists applied themselves to dyeing and bleaching. Bancroft's treatise went through two editions, and works of the Frenchmen Hellot, Macquer, and Berthollet were translated into English. The rapid, widespread translation of works on dyeing is another indication of its importance in late eighteenth-century Europe.[56]

[53] Hellot, "Théorie chimique de la teinture des étoffes" (1740–1), p. 94. *L'Art de la teinture des laines et des étoffes de laine, en grand et petit teint* (Paris, 1750); which was translated in *The Art of Dying Wool, Silk and Cotton. Translated from the French of M. Hellot, M. Macquer, and M. Le Pileur d'Apligny* (London, 1789); and also into German (1751), Spanish (1752), and Italian (1791).

[54] Macquer, *Art de la teinture en soie* (Paris, 1763); translated into German (1764), Spanish (1771), English (1789), and Dutch (1791); and "Examen chymique du bleu de Prusse" (1752). See Coleby, *Chemical Studies*, pp. 52–9, 85–96.

[55] Berthollet, "Mémoire sur l'acide Prussique" (1787), p. 148. See also Scheele, "Experiments concerning the colouring principle in Prussian blue" (1782–3), *Collected Papers*, pp. 238–55. On Berthollet's achievements during his tenure as director of dyeing, see Brunello, *Art of Dyeing*, pp. 258–61; Michelle Sadoun-Goupil, "Science pure et science apliquée"; and *Le chimiste Claude-Louis Berthollet*, pp. 21–5, 138–9; Gillispie, *Science and Polity in France*, pp. 408–13; and Barbara Whitney Keyser, "Between science and craft"; see also note 64.

[56] For the British exploitation of Berthollet's development of chlorine bleaching, which involved the active participation of scientists such as James Watt, see Clow, *Chemical*

Granted the importance of dyeing for eighteenth-century industry and chemistry, one may ask what significance it had for the theory of colored bodies. Chemists, with increasing success, were themselves altering the colors of bodies and studying those changes, whether caused by chemicals or by light itself. Consequently, whatever lingering doubts there may have been as to whether color was a physical property of bodies, they quite naturally interpreted the causes of color and color changes as chemical in nature, and not mechanical. For Newton, or an adherent to his theory of colored bodies, the color of a body or its change is due to the structure or arrangement of the corpuscles and the vibrations set up by the light particles. For chemists the colors were caused by the different affinities that the substances had for light. By about 1775, mechanical explanations in the mode of a Boyle that invoked the size, shape, arrangement, and motion of corpuscles were on methodological grounds no longer acceptable to chemists. Both Berthollet and Bancroft all but mocked the naïveté of Newton's mechanical theory of color. A remark by Chaptal on Hellot's explanation of dyeing by means of the size of the pores of the cloth and atoms of the dye well illustrates this change of style: "By a curious aberration of the human mind, Hellot and Macquer attributed the results of the process of dyeing to mechanical causes at the very moment when analysis was beginning to recognize affinity as the decisive factor in chemical actions."[57]

Venel's contrast of the physicist's and the chemist's conceptions of color in the *Encyclopédie* in 1753, I believe, typified that of most chemists even at the end of the century: "Color considered in the colored body is, for the physicist, a certain disposition in the surface of the body that makes it fit to return such and such a light ray; but for the chemist the greenness of a plant is inherent in a certain green resinous

Revolution, Ch. 9; and Musson and Robinson, *Industrial Revolution*, Ch. 8. The latter also treats the growing role of scientists in the dye industry (Ch. 9), but see, too, Clow, pp. 200, 217. See, for example, the paper by Thomas Henry, "Considerations relative to the nature of wool, silk, and cotton" (1790). Berthollet, with Nicolas Desmarets, annotated and reviewed the French translation of Carl Wilhelm Pörner's *Chymische Versuche und Bemerkungen zum Nutzen der Färbekunst* (Leipzig, 1772); *Instruction sur l'art de la teinture et particulièrement sur la teinture des laines, par M. Poerner.* Ouvrage traduit de l'allemand par M. C***. Revu et augmenté de notes par MM. Desmarets et Berthollet (Paris, 1791). For evidence of the demand for works on dyeing and their translation see notes 53, 54; Ch. 5, note 53; Ch. 7, note 8; and Ch. 8, note 21.

[57] Quoted without reference in Wescher, "Great Masters," p. 634. Because my aim here is to illustrate the changed attitude to mechanical models, my point is only reinforced by Macquer's later abandonment of his mechanical concept of dyeing; see Coleby, *Macquer*, pp. 91–3.

body, which he is able to separate from this plant."[58] A science whose conceptual foundation was quality-bearing substances would naturally associate the quality of color with a substance. To chemists, the ultimate proof of their belief was that they had succeeded in isolating and preparing these coloring principles in their vats and beakers. Thus, in France, during the last quarter of the century there occurred a conjunction of three developments that led to direct conflict between a physical and a chemical interpretation of color: conclusive evidence from photochemistry that light enters into chemical reactions; a rapid growth in dye and bleaching chemistry; and the maturity of chemistry as a physical science with chemists asserting their independence. Britain saw a similar conjunction but not quite as strongly focused as in France.

Lavoisier's introduction of oxygen in place of phlogiston was just one aspect of the broad sweep of the Chemical Revolution instituted by him. For our purposes, however, it is the most important one. Although the interpretation of photochemical reactions now became somewhat more complex, the idea of the chemical nature of light was carried over into the new chemistry unchanged. Lavoisier explained the combustion and calcination of bodies as a combination with oxygen from the air rather than as a loss of phlogiston from the body. According to him, a gas such as oxygen consisted of the "base" of the gas plus caloric (heat or the matter of fire), which rendered the gas elastic. Thus, in combustion and calcination, a body combined with oxygen from the air and released caloric as heat and light. Lavoisier's caloric was very much like the old phlogiston, transferred from bodies to the air, insofar as it was a materialization of heat and light and capable of chemically combining with bodies. In the table of "Simple substances belonging to all the kingdoms of nature, which may be considered as *the elements of bodies*" in his *Traité élémentaire de chimie*, pride of place went to light, followed by caloric. He thought it probable the two were identical and certain that they were related:

> We are unable to determine whether light be a modification of caloric, or if caloric be, on the contrary, a modification of light. . . . We therefore distinguish light from caloric; though we do not therefore deny that they have certain qualities in common, and that, in certain circumstances, they combine

[58] Venel, "Chymie," p. 419; compare Crosland, "Development," pp. 409–10.

with other bodies almost in the same manner, and produce, in part, the same effects.[59]

The relation of caloric and light (as well as electricity) would be a center of debate well into the nineteenth century. Lavoisier granted that the combination of light with bodies was still little known, but noted that Berthollet's experiments (soon to be examined) show that it has a "great affinity with oxygen" and others that it combines with plants.[60]

The first phase of the Lavoisian revolution, insofar as coloration is concerned, was to attack the phlogiston interpretation. In his classic "Reflections on phlogiston," which was read to the Academy of Sciences in the summer of 1785, Lavoisier launched his full-scale attack on phlogiston chemistry. He devoted a small section to its explanation of coloration, which he judged to be self-contradictory. One of Lavoisier's arguments against it was based on the colors of metallic calxes (or oxides, as they were called in the new chemistry) and was apparently drawn from Fourcroy.[61] It is quite possible that he had in mind Rouelle's lectures, which he had attended, or Opoix's theory, which was presented to the academy the preceding decade. The phlogistonists "teach us that phlogiston is the principle of colors, and yet in proportion that metallic calxes are more deprived of it they become more colored." For instance, the calx of iron is initially yellow, but as it loses phlogiston it becomes red and then brown; and the calx of mercury is red, whereas those of copper are green and blue. Thus, if phlogiston is the principle of colors, these calxes must contain phlogiston, but it a fundamental doctrine of that theory that calxes contain no phlogiston. Although it is true that some calxes are white, Lavoisier appealed to Newton's theory of light and color and charged that "the partisans of Stahl's doctrine" err in considering white to be the absence of all color (and phlogiston), when it is actually a combination of all colors. Therefore, he concluded, "One must admit that all metallic calxes contain some phlogiston, since some unite all colors and others present particular ones."[62]

The antiphlogistonists next had to propose an alternative expla-

[59] Lavoisier, *Elements of Chemistry*, Pt. I, Ch. 1, p. 6; Pt. II, Introduction, p. 175; italics added.

[60] Ibid., Pt. II, Sect. III, p. 18[3]. See Fox, *The Caloric Theory of Gases*.

[61] See W. A. Smeaton, *Fourcroy: Chemist and Revolutionary*, p. 97, for Fourcroy's similar criticism of Macquer's version of the phlogiston theory in his *Leçons élémentaires d'histoire naturelle et de chimie* (Paris, 1782).

[62] Lavoisier, "Réflexions sur le phlogistique, pour servir de suite à la théorie de la combustion et de la calcination, publiée en 1777," *Oeuvres*, 2:623–55, on p. 638.

nation of color, and they quite naturally turned to oxygen, or antiphlogiston. In their zeal to embrace Lavoisier's new theory, his supporters, Berthollet, Chaptal, and Fourcroy, initially tended to overemphasize oxygen. Lavoisier himself, as far as I can determine, did not participate in the effort to develop a new chemical theory of color. Berthollet, the newly installed director of dyeing and administrator of the renowned Gobelins tapestry works, was the most active in the period between 1785 and 1791, when his *Eléments de l'art de la teinture* appeared. He wrote a series of important papers on dephlogisticated marine acid (chlorine), its use in bleaching, and its effects on the coloring parts; a paper on an analysis of Prussic acid and Prussian blue; and one on general considerations of the chemical influence of light, which reinterpreted with oxygen many of the phenomena treated by Senebier, photosynthesis, etiolation, and the blackening of silver chloride. Berthollet's explanation of the green color of plants shows how the new account differs from that with phlogiston, which was held to enter the plant directly from light. When light shines on a plant, it frees the oxygen from the water contained in the plant because light has an affinity for oxygen. In darkness, where there is no light to free the oxygen, the oxygen combines with the coloring parts and renders them colorless.[63] Dephlogisticated marine acid bleaches substances in the same way; the oxygen of the acid (as Berthollet believed) combines with the coloring parts and effaces the color.[64] In his paper "On the influence of light," he demonstrated the affinity of light and oxygen by showing that in sunlight dephlogisticated marine acid and nitric acid give up oxygen; both also change color in the process, the former from yellow to colorless and the latter from white to yellow.[65]

In 1790, Fourcroy, utilizing the work of Berthollet and Scheele, deduced some general rules "on the coloration of vegetable matter by vital air [oxygen]." These further emphasized the role of oxygen: "1. That oxygen combined with vegetable substances changes their

[63] Berthollet, "De l'influence de la lumière" (1786), p. 84; see Kottler, "Jean Senebier," pp. 261–3.

[64] Berthollet, "Mémoire sur l'acide marin déphlogistique" (1785), p. 293. In the new terminology adopted in his later papers dephlogisticated marine acid became oxygenated muriatic acid; Davy proved in 1810 that it is actually the element chlorine. Berthollet's other papers on chlorine bleaching in this period are "Description du blanchîment" (1789), "Additions à la description du blanchîment" (1790), and "Mémoire sur l'action que l'acide muriatique oxigéné exerce sur les parties colorantes" (1790).

[65] Berthollet, "De l'influence de la lumière," pp. 81–3.

color. 2. That the proportions of this principle vary the hues of colored vegetable matters. . . . 6. That this complete saturation [of the coloring parts with oxygen] most often gives yellow colors."[66] Both Fourcroy and Berthollet took a purely chemical approach to coloration and were concerned only with the substances that cause color.

Chaptal, in an unlikely source, a memoir on Roquefort cheese a year earlier, utilized an optical explanation of coloration based on oxygen in order to interpret the changing colors of the rind of the cheese. He makes the extravagant claim that all chemical changes are due to various combinations of the principles of bodies with oxygen,

> and the changes of color which are almost an inevitable consequence of it admit no other cause than the fixation and condensation of this gas with these same principles. In being concentrated more or less and in a greater or lesser quantity, it [the oxygen] necessarily acquires different degrees of density which makes it fit to refract this or that ray according to the flexibility of each.[67]

Chaptal then proceeds to fall into the same trap as Opoix by confusing reflection, absorption, and refraction and focusing on the reflected ray. Because, he argues, the blue is the weakest and most flexible ray, it is the first ray reflected when oxygen is fixed in a body. A black body absorbs all rays, so that in this explanation the added oxygen repels rather than attracts the blue ray. Moreover, because his explanation depends on only one variable (oxygen density), color changes must all be in a regular sequence from blue to red.

In the following year Chaptal largely resolved these problems. He now recognized that he had to explain absorption and introduced "the greater or less affinity of the several rays with various bodies" as the cause of the absorption or "combination" of some rays and reflection of others.[68] Moreover, he no longer completely denied a physical theory of coloration and allowed that the disposition of the pores (Nollet's explanation) could also contribute to coloration.[69] By the next decade the excessive concentration on the role of oxygen in coloration would also generally abate. Indeed, by 1807, Chaptal declared: "As we are still little enlightened on the cause of the coloration of bodies, we will restrict ourselves to relating some facts."[70]

[66] Fourcroy, "Mémoire sur la coloration des matières végétables par l'air vital," pp. 90–1.
[67] Chaptal, "Observations sur les caves et le fromage de Roquefort" (1789), p. 57.
[68] Chaptal, *Elements of Chemistry* (1791), 1:86; a translation of his *Elémens de chimie*, 3 vols. (Montpellier, 1790).
[69] Chaptal, *Elements of Chemistry*, 1:86, 3:144.
[70] Chaptal, *Chimie appliquée aux arts*, 4:405.

In this first phase of the chemists' revolt against Newton's theory of colored bodies, their principal concern was to establish a chemical theory of light and coloration by showing that the color of bodies depends on their chemical composition. As chemists, they devoted little attention to optics – the composition of the colors of bodies, or the physical process of coloration – nor did they wield the principal instrument of eighteenth-century optics, the prism. In the following decade this would change as some would apply the concept of chemical affinities to the optical process of absorption and some would pick up a prism. Indeed, in Britain, the optical investigation of coloration had already begun. Before we turn to the renewed attack on Newton's theory of colored bodies, which was launched by Berthollet in 1791, let us cross the Channel to study these new developments and the very different milieu in which they occurred.

7

THE CHEMICAL PHILOSOPHY IN BRITAIN: 1785–1815

In 1776, when Opoix launched his attack on Newton's theory of colored bodies in France, in Britain, Priestley had just enthusiastically endorsed it and Delaval was carrying out research to confirm it. Their work shows how chemistry and natural philosophy were part of a common, Newtonian tradition in a way that was not true in France, where natural philosophy, as exemplified by Nollet, Dutour, and Lamarck, occupied an ever-diminishing position after the third quarter of the century and was being replaced by physics. The Newtonian natural philosophical tradition, or rather traditions, was rich and diverse. It included the classical, dynamical view of corpuscles with forces acting at a distance; Boscovichean atomism with point masses possessing attractive and repulsive forces at various distances; and a chemical tradition of active aethereal mediums, fire, solar matter, light, or phlogiston. Because the Newtonian tradition encompassed such a diversity of approaches, and chemistry was an integral, and perhaps the dominant, component of British science, the attack on Newton's theory of colored bodies was not nearly so strident or part of a broader ideological struggle between chemistry and physics as in France.

In 1785, Delaval, who had done so much in the previous two decades to establish Newton's theory, switched his allegiance to a chemical theory of colored bodies. His new work also contained an experimental refutation of Newton's phenomenological theory that showed that colored bodies do not reflect colored light. He was the first (and for another fifteen years the only) advocate of the chemical theory to undertake an experimental investigation of the optical cause of coloration. It completely undercut the physical basis for Newton's analogical theory, although its significance was not widely recognized. Almost a decade later, Bancroft presented a thoroughly argued case against Newton's physical theory (and Delaval's earlier work in support of it) and proposed that coloration is due to the elective affinities of bodies for some colors, which are selectively absorbed. Though he carried out no experiments on absorption, because of the prevailing chemical philosophy his suggestion was rapidly adopted and, together with Delaval's experimental work, helped to put Newton's theory into eclipse.

In a nation purportedly peopled by staunch Newtonians, it may come as a surprise that no one stepped forth to defend Newton's theory. Because of the prevailing chemical philosophy, most British scientists were receptive to the chemistry of light and unsympathetic to Newton's mechanical theory. Of the few quantitative physicists who might have been expected to defend Newton's theory, John Robison and John Leslie were adherents of the chemical view, whereas Henry Brougham and Young had lost confidence in it because of negative optical evidence. Brougham and Young were untouched by the chemistry of light and started out as supporters of Newton's theory.

7.1. THE CHEMISTRY OF LIGHT AND COLOR

Delaval began his "Experimental inquiry into the cause of the permanent colours of opake bodies" with a careful analysis of Newton's account of his phenomenological theory of colored bodies in the *Opticks* (Bk. I, Pt. II, Prop. X). He immediately focused on one of its principal difficulties – namely, it requires that all bodies be one color by reflection and the complement by transmission, whereas most transparent bodies are the same color whether seen by reflection or transmission. Newton argued that this was an artifact: If the rear surface of a transparent body were blackened to eliminate the return of the transmitted light, then the light reflected from the "tinging" particles would be seen alone and be "apt to vary" from the transmitted light. After Delaval performed the experiment proposed by Newton, he reported:

> It will appear from the experiments . . . that in Transparent Coloured Substances, the Colouring Matter *does not reflect any light;* . . . by intercepting the light which was transmitted . . . they do not vary from their former colour, to any other colour, but become entirely *black.*
>
> As the incapacity of the Colouring Particles of Transparent Bodies to reflect

light, is deduced from experiments which are very numerous, and whose results are constant and invariable, it may be held as a general law. . . .

This law will appear still more extensive, if it be considered that, for the most part, the tinging Particles of liquors, or other Transparent Substances, are extracted from Opake Bodies; that the Opake Bodies owe their colours to those particles in like manner as the Transparent Substances do.[1]

Delaval's experiment, though qualitative, was carefully designed and executed and represents the best of "experimental philosophy." He took a vial (Figure 7.1) with a round neck and rectangular base and thoroughly covered it with black varnish except for the neck *A* and part of one side of the base *B*. To exclude scattered, transmitted light, he took care to keep the glass "perfectly clean" and not allow sediments to form on the faces. The vial was filled with various colored fluids and placed so that light coming from a window simultaneously struck the neck and a fully varnished side while the uncovered side was perpendicular to the window. In this way the fluid in the neck would be seen by transmitted light, which was excluded from the base. After examining sixty-eight fluids he found the fluid in the neck to be always colored and that in the base to be colorless or black and thus to reflect no light. Contrary to Newton's claim, Delaval declared "Transparent Coloured Liquors do not yield any colour by reflection, but by transmission only." He carried out a number of other experiments to confirm the result, including an observation of candle light reflected from colored glass plates: The reflected image was "*not in the least tinged by the coloured glass.*"[2] Delaval's observations of the nature of the reflected and transmitted light in colored bodies were the most significant part of his paper. His persuasive series of experiments corrected Newton's erroneous interpretation and restored the seventeenth-century understanding of the phenomenon. That Delaval was equally adept in chemistry and experimental physics, I believe, helps to explain why he was one of the very few chemical critics of Newton's theory to undertake an optical investigation of the cause of coloration.

In the remainder of his long paper, Delaval set forth his own physico-chemical explanation of the cause of coloration. His experiments

[1] "An experimental inquiry into the cause of the permanent colours of opake bodies" (1789), pp. 162–3; read 19 May 1784; also published in 1785 as a book with the same title. I will cite the paper. In contrast to his earlier book (Ch. 5, note 53), this one was translated only into German (1788), though it was certainly known in France and Italy. Newton's phenomological theory and his experiment are described in §3.2.

[2] Ibid., pp. 171, 174.

Figure 7.1 To test whether transparent colored liquids reflect colored light, Delaval placed them in a glass vial covered with black varnish except for the neck and part of one side. (From *Memoirs of the Literary and Philosophical Society of Manchester*, 2 [1789]:164.)

had established that colored light is not reflected from the surface of a colored body but must be transmitted within it and then be reflected. He interpreted this as showing that the coloring particles have no power to reflect light (an argument that, incidentally, invokes transduction exactly as Newton had) but, rather, one to transmit it. When he chemically separated the coloring matter from plants, he found that a white, earthy matter remained and attributed a reflecting power to it.[3] Thus he explained coloration by two sets of corpuscles: transparent coloring particles that transmit light and opaque white particles that reflect colored light within the body. The gap in his explanation, soon to be rectified by others, is that it altogether ignores absorbed light. Delaval related a series of observations and experiments in order to show, just as his Continental colleagues did, that the "Colouring Particles of plants consist principally of inflammable matter"; and that the "action and properties of phlogiston, and of the sun's light, are so exactly similar, that the identity of those subtile principles can scarcely be doubted."[4]

In an attempt to present nearly as comprehensive a theory as Newton, Delaval applied these principles to the colors of animal substances, minerals, metals, and Newton's archetypal phenomena (called "semipellucid colored substances" by Delaval), but he was compelled to introduce ad hoc explanations along the way. Near the end of his paper he tried to salvage his two earlier papers (which showed that color changes depend on the size and density of the corpuscles) by modifying Newton's theory:

> If the experiments, which have been made with thin Transparent Colourless Substances, and those which have been performed with Coloured Media, be jointly considered, it will appear that the effects, which Sir Isaac Newton has attributed to Reflected Light, should have been applied to Transmitted Light. And the observation of this great philosopher, relative to the connection between the colours of natural bodies, and the size and density of their particles, if thus modified, will coincide with his own experiments, and all the others which have been made subsequent to them.

He does not explain how this revised theory can be reconciled with phlogiston as a cause of color and yet another cause ("the greater, or less obstruction which the rays meet with" in passing through the pores of bodies) that he introduces on the next page.[5] Delaval's work was well known and well received in Britain, but before we turn to its reception let us examine Bancroft's contributions to the debate over Newton's theory.

Edward Bancroft, who was born in Massachusetts in 1744 and died in England in 1821, is undoubtedly the most fascinating character in

[3] Ibid., pp. 185, 245.
[4] Ibid., pp. 186, 187–8. Chemists were beginning to adopt the oxygen theory at about the time Delaval read his paper, but phlogiston theory continued to be used in Britain until the turn of the century; Partington, *History of Chemistry*, 3:489–90.
[5] Delaval, "An experimental inquiry," pp. 248, 249.

our story: physician and innovative dye chemist and also spy, war profiteer, and speculator.[6] The young Bancroft was tutored by Silas Deane, who later became a member of the Continental Congress and American commissioner (ambassador) to France. Bancroft was afterward apprenticed to a physician in Connecticut, but at the age of eighteen he ran away to Surinam, where he practiced surgery. In 1767, he emigrated to England and enrolled as a physician's pupil at St. Bartholomew's Hospital in London, and he later received an M.D. from Aberdeen University. Two years after arriving in England the twenty-five-year-old published two books: an important account of the natural history of Dutch Guiana that included scientific and medical observations and a political tract on the growing controversy between Great Britain and the American colonies. Bancroft became a friend of Benjamin Franklin and Priestley and entered the London scientific community. On the basis of his book on the natural history of Guiana and three unpublished essays on dyeing, he was elected a fellow of the Royal Society in 1773. Bancroft's notoriety derives from his life as a double spy during the American Revolution, in which he managed simultaneously to serve and betray both his native and adopted countries and evade detection until the publication of British diplomatic papers in the late nineteenth century. He has also been accused of using his expertise with poisons to murder Deane, his former tutor and fellow rogue.[7]

Bancroft enters our story with his discovery that the inner bark of American black oak could be commercially used as a yellow dye, which he called quercitron and which was still used extensively until the beginning of this century. In 1779, Bancroft received a patent (monopoly) for the importation and sale of the bark. The patent was extended in 1785, but a second extension was defeated in the House of Lords in 1799. Bancroft, in the manner of the French chemists, turned to the scientific study of dyeing in 1769. In 1794 he succeeded in bringing out the first volume of his *Experimental Researches concerning the Philosophy of Permanent Colours*. He intended to follow this with a second volume, but because of the financial difficulties caused

[6] For biographical information, see the fascinating paper by Julian P. Boyd, "Silas Deane: Death by a kindly teacher of treason?"; and Godfrey Tryggve Anderson and Dennis Kent Anderson, "Edward Bancroft. M.D." See also Arthur S. McNalty, "Edward Bancroft"; C. A. Browne, "A sketch of the life and chemical theories of Dr. Edward Bancroft"; and Sidney M. Edelstein, "Historical notes on the wet-processing industry."

[7] Boyd, "Silas Deane," bases his case entirely on motivation and circumstantial evidence.

by the loss of his bark monopoly, it was published only with the second edition in 1813.[8] His book, highly regarded by historians of dyeing, appeared in an American edition in 1814 and in German translations in 1797 and 1817–18, the latter being reprinted as late as 1834.

Bancroft devoted the long introductory chapter of his *Experimental Researches* to attacking Newton's theory of colored bodies and presenting the case that their colors are chemical in origin. His brief account of Newton's theory describes its general features but exhibits some of the misunderstandings common among chemists like Berthollet. He did not realize that the various orders of the same color, say, a blue, are in fact different colors because their composition differs.[9] Only if this point is grasped can the predictions of Newton's theory be recognized and put to an experimental test. Most of his direct criticism of Newton's theory is actually aimed at Delaval's conflation (which follows directly from Newton) of the mechanical and the chemical in his first two papers (§5.2). At the conclusion of his vigorous argument against Delaval's "chimeras," Bancroft vents the chemists' frustration with this sort of mechanical theory:

> The common sense and experience of mankind, if fairly consulted, will condemn and revolt at the idea, of making the colours of bodies depend on their weight, or the size of their particles; for it certainly never has been observed, that the heaviest substances were red, or the lightest violet-coloured, or that bodies equally heavy were all of the same colour. Different parcels of indigo, for instance, vary extremely as to specific gravity, without any variation of colour; a fact which is not only at variance with Mr. Delaval's hypothesis, but which renders it easy to find samples of indigo, of exactly the same specific gravity as the colouring matter of cochineal (exhibited in what is called *carmine*,) which of all colours is the farthest removed from that of indigo: and if Mr. Delaval should allege, that, though agreeing in weight, they differ as to the sizes of their respective particles, let him correct this difference by the only means suited to do it, without doing more; I mean by simple mechanical division, or grinding. Let this be employed upon either of the substances in question, whose particles he may suppose too large, as long as he shall think proper, and let us then see whether he can thereby render the colour of indigo red, or that of cochineal blue or violet.[10]

This outburst struck a chord among the chemical opposition to Newton's theory. Berthollet cited and elaborated upon it in the second

[8] Bancroft, *Experimental Researches* (1813), 1:v–vi. For the portions that concern us the two editions are largely the same, but I will indicate significant differences.
[9] Ibid., pp. 7–8; Berthollet, *Eléments* (1791), 1:3.
[10] Bancroft, *Experimental Researches* (1813), 1:29–30.

edition of his *Eléments*, and Hassenfratz likewise appealed to it. It was easy for Haüy to reply that the particles of indigo and carmine are already so fine that ordinary means of grinding them cannot further alter their size.[11] Arguments of this sort prove neither view. Even before Newton's day both physicists and chemists knew that the sizes and arrangements of the elementary constituents of matter were unknown. But such arguments do reveal the underlying assumptions and attitudes of their proponents.

In turning to establish a chemical theory of coloration, Bancroft first dispatches the older phlogiston theories by showing their inconsistency. He observes that most modern chemists have now substituted oxygen and, following Newton, assign it a great refractive power.[12] Then he clarifies a misunderstanding that had previously beset the chemical-affinity approach to coloration, the confusion between refraction and absorption, and recognizes the latter as an independent physical process. Although it is true that glass prisms and other transparent colorless substances *separate* the rays of different color by their different degrees of refrangibility, "yet the *permanent* colours of different bodies, or substances *are not,* as I believe, *produced by mere refraction.*" Their colors, he explains,

> evidently depend on other properties, which determine, or occasion the reflection or transmission of some particular sort or sorts of rays, and an absorption or disappearance of the rest; and these I conceive to be certain *affinities,* or *elective attractions,* existing in or between the differently coloured matters and the particular sorts of rays of light so absorbed or made *latent.*[13]

Bancroft clearly recognizes that it is the missing and not the reflected or refracted rays that must be explained.

Like his Continental contemporaries, especially Senebier, whom he freely cites, Bancroft describes virtually all known photochemical reactions. His aim is to show the predominant role of oxygen "in producing and changing those affinities, or elective attractions, from which the permanent colours of different substances arise."[14] However, he takes a more moderate position than the French, particularly Berthollet, whom he criticizes:

[11] C.-L. and A.-M. Berthollet, *Elements* (1824), 1:35–6; Hassenfratz, "Mémoire sur la colorisation des corps" (1808), p. 165; and Haüy, *Traité élémentaire de physique* (1806), 2:261–2.
[12] Bancroft, *Experimental Researches* (1813), 1:31–3.
[13] Ibid., p. 34.
[14] Ibid., p. 35.

> Though most of the changes of colour, in permanently coloured bodies, evidently depend on changes in their respective portions of oxygene, I am far from thinking that this cause operates exclusively in all cases, or that chymical knowledge is yet far enough advanced to justify even an attempt towards a complete hypothesis respecting these most abstruse and most interesting phenomena.[15]

Bancroft's aim was in the first place to propose an alternative explanation of the colors of bodies in terms of the chemical composition of bodies and light. Although he could not "assign a reason or cause for these affinities, and their connection with particular proportions of oxygene," by identifying elective affinities with the optical process of selective absorption he converted the chemical concept of affinities into a meaningful physical cause of coloration.[16] This way of conceiving affinities would soon be applied to optical investigations of absorption. In the second place Bancroft wanted to undermine the plausibility of Newton's mechanical theory, though he was more successful in discrediting Delaval's interpretation of it than Newton's. He did not, however, refute the theory; he rejected and undermined it. He presented no observations or experiments that were inconsistent with Newton's theory. Chemistry was (and would long be) incapable of determining the size and density of the coloring particles. Physical, in particular, optical, evidence would be required to refute the theory, and Bancroft, like most chemists, was simply not sufficiently adept or interested in physics to undertake such an enterprise.

In Chapter III of his *Experimental Researches*, Bancroft introduces the concept of "colouring matter," which he defines as "a substance which possesses, or acquires a power of acting upon the rays of light, so as either to absorb them all, and produce the sensation of black; or only to absorb particular rays, and transmit or reflect others, and thereby produce the perception of that particular colour, which belongs to the ray or rays so transmitted or reflected."[17] This leads him to consider the reflection of rays from coloring matter and thence to Newton's phenomenological theory and Delaval's refutation of it. Bancroft endorsed Delaval's refutation, and his extensive quotation from Delaval's paper helped to diffuse knowledge of it.[18] However,

[15] Ibid. (1794), p. 57; though this passage was revised in the 1813 edition, 1:82, it expresses the same idea. For his criticism of Berthollet, see (1813), 1:55–82.
[16] Ibid. (1813), 1:82.
[17] Ibid., p. 111.
[18] Ibid., pp. 112–14. He also accepted Delaval's idea that colored bodies consist of transparent coloring and opaque reflecting corpuscles, pp. 114–15.

it again reveals his relatively limited grasp of optical theory, for had he recognized that Delaval's investigation eliminated the physical basis for Newton's analogical theory, he would have presented it in his first chapter, when he was attempting to refute that theory. In this chapter Bancroft also explains the distinction between simple and compound colors. Confusing the concept of primary or simple colors from pigment mixing with that from Newton's theory of light, he held that the colors red, yellow, and blue, whether in light or bodies, are always simple. But he also believed that very many other colors of bodies are simple, for instance, all metallic oxides, vegetable green, and Tyrolean purple.[19] In fact, he does not offer a single example of a compound color here.

Delaval's investigation of colored bodies was widely known and disseminated in England, though its negative implication for Newton's analogical theory was not grasped. The volume of the *Memoirs of the Literary and Philosophical Society* with Delaval's paper was reviewed anonymously in the *Monthly Review* in 1786 by Alexander Chisholm, who had served for thirty years as the chemical assistant and collaborator of William Lewis and, after Lewis's death in 1781, of Josiah Wedgwood. Chisholm recognized the "importance" of the paper and that "it establishes . . . a new theory of colours; and ascertains several properties of coloured matter, very different from those which have been hitherto thought to prevail. It proves that colouring particles, universally, *transmit* light, but never *reflect* any."[20] The long review is mostly devoted to Delaval's chemistry and does not specify Newton's theory as that which "hitherto prevailed."

Upon the recommendation of Chisholm and Wedgwood, John Leslie, who was then tutor to Wedgwood's sons, became associated with the *Monthly Review* and was assigned Bancroft's *Philosophy of Permanent Colours*. Best known for his studies of radiant heat, Leslie later became professor of mathematics and then of natural philosophy at the Uni-

[19] Ibid., pp. 115–17. Berthollet held similar views, see Ch. 8, note 34.

[20] [Chisholm], "[Review of] *Memoirs of the Literary and Philosophical Society of Manchester*, Vol. II" (1786), p. 357. The reviewer is identified by Nangle, *The Monthly Review, First Series*, pp. 9, 157. See also the anonymous "[Review of] *Memoirs of the Literary and Philosophical Society*," pp. 345–8. Thomas Henry in his paper on dyeing, "Considerations relative to the nature of wool," in the volume of the *Memoirs* following that with Delaval's paper, frequently cites him and also adopted the chemistry of light. On Chisholm see Eliza Meteyard, *The Life of Josiah Wedgwood*, 2:465–6; Schofield, "Josiah Wedgwood, industrial chemist," pp. 182, 186; Gibbs, "William Lewis," p. 128; and Sivin, "William Lewis (1708–1781)," p. 66.

versity of Edinburgh. His review in 1795 contains the clearest explanation of the process of coloration in Britain until those of Brewster and Herschel in the 1820s. Leslie declared Book II of Newton's *Opticks* to contain two "essential mistakes," his theories of colored bodies and fits. He considered these to be "the fruits of declining age" in which Newton "gave way to the indulgence of fancy and to the prejudices of the mechanical philosophy then in vogue." Turning to Bancroft's explanation of colored bodies, he quoted his account of absorption by "*affinities or elective attractions*" and then endorsed it:

> This position, though neither fully conceived nor precisely expressed, is in the main correct. Of the particles of light that enter a body, some proceed uniformly without obstruction; others, which chance to approach within certain limits of the corporeal atoms, being either attracted or repelled, according to the peculiar relations and forces subsisting between the specific rays and the permeable matter, are absorbed or dispersed. It is this dispersed light that causes the sensation of colour. It will be deemed paradoxical, we fear, to assert that the permanent colours of bodies are not occasioned by the rays *reflected* from the surface, but from those *transmitted* from the *internal mass*: yet this opinion might be substantiated by conclusive arguments.[21]

Although Leslie does not cite Delaval by name, he certainly knew of his refutation of Newton's phenomenological theory from Bancroft and probably also from Chisholm. By combining Delaval's results with Bancroft's concept of selective absorption, he proposed an alternative to Newton's theory. Leslie argued that an "evident consequence" of his explanation is that color must depend on a body's chemical composition, "on the nature of which the discriminating *affinities to particular sorts of light* depend."[22] Citing Berthollet and Bancroft, he considered oxygen to be the most important element in altering the affinities for light. Although he did not further pursue "substantiating" his account, this early episode explains why Leslie, one of the few quantitative British physicists, did not defend Newton's physical theory: He was convinced it was erroneous. Because Leslie's expla-

[21] [Leslie], "[Review of] *Experimental Researches*" (1795), pp. 290, 291; Leslie is identified by Nangle, *The Monthly Review, Second Series*, pp. 37–8, 83. It is fascinating to note that Leslie, Chisholm, his former employer, Lewis, and Wedgwood's son, Tom, were all involved in photochemical experiments. Around 1800 Tom Wedgwood, later assisted by Humphry Davy, made primitive "photographs." For the shadowy history of these experiments and the relation of these individuals to one another, see Helmut Gernsheim, *The Origins of Photography*, pp. 21, 24–8.

[22] Leslie, "[Review]," pp. 291–2; italics added. See also the anonymous "[Review of] *Experimental researches, &c.*" in the Genevan journal *Bibliothèque Britannique* (1796).

nation of coloration was so brief and in a book review, it was without direct influence on subsequent research.[23]

In 1792, when Bancroft's book was already in press, Leslie's older friend, the natural philosopher and eminent uniformitarian geologist James Hutton, had on more general grounds proposed selective absorption as the cause of coloration. In the seventh of his *Dissertations on Different Subjects in Natural Philosophy*, on the "solar substance in the composition of bodies," Hutton argued on phenomenological grounds – as Newton had over a century earlier in his *Optical Lectures* – that transparent colored bodies must absorb rays of certain colors. He held that "phlogistic matter" causes absorption by "an elective action that takes place, between certain substances and particular species of light."[24] Unlike Bancroft and the Continental chemists, Hutton did not present any empirical evidence, particularly photochemical reactions, to support his view. It was, rather, just one aspect of his broader chemical philosophy in which heat, light, and electricity are different manifestations of the "solar substance." The circulation of phlogiston (which was the matter of heat and light combined with bodies) was necessary for the maintenance of the operations of nature. Hutton synthesized Boerhaave's fire, Rouelle's phlogiston, and Newton's repulsive forces into a decidedly chemical natural philosophy. Although his writing is notoriously opaque and prolix, he was an

[23] The results of Delaval's investigation were further propagated through the article "Chromatics" in the *Encyclopaedia Britannica* (1797). The article was apparently written by Charles Hutton, an applied mathematician who taught at the Royal Military Academy in Woolwich, inasmuch as an abbreviated version of it appeared *earlier* in his *Mathematical and Philosophical Dictionary* (1795), 1:280. Hutton presented a superficial account of Newton's theory of colored bodies, together with a summary of Delaval's first two papers in support of it. He ostensibly rejected Newton's theory because he thought the colors of bodies would vary too much with the angle of viewing them (p. 728), though he did not present Newton's solution to this problem. In fact he was an adherent of the chemistry of light. Hutton devoted close to half the article to an account of Delaval's third paper on "permanent colours," which he judged to "be of the utmost consequence in the arts of colour-making and dyeing" (p. 730). He was not uncritical, but his disagreements with Delaval were largely over the chemistry of light. He rejected Delaval's identification of light and phlogiston (p. 733) and noted his failure to assign a role to "the element of fire," which "has a considerable share in the production of colours" (p. 736). The article judiciously concluded "that the theory of colours seems not yet to be determined with certainty; and very formidable, perhaps unanswerable, objections might be brought against every hypothesis on this subject that hath been invented" (p. 738).

[24] Hutton, *Dissertations on Different Subjects in Natural Philosophy*, p. 599. On Hutton's natural philosophy see Schofield, *Mechanism and Materialism*, pp. 273–6; Heimann and McGuire, "Newtonian forces," pp. 281–95; Heimann, " 'Nature is a perpetual worker'," pp. 17–22; and "Ether and imponderables," pp. 76–8; Ziemacki, "Humphry Davy," pp. 104–7; and Cantor, *Optics after Newton*, pp. 58–9.

important member of the burgeoning Scottish scientific community. For our purpose, Hutton's views are significant insofar as they characterize the chemical outlook of British natural philosophy and, more specifically, how chemical elective affinities became associated with selective absorption in coloration. Hutton held that light was an imponderable fluid rather than discrete independent particles, but this is of no consequence for the heurestic value of the affinity conception of absorption, as it was a purely qualitative model that was used to interpret phenomena and, for some, to suggest experiments.

Four years later an anonymous paper by C.A. (generally known to be Bartholomew Parr, a physician educated at Edinburgh) appeared in a collection of *Essays by a Society of Gentlemen at Exeter.* Its title, "Observations on Light, particularly on its combination and separation as a chemical principle," shows its orientation. Although he says nothing about Newton's theory, his discussion of color changes shows him to be an advocate of a chemical theory: "The doctrine of colours is still very obscure; and we can enlarge no farther on the subject than to connect it, with the principal position which pervades this essay, that light and oxygen mutually separate each other."[25] Parr's paper stirred up a discussion in *Nicholson's Journal*, and "a correspondent," who generally supported Parr's views, represented it as that of "the chemical philosopher of the present day."[26]

A final example from the *Philosophical Magazine* of 1800 will suffice to show how widespread the chemistry of light had become in Britain. William Blackburne (also an Edinburgh-educated physician) had promised to publish a theory of heat and light, but as he was pressed for time, he sent in just the results of his investigations on light. He held that light was a combination of caloric and oxygen. Though fire and phlogiston were now gone, the underlying approach of the chemistry of light had changed little in a quarter of a century, as we can see from his conclusions on color:

> 3dly, The phenomena of colours are to be ascribed to the different qualities of light, as containing caloric and oxygen in different proportions. The different proportions manifest themselves in the circumstances both of the decomposition and the formation of variously coloured light.

[25] [Parr], "Observations on light" (1796), p. 512. For the identification of C. A. as Parr, see the anonymous letter, "On Dr. Parr's theory of light and heat" (1799). A footnote on p. 547 refers to the paper as an "Essay upon Light, in a miscellaneous volume, published by a society at Exeter." George Pearson also refers to "Dr. Parr" in his "Experiments and observations" (1797), p. 351.

[26] "On Dr. Parr's theory," p. 547.

4thly, The separation of light by the prism is to be regarded as a *chemical* decomposition, not a *physical* or *mechanical* division of light.

5thly, The changes which take place in the colours of different substances, as of plants in the process of vegetation, of metals in that of oxydation, are owing to correspondent changes which these substances undergo in their chemical action upon light.

6thly, The evanescence, or, as it is frequently termed, the *absorption*, of light, is to be referred to the complete resolution of this compound into its constituent parts.[27]

Because Blackburne considers all production of color to be a chemical decomposition and rejects all physical or mechanical causes, he necessarily rejects Newton's theory of colored bodies.

7.2. THE PHYSICISTS' DEFECTION

Let us return to the mainstream of natural philosophy and examine the attitude toward Newton's theory held by the small British community of quantitative physicists. In particular I will evaluate the relative roles of the chemical philosophy and optical evidence in forming their views. John Robison was professor of natural philosophy at the University of Edinburgh from 1773 to 1805 and introduced Boscovichean ideas to Scottish natural philosophy. However, he was also Black's successor as lecturer in chemistry at the University of Glasgow from 1766 to 1770 and the editor of his chemical lectures. In his one published optical paper in 1790, on refraction in moving media, he showed insight and skill in his handling of Newtonian optics utilizing short-range forces and corpuscles.[28] Robison wrote a number of articles for the third edition of the *Encylopaedia Britannica* (1797). In the article "Optics" he gave a more general presentation of Newton's mechanical account of refraction, but now chemical forces entered.

Newton had assumed that chromatic dispersion is an innate property of light and is proportional to the mean refraction, which depends on the nature of the refracting body. As a consequence, he suggested in the *Opticks* that it would be impossible to eliminate the chromatic

[27] Blackburne, "Communication . . . respecting caloric, light, and colours," pp. 334–5; italics added.

[28] Robison, "On the motion of light." On Robison's optics see Kurt Møller Pedersen, "Roger Joseph Boscovich and John Robison on terrestrial aberration"; and Cantor, *Optics after Newton*, pp. 75–6, 212–16. On his natural philosophy see Schofield, *Mechanism and Materialism*, pp. 277–8, 280–2; Richard Olson, *Scottish Philosophy and British Physics*, pp. 157–61; and Crosbie Smith, " 'Mechanical philosophy' and the emergence of physics in Britain."

aberration of lenses. John Dollond's discovery in 1759 that dispersion is not always proportional to the mean refraction led to the construction of alchromatic lenses and significant improvements in optical instruments. It was also significant in showing that even Newton could make serious errors in optics, and, to the more sophisticated, in strikingly indicating that the force of refraction was not mechanical. Unlike the gravitational force, it depended on the nature of the refracting substance and it did not act on all light particles alike; that is, it was a selective force.[29] Robison, like so many others, drew the obvious conclusion that optical forces are chemical. Newton, he explained,

> supposed that the action of all bodies was similar on the different kinds of light, that is, that the specific velocities of the differently coloured rays had a determined proportion to each other. This was gratuitous; and it might have been doubted by him who had observed the analogy between the chemical actions of bodies by elective attractions and repulsions, and the similar actions on light. . . . In like manner, we might expect not only that some bodies would attract light in general more than others, but also might differ in the proportion of their actions on the different kinds of light, and this so much, that some might even attract the red more than the violet. The late discoveries in chemistry show us some very distinct proofs, that light is not exempted from the laws of chemical action, and that it is susceptible of chemical combination.[30]

Inasmuch as Robison had adopted the chemistry of light, it is highly unlikely that he would have supported Newton's theory of colored bodies, though he never commented on it insofar as I can determine.

The young Henry Brougham, educated at the University of Edinburgh and later lord chancellor of England, imagined himself as a restorer and promoter of Newtonian optics, yet he rejected many of its central tenets. Brougham based his optics on corpuscles and short-range forces, but, like so many Scottish natural philosophers, he adopted a Boscovichean scheme of alternating attractive and repulsive

[29] On a deeper level Newton's "gravitational" model of refraction leads to the same conclusion. To explain the existence of dispersion the force of refraction must depend on the mass or the nature of light particles of different color, which is totally unlike the gravitational force: All projectiles, whatever their mass or nature, fall at the same rate. Observation shows dispersion cannot depend on velocity; see Bechler, "Newton's search for a mechanistic model." Newton's investigation of the relation of the refractive force and the density of refracting substances also showed that the force was not mechanical; see Ch. 6, note 13.

[30] [Robison], "Optics," p. 286. John Playfair attributes this unsigned article to Robison; "Biographical account of the late John Robison" (1815), p. 521.

forces. His principal innovation was his "discovery" that rays of different color are reflected at different angles, which depend on the size of the light particles. He then applied his new concept of "different reflexibility" to explain a number of phenomena, including the colors of thick plates and thin films. He now imagined

> that upon the principles which have been laid down, the colours of natural bodies may be explained. The celebrated discovery of Newton, that these depend on the thickness of their parts, is degraded by a comparison with his hypothesis of the fits of rays and waves of aether. Delighted and astonished by the former, we gladly turn from the latter; and unwilling to involve in the smoke of unintelligible theory so fair a fabric, founded on strict induction, we wish to find some continuation of experiments and observations which may relieve us from the necessity of the supposition.[31]

Brougham, on the one hand, must be counted as one of the few British supporters of Newton's "fair" theory. On the other hand, he rejected his explanation of thin films so that he had to emend that theory. The corpuscles of bodies, which for Newton were like transparent thin films, became for Brougham "specular and spherical" parts like the heads of pins.[32] Brougham's theory of coloration was strictly physical and singularly untouched by the chemistry of light. Indeed, when some years later he reviewed Venturi's study of optical absorption in the *Edinburgh Review*, he could only reluctantly accept selective absorption as a mode of coloration because of Venturi's chemical interpretation. "It cannot be denied," Brougham allowed, "that such an account of the phenomena is much more simple and consistent with various analogies, than the one which it is intended to displace," that is, Newton's. He also granted that "if light is a material substance . . . it is reasonable to think that, like other substances, it should be capable of entering into various unions, according to the laws of chemical affinity and corpuscular attraction."[33] Yet he found it impossible to imagine that so much light could be chem-

[31] Brougham, "Experiments and observations on the inflection, reflection, and colours of light" (1796), pp. 271–2. On Brougham's optics see Olson, *Scottish Philosophy*, pp. 219–24; Henry John Steffens, *Development of Newtonian Optics*, pp. 86–92; Cantor, *Optics after Newton*, pp. 78–80; and Kipnis, *History of the Principle of Interference*, pp. 71–2.

[32] Brougham, "Experiments and observations," pp. 273–4.

[33] [Brougham], "[Review of] *Indagine fisica sui colori*" (1805), p. 38; his authorship is evident from the discussion of his own earlier work, e.g., on pp. 23–30. I suspect that Venturi sent a presentation copy of his book to Brougham, since he had discussed his papers in it; see §8.1 for a further account of Brougham's response to the evidence of absorption spectroscopy.

ically combined with and retained in bodies. Brougham's resistance to the chemistry of light is anomalous for an emission theorist in this period, especially one from Scotland.

Thomas Young, Britain's leading physicist at the turn of the century, represents an important transition with regard to the fate of Newton's theory, for he introduced the experimental study of absorption spectra to Britain. We might expect Young, as a proponent of the wave theory, to have rejected Newton's theory of colored bodies, but there is in principle no conflict between the two; Newton's theory was formulated independently of a theory of light directly from the analogy between the phenomena of thin plates and colored bodies. Young in fact began as an enthusiastic supporter of the theory and became increasingly skeptical of it, but he was unable to find a suitable physical alternative.

Young's investigation of Newton's theory of colored bodies was brought about by an announcement by William Hyde Wollaston at a meeting of the Royal Society on 24 June 1802 that when he observed luminous sources such as the sun through a very narrow slit (1/20-inch as compared to Newton's 1/4-inch hole) with a fine prism free from veins, their spectra were divided into a small number of uniformly colored bands separated by dark lines and spaces.[34] The lines quickly passed from public attention. Though Wollaston described them carefully, he was mainly interested in the four apparently pure colors that he observed in sunlight in order to resolve the question of the number of simple spectral colors. Young observed Wollaston's dark lines, but he focused his attention on the colors and dark bands, which he thought he could explain by Newton's theory of colored bodies. When Fraunhofer rediscovered the dark lines in the next decade, they would come to play a significant role in promoting the study of absorption and coloration (§9.1).

Young immediately seized on Wollaston's discovery to redetermine the primary colors for his theory of color vision and substituted new ones. One week later, at the next meeting of the Royal Society on 1 July, he included his own observations of Wollaston's spectra in his second paper on the wave theory of light and reported his new primaries. Young was familiar with Newton's entire optical oeuvre, in particular his investigations of the colors of thin films and of natural bodies. At this time probably he alone in Britain understood that the

[34] Wollaston, "A method of examining refractive and dispersive powers," pp. 378–80; see also David Hargreave, "Thomas Young's theory of color vision," pp. 158–78.

colors of thin films are compound and that their spectra consist of colored bands with intervening dark spaces (§2.4). By invoking Newton's theory of colored bodies, he was able to explain the five colored bands and the dark spaces that Wollaston discovered in the blue part of the flame of a candle:

The light transmitted through every part of a thin plate, is divided in a similar manner into distinct portions, increasing in number with the thickness of the plate, until they become too minute to be visible. At the thickness corresponding to the ninth or tenth portion of red light, the number of portions of different colours is five; and their proportions, as exhibited by refraction, are nearly the same as in the light of a candle, the violet being the broadest. We have only to suppose each particle of tallow to be, at its first evaporation, of such dimensions as to produce the same effect as the thin plate of air at this point, where it is about $\frac{1}{10000}$ of an inch in thickness, and to reflect, or perhaps rather to transmit, the mixed light produced by the incipient combustion around it, and we shall have a light completely resembling that which Dr. Wollaston has observed.[35]

The qualitative resemblance between the form of the spectra predicted by Newton's theory and those observed from colored bodies is so striking that it seems at first glance to confirm the theory. It is only when it is subjected to more careful scrutiny that discrepancies become apparent. By interpreting Wollaston's spectrum of a flame as being that of a colored body (particles of evaporated tallow), Young was the first in Britain to compare the observed spectrum of a colored body with that of a thin film; Venturi had carried out such a test three years earlier in Italy. Like so many mathematical physicists, Young was very much taken by the theory:

Experiments upon this subject might tend greatly to establish the Newtonian opinion, that the colours of all natural bodies are similar in their origin to those of thin plates; an opinion which appears to do the highest honour to the sagacity of its author, and indeed to form a very considerable step in our advances toward an acquaintance with the intimate constitution and arrangement of material substances.[36]

Nearly a year and a half later, in the last of his trio of papers on the wave theory, Young's enthusiasm had noticeably waned. In the

[35] Young, "An account of some cases of the production of colours," pp. 395–6.
[36] Ibid., p. 396. Five years later, in his *Course of Lectures* (1:786, Plate 29, Figure 420), Young published a colored illustration of the spectrum of a candle's flame, but he no longer interpreted it as being that of a colored body, p. 438.

interim he had put the theory to a serious test by observing (true) absorption spectra of various colored bodies and comparing them in detail with those predicted by Newton's theory. For example, when he viewed a blue glass through a prism, he saw a spectrum divided into seven segments: two red, a yellowish green, green, blue, bluish violet, and violet. According to Young, the eleventh order of red (or eighteenth of violet) reflected from a thin film of air 1/6840 inch thick "agrees very nearly" with the observed division; if the film were of a denser metallic oxide, it would be about 1/15,000 inch thick. "But it must be confessed," Young cautioned, "that there are strong reasons for believing the colouring particles of natural bodies in general to be incomparably smaller than this; and it is probable that the analogy, suggested by Newton, is somewhat less close than he imagined." He then enumerated more damaging optical evidence.

Young now realized that the reflected light of such high orders as the eleventh and eighteenth appears "very nearly white." Moreover, after observing the candle's flame again, he found that it did not agree as well as he had reported in his previous paper: "In the blue light of a candle, expanded by the prism, the portions of each colour appear to be narrower, and the intervening dark spaces wider, than in the analogous spectrum derived from the light reflected from a thin plate."[37] He also casually reported a few absorption spectra of colored glasses, for instance, one was red with a little green. Such a combination is not possible with a thin film according to Newton's nomograph. However, without an explanation of how the composition of the spectra of thin films are determined (later provided by Hassenfratz; §8.2) there were, at best, only a few people in the world who appreciated this, and almost certainly none in Britain. Young, in his characteristically laconic and ultimately obscure way, has all but refuted Newton's theory of colored bodies.

Young's observations of absorption spectra cannot strictly be called absorption spectra, as he did not in fact admit the concept of selective absorption, or the total extinction of some colors. Of course, he described spectra of colored glasses and thin films as having some colors "intercepted," but he never suggested that these rays were physically absorbed and destroyed (and indeed they are not in thin films). As a wave theorist he could not accept that light was attracted to and combined with bodies, as emission theorists in the chemical tradition

[37] Young, "Experiments and calculations," pp. 14, 15; read 24 November 1803. In the reprint of this paper in his *Course of Lectures* Young inserted a sentence (quoted at note 39) immediately following this one.

could so readily do. In his *Lectures on Natural Philosophy*, Young dismissed the chemically conceived elective attractions used in the emission theory to explain dispersion as only a name without any ulterior mechanical cause: "But an elective attraction of this kind is a property foreign to mechanical philosophy, and when we use the term in chemistry, we only confess our incapacity to assign a mechanical cause for the effect, and refer to an analogy with other facts, of which the intimate nature is perfectly unknown to us."[38]

Once Young realized the inadequacy of Newton's theory of colored bodies, he tentatively suggested an explanation of "absorption" spectra that did not involve the combination of light particles with bodies. In the reprint of his third paper on the wave theory that he appended to his *Lectures* he inserted one sentence after he reported the disagreement between the observed and predicted spectral bands of a flame: "Perhaps their origin may have some resemblance to that of the different harmonics of a single vibrating substance."[39] Young did not pursue this idea, which he apparently drew from Euler's theory of color, but it shows how wave theorists would have to develop alternative conceptions of absorption that depended on energy transfer rather than the capture of particles. In the text of his *Lectures*, Young's attitude toward Newton's theory remained ambiguous. He was still impressed with the way in which the spectra of thin films appeared so much like those of colored bodies, in accordance with Newton's theory, but he considered it "impossible to suppose the production of natural colors *perfectly identical* with those of thin plates."[40] His major objection was that the corpuscles of matter were much smaller than the theory predicted. With his equivocal assessment of Newton's theory, Young was scarcely in a position to defend it.

[38] Young, *Course of Lectures*, 1:463. Young delivered his *Lectures* as professor of natural philosophy at the Royal Institution, whose founder, Count Rumford (Benjamin Thomson), arranged for Young's appointment. Rumford not only advocated the idea that heat was a motion (which was supported by Young), in opposition to the contemporary caloric or material theory, but also a wave theory of light; and he was as opposed to the affinity of light as Young was. In 1812, Rumford vigorously attacked the emission theory and its "fruitless" chemistry of light: "As long as the doctrine which supposes light to be substance emitted by luminous bodies continues to be believed and universally taught, a great deal of time will no doubt continue to be employed in useless researches concerning its supposed affinities and combinations" (Thomson, "An inquiry concerning the source of the light which is manifested in the combustion of inflammable bodies," *The Collected Works*, 4:229–50, on pp. 244–5).

[39] Young, *Course of Lectures*, 2:647.

[40] Ibid., 1:469; italics added.

If Young's tests of Newton's theory and doubts about it had any influence, they would only have reinforced the prevailing chemical views already opposed to it. By 1807, Newton's theory of colored bodies was eclipsed by the combined forces of the chemical philosophy and optical evidence. Two of the initial supporters of the theory that I have identified in this period, the physicists Brougham and Young, were untouched by – indeed, opposed to – the chemical interpretation. They were persuaded of the problems with Newton's theory by the optical evidence of absorption: Young by his own experiments and Brougham by Venturi's. The chemist and natural philosopher Delaval switched to a chemical theory with the support of his own experiments on the color, or lack of color, of reflected light. Leslie proposed that colors are due to absorbed light under the joint influence of Delaval's experiments and Bancroft's cogent chemical arguments. Delaval's third paper, although widely known in Britain, was not as crucial in overthrowing Newton's analogical theory as might be expected. Leslie alone at this time in Britain understood that if no colored light is reflected from bodies, then the analogy to thin films, which specularly reflect light, is invalid. Even as astute a physicist as Thomas Young, who recognized that Delaval showed "very satisfactorily" that bodies are colored only by transmitted light, "and not as Newton supposed," did not relate this to Newton's analogical theory.[41] The greatest influence of Delaval's experiments in his third paper was in more generally discrediting Newton's views on colored bodies, while simultaneously advocating the chemistry of light. Young's (and Venturi's) observations of absorption spectra would be totally forgotten until the 1820s.

An incident at the Royal Society in 1807 well illustrates the standing of Newton's theory in this period. The astronomer William Herschel had completed the reading of the first of three papers in which he attempted to refute Newton's explanation of the colors of thin films and his theory of fits. In his conclusion, he argued (erroneously) that the theory of colored bodies, "which Sir I. Newton has founded upon the existence of fits of easy reflection and easy transmission," must

[41] Ibid., 1:481; this remark is in the chapter on the history of optics and not in the earlier one on the nature of light and colors, where he treated Newton's theory of colored bodies. See also Young's abstract of Delaval's paper in the appended bibliography, 2:321. The import of Delaval's experiment appears to have been more widely grasped on the Continent – Venturi, Berthollet, and Hassenfratz cite it – but it was evidently insufficiently convincing to advocates of Newton's theory, such as Haüy and Biot, to compell them to abandon it; see Ch. 8, notes 5, 25, and 50.

therefore "remain unsupported."[42] The account of this Royal Society meeting in the *Philosophical Magazine* observed: "It may not be improper to remark here, that many of the author's observations on Newton's theory of colours have been anticipated both by Dr. Bancroft and by an anonymous writer on that subject"; a footnote referred to a review of Jordan's anonymous work, *An Account of the Irides or Coronae.*[43] Thus Herschel's attack on Newton's theories of fits and colored bodies was in 1807 considered unoriginal and Bancroft's thirteen-year-old critique of the latter was easily recalled.

Inasmuch as we have uncovered a strong correlation between a rejection of Newton's theory with a belief in the chemistry of light, we should not be surprised to find Herschel subscribing to that belief. In one of his epoch-making series of papers in 1800 demonstrating the existence of infrared rays ("invisible rays of heat"), he called for the investigation of the "chemical properties of the prismatic colours," for light "is the most subtle of all the active principles." He then suggested:

> If the chemical properties of colours also, when ascertained, should be such that an acid principle, for instance . . . may reside only in one of the colours, while others may prove to be differently invested, it will follow, that bodies may be variously affected by light, according as they *imbibe and retain*, or transmit and reflect, the different colours of which it is composed.[44]

Chemists themselves, of course, taught a chemical theory of color. Thomas Thomson, in 1804, lecturer in chemistry at the University of Edinburgh and afterward professor of chemistry at Glasgow, explained in his influential textbook, *A System of Chemistry*, that light combines with bodies "and constitutes one of their component parts." He then clearly articulated the nature of coloration by the chemistry of light:

> Almost all bodies have the property of absorbing light. . . . But they by no means absorb all the rays indiscriminately; some absorb one coloured ray,

[42] W. Herschel, "Experiments for investigating the cause of the coloured concentric rings," p. 232.

[43] "Proceedings of the learned societies. Royal Society of London," *Philosophical Magazine*, 27(1807):84–5. The favorable review of Jordan's book that is cited is in ibid., 8(1800):78–80. On Jordan see Appendix 1; and on the furor over Herschel's papers see J.L.E. Dreyer's comments in W. Herschel, *Scientific Papers*, 1:lvii–lviii.

[44] W. Herschel, "Investigation of the powers of the prismatic colours," pp. 270–1; italics added. See his early paper, read to the Bath Philosophical Society in 1780, "Observations on Dr. Priestley's desideratum – 'What becomes of light?' " in Herschel, *Scientific Papers*, 1:lxviii–lxxviii.

others another, while they reflect the rest. This is the cause of the different colours of bodies. . . . The different colours of bodies, then, depend upon the affinity of each for particular rays, and its want of affinity for the others.[45]

Nonetheless, Newton's theory was only in eclipse and could not yet be counted out. The most important, and apparently the sole, advocate of the theory in Britain was a Frenchman, Haüy, whose physics textbook taught Newton's theory of colored bodies. Olinthus Gregory, the translator of Haüy's book, was a mathematical popularizer and teacher of mathematics at the Royal Military Academy in Woolwich. To Haüy's account of Newton's theory Gregory added a note that "several philosophers have recently contemplated the doctrine of colours in connection with chemistry, and the art of dyeing."[46] He then related that Berthollet rejected Newton's analogy to the colors of thin plates. Still, two young adherents to Newton's theory remained on the sidelines: the physicists John Herschel (William's son) and David Brewster. By the 1820s, only supporters of Newton's theory would still devote themselves to refuting it.

[45] Thomson, *A System of Chemistry*, Pt. I, Bk. I, Div. II, Ch. 1, §13, 1:289; the first edition of 1802 is the same. For Fresnel's response to Thomson see Ch. 9, note 63.

[46] Haüy, *An Elementary Treatise on Natural Philosophy*, 2:253. This translation is based on the first edition of the *Traité élémentaire de physique* (1803) and therefore does not contain the reply to Berthollet's attack that Haüy added to the second edition of 1806 (see §8.2).

8

DEBATE AND ABSORPTION SPECTROSCOPY IN FRANCE: 1791–1816

France, in the decades around the turn of the century, presents a contrasting picture to the predominance of the chemical view in Britain. The mathematical crystallographer Haüy and the physicist Biot staunchly defended Newton's physical theory of colored bodies against the renewed chemical offensive lead by Berthollet. Hitherto, physicists had in public chosen to ignore the chemists' arguments. The most significant new weapon in the chemists' arsenal, which was otherwise becoming depleted, was the experimental study of absorption spectra, which grew out of their concept of elective attractions or affinities and their recognition that such observations would allow Newton's theory of colored bodies to be tested. Between 1799 and 1807, the physicist Venturi, the chemist-cum-physicist Hassenfratz, and the military engineer Prieur de la Côte-d'Or carried out studies of selective absorption and proposed it as the most general cause of the colors of bodies. The lack of agreement between a number of observed absorption spectra with those predicted by Newton's theory, they declared, refuted that theory. An essential step in the development of absorption spectroscopy and the testing of Newton's theory was the recognition that the colors of most bodies are compound. Otherwise, there would have been no motivation to study the composition of their colors by separating them with a prism. The compound nature of the colors of natural bodies became generally recognized by 1805, when Haüy corrected Berthollet's misunderstanding on this point, which was especially common among chemists.

After these pioneering studies, no further experimental research on selective absorption was undertaken in France, or elsewhere, insofar as I can determine, until absorption spectroscopy was independently reinvented at the end of the next decade. I will later propose some reasons for this abortive birth of absorption spectroscopy. One key insight is provided by the principal objection that Haüy and Biot brought against selective absorption by affinities as a cause of color. They held that it was not lawlike, and affinities had to be arbitrarily assumed for each substance and spectral color, whereas Newton's theory made specific predictions about the composition of the color of any body according to the well-established laws for the colors of thin films. Thus French physicists, who required that nature obey mathematical laws, considered absorption to be too capricious to be an independent physical process; and because they did not consider it to be the cause of color, why study such a disorderly phenomenon? French chemists, on the other hand, accepted affinities as the cause of coloration, rejected Newton's theory, and were not inclined to study selective absorption because such optical experiments were not their business but that of physicists. Moreover, a chemistry based on traditional affinities was rapidly losing its viability.

If the work of Hassenfratz and Prieur was not pursued, at least it was known and discussed in the French scientific community. The research of the Italian physicist Venturi, however, was totally ignored. Because Venturi had sojourned in Paris immediately before he undertook his research on the cause of coloration, I consider his work part of the French story and thus begin with a detour to Italy.

8.1. AN ITALIAN INTERLUDE: VENTURI

Venturi's paper of 1799, "Indagine fisica sui colori" (Physical research on colors), was a thorough analysis of selective absorption and the coloration of bodies that was not to be surpassed in comprehensiveness and reliability until the 1820s. Despite winning a prize from the Italian Society of Science and republication in 1801 as a presumably more accessible book, it was almost totally ignored. The sole notice that I have found of it in this period is a long review by Brougham in the *Edinburgh Review* in 1805. Thereafter it fell totally from view. It is especially surprising that his work was ignored in France, where coloration and Newton's theory were then a central topic, for Venturi had just spent two years there.

Venturi arrived in Paris in 1796 on a diplomatic mission as the secretary to the count of San-Romano. After the French invasion of Italy, the count negotiated with the Directory to preserve the independence of Modena. When the negotiations failed, Venturi, who had been professor of experimental physics at the University of Modena, remained in Paris and entered into the center of Parisian science. He associated with Berthollet, Biot, Chaptal, Fourcroy, Haüy, and Laplace, among others; participated in the sessions of the institute; and published a number of memoirs in French journals, including his most widely known work, on hydrodynamics. In 1800, he was appointed professor of physics at the University of Pavia, but the following year he became ambassador of the Cisalpine Republic to the Helvetic Confederation in Bern, where he remained for twelve years. Earlier, in a paper published in 1786, he exhibited a sophisticated knowledge of optics and a broad command of the literature, having read not only Newton's *Opticks* but also, which was relatively unusual, his *Optical Lectures.*[1] While in Paris he pursued chemistry and mineralogy, and judging from the company he kept it was no doubt during this sojourn that he was drawn to the study of coloration.

Perhaps the most striking feature of Venturi's paper is the lack of polemics in such a contentious subject. All the usual chemical topics are missing, such as the invocation of the chemical effects of light and the criticism of Newton's and Delaval's arguments on the change of the size and density of corpuscles by the action of acids and alkalies. To clear the way for a proper assessment of Newton's theory of colored bodies, Venturi opened the "Indagine" by refuting Dutour's attempt to reduce the colors of thin films to ordinary refractive dispersion (§5.2). If Dutour's claim were true, it would completely subvert Newton's demonstration of the periodicity of light, which Venturi accepted with the theory of fits.

Beginning his investigation of coloration itself in Chapter 2, Venturi insisted that to investigate the colors of bodies properly we must know the composition of their colors. To carry this out, he at first adopted Dutour's method ("following the example of others") of viewing through a prism narrow strips painted with various colored substances, but he soon recognized that because of all the white light reflected from the upper surface, this method was "unreliable and fruitless." The technique that he then developed was quite simple and natural and with some variation that used by subsequent investigators. He inserted a transparent colored substance (see Hassenfratz's Figure 1 in my Figure 8.3) into a narrow beam of sunlight that

[1] Venturi, "Considerazioni ottiche" (1786); see Newton, *Optical Papers*, 1:278. For Venturi's biography and bibliography, see Girolamo Tiraboschi, *Notizie biografiche*, 3:187–369, 486; 4:vi–vii, 5:xxxix–xlii.

passed through a small hole in a window shutter and then through a prism placed at minimum deviation, the classical Newtonian arrangement. The spectrum, which is either projected onto a screen or viewed directly, shows the composition of the light transmitted by the colored substance and, by the missing colors, those that have been absorbed. Because of the different rates of absorption for different colors, the spectrum varies as the thickness or density of the substance is altered.

Venturi introduced three variants on this arrangement: First, he placed a second identical hole next to the first one. When the colored substance was inserted before the first hole, and both holes were viewed through a prism, he saw two adjacent spectra and could immediately determine which colors were missing. Second, using a single hole and following Newton's procedure in the *Opticks* (Bk. I, Pt. I, Expt. 11), he inserted directly in front of the prism a lens that cast an image on a screen placed at its focal point, 8 feet away. This method is closest to a spectroscope. Third, in the preceding arrangement, instead of placing the colored substance in front of the hole, he placed it before his eye and viewed the solar spectrum cast onto the screen, which had a scale on it so that he could readily determine the missing colors. This was the easiest way, but generally he verified the results by all three methods.[2]

Venturi examined about fifty transparent colored liquids and described thirty different sequences of spectral colors. His immediate aim was to describe selective absorption without any physical interpretation and to show how the transmitted colors vary with absorption. He carefully purified and filtered a variety of colored liquids and placed them in vials with flat rectangular faces. For each substance Venturi described five spectra: *A* represents the colors transmitted

[2] Venturi, "Indagine fisica sui colori," pp. 709–11; read 15 December 1799. The first four chapters of his book *Indagine fisica sui colori* (1801) are a reprint of the paper, but a new chapter on physiological colors, which I will not consider, was added. I shall cite the paper. Paolo S. Palladino translated Venturi's Italian for me. In his description of the observed spectral colors, Venturi corrected for the nature of the glass composing his prisms and the variation of the space occupied by each color by "normalizing" his spectra. He identified the colors by the *position* they occupied with respect to Newton's division of the spectrum rather than by the perceived *color*. In Newton's (musical) division of the spectrum (*Opticks*, Bk. I, Pt. II, Prop. III, Expt. 7; and *Optical Papers*, 1:544–7), if the spectrum's length is 120 units, as Venturi assumes, then red occupies 15 units, orange 9, yellow 16, green 20, blue 20, indigo 13⅓, and violet 26⅔. Thus, it is essential to note that when Venturi refers to the color at the border of red and orange, he is indicating only that it is $^{15}/_{120}$th from the red end of the spectrum, and not its perceived color.

(34)

E. Rosso, Dorè, Giallo, Verde, Azzurro.

XII. Indaco macinato con olio e schiacciato fra due vetri piani.

A. Verde, la metà bassa dell' Azzurro.

B. Giallo, Verde, Azzurro.

C. Dorè, Giallo, Verde, Azzurro, Indaco.

D. La terza parte più alta del Rosso; Dorè, Giallo, Verde, Azzurro, Indaco; la metà inferiore del Paonazzo.

E. Manca solamente l'estremità inferiore del Rosso.

XIII. Indaco disciolto nell' Acido Sulfurico (*Azzurro di Sassonia*).

A. L' estremo più basso del Rosso, la terza parte più alta del Verde, e i due terzi più bassi dell' Azzurro.

B. La terza parte più bassa del Rosso, la metà alta del Verde; l' Azzurro.

C. La metà bassa del Rosso; Verde, Azzurro, Indaco.

D. I due terzi bassi del Rosso; Verde, Azzurro, Indaco, la metà bassa del Paonazzo.

E. Manca il solo Dorè.

XIV. Indaco e Potassa disciolti in Acqua.

A. Giallo, la terza parte più bassa del Verde.

B. Dorè, Giallo, Verde.

Figure 8.1. Venturi's description of absorption spectra at five thicknesses of various preparations of indigo (translated in the text). (From *Indagine fisica sui colori* [1801].) For someone who never actually saw an absorption spectrum it would be difficult to form an image of such spectra from this verbal account. In contrast, Hassenfratz's and Brewster's illustrations (Figure 8.3 and Plate 1) directly represent the appearance of the spectra.

through the thickest or densest quantity beyond which the body becomes opaque; *E*, those colors transmitted at the rarest or thinnest quantity above which all seven colors pass through; and *B*, *C*, and *D*, the colors transmitted at three intermediate densities. Because he found that changes in density and thickness are equivalent, he varied both. His observations for two preparations of indigo (see Figure 8.1) are typical; "lower" means toward the red end of the spectrum and "upper" toward the violet end:

XII. Indigo ground with oil and pressed between two glass plates.
- A. Green, lower half of Blue
- B. Yellow, Green, Blue
- C. Orange, Yellow, Green, Blue, Indigo
- D. The upper third of Red, Orange, Yellow, Green, Blue, Indigo, the lower half of Violet
- E. Only the lower extremity of Red is missing

XIII. Indigo dissolved in sulphuric acid (*Saxony blue*)
- A. The lower extremity of Red, the upper third of Green, the lower two-thirds of Blue
- B. The lower third of Red, the upper half of Green, Blue
- C. The lower half of Red, Green, Blue, Indigo
- D. The lower two-thirds of Red, Green, Blue, Indigo, the lower half of Violet
- E. Orange alone is missing[3]

Thus, in XII, absorption proceeds from both ends of the spectrum until only some green and blue remain, whereas in XIII it proceeds both from the middle at orange and also the violet end until only some red and some green and blue remain.

Venturi's verbal descriptions of absorption spectra contain a great deal of information about the stages of absorption (often lacking from later studies), but it is difficult to form a picture of the phenomenon from this mode of representation. In fact, Brougham could form none at all:

> When, for example, he [Venturi] viewed the spectrum made by light, which passed through a red solution, and saw no green in it . . . we should wish to know what colour occupied the middle part of the spectrum. Indeed, we are left to guess what was the form and appearance of the prismatic spectrum in all these experiments. We are told that one solution passed all but yellow, another all but blue. . . . We should wish to know what kind of oblong spectra were exhibited in the two cases now cited.[4]

For more than a century the world had viewed the Newtonian spectrum with its continuous succession of colors, and Brougham could

[3] Venturi, "Indagine fisica," pp. 715–16.
[4] [Brougham], "[Review of] *Indagine fisica*," p. 32. For the identification of Brougham, see Ch. 7, note 33.

not imagine a spectrum with parts altogether missing, with neither color nor light. His demand to know what such spectra looked like was a perfectly reasonable one, and both Hassenfratz and Brewster later published illustrations of the observed absorption spectra (Figure 8.3 and Plate 1).

Venturi maintained that Newton's theory of colored bodies was an overgeneralization, for there were not one but four distinct causes of coloration. The first three need not detain us: refraction, diffraction, and the periodic colors of thin films. The fourth cause, responsible for the colors of most bodies, is transmission with selective absorption. Before explaining this new cause, Venturi shows that Newton's theory cannot explain the colors of bodies. First, he appeals to Delaval's third paper ("which is preferable to the earlier one already translated into Italian"). Delaval showed that no colored light is reflected from transparent colored bodies, and that they acquire their color from light transmitted into the body and then returned outward.[5] The colors of such bodies cannot be explained by Newton's theory, for in thin films the transmitted and reflected colors are always complements.

In the second place, he notes that "the series of different sorts of rays" that were found in his absorption spectra "do not correspond in form or progression to the series and progression that Newton assigned to the various thicknesses and refractive powers of his thin plates." For example, it is impossible for a thin film to transmit red, orange, and blue as in density *C* of gold oxide (case II) or to transmit the upper third of red, the upper half of green, and blue as in density *B* of Saxony blue (case XIII). "These and many other similar bodies," Venturi declared, "transmit several sorts of rays with such interposed breaks in the prismatic order as Newton's theory does not in any way allow."[6] Nor do thin films allow the passage of orange alone as in cobalt ink (case X), nor just yellow and green as in emerald glass (case V) and a solution of indigo and potash (case XIV). Other than for a footnote referring to Newton's nomograph, Venturi (like Young three years later) did not explain how Newton's theory predicts the appearance of particular combinations of colors. This would have scarcely been evident to any reader. Hassenfratz judged it necessary to devote a significant part of his memoir to deriving the color sequences of Newton's theory and comparing them with his observed

[5] Venturi, "Indagine fisica," p. 729; see also p. 745.
[6] Ibid., p. 731.

spectra. Venturi's principal aim, however, in contrast to Hassenfratz's, was not to refute Newton's theory but, rather, to establish selective absorption as a cause of coloration.[7]

Having eliminated Newton's theory, Venturi explains that most bodies are colored by transmission and selective absorption. Just as black bodies extinguish transmitted rays of all colors gradually (because they become transparent when sufficiently thin), so the same process occurs in colored bodies: "*They suffocate and extinguish certain determinate species of rays more quickly and more easily than others.*"[8] Venturi then immediately turned to the intersection of physics and chemistry and examined the relation between absorption and chemical changes in the body. To study this he cast sunlight through various colored filters (the transparent liquids whose absorption spectra he had already determined) into a solution of chlorophyl (green extract of clover or *trifoglio* in alcohol, whose absorption spectrum was also known) and observed the discoloration that occurred in a given time with different sorts of rays. From this and a similar experiment, as well as others of Senebier – most likely the inspiration for his investigation – and Henri Alexandre Tessier, he inferred the general rule (*massima generale*) that "*A given species of rays acts to induce change only in those bodies in which it is neither transmitted nor reflected,*" that is, only in those in which it is absorbed or extinguished.[9] That Venturi proposed this fundamental law of photochemistry decades before Theodor von Grotthuss (1818) and John William Draper (1841), after both of whom the law is named, is not here my concern.[10] Rather, it is the

[7] Venturi's final argument against Newton's theory (ibid.) involves a mistake in transduction. He correctly observes that when the corpuscles' density or size changes, the colors should either ascend to higher orders or descend to lower ones in Newton's nomograph. Yet in many substances (e.g., case XII above) they do both simultaneously. In his experiments, however, Venturi was changing not the density or thickness of the corpuscles – to which alone Newton's theory applies – but that of the whole fluid.

[8] Ibid., p. 734; Venturi's italics.

[9] "*. . . un dato Genere di Raggi non agisce ad indurre cangiamento se non in que' Corpi, dai quali non è trasmesso nè riflettuto*" (ibid., p. 736).

[10] Grotthuss, "Auszug aus vier Abhandlungen physikalisch-chemischen Inhalts," pp. 58–9; and Draper, "On some analogies between the phaenomena of the chemical rays," p. 195; see Eder, *History of Photography*, pp. 110, 166–8. It is not surprising that the law had multiple discoverers, for it is so obvious. Brougham thought so. After describing Venturi's result, he remarked that "though we willingly acknowledge the elegance of the above experiment . . . we do not apprehend that there is great originality in the conclusion which our author draws from it, viz. that those rays only discolour a body which are absorbed by it. It would be difficult, upon the principle of absorption, to suppose that those rays which pass freely, alter the internal structure of the body" ("[Reveiw of] *Indagine fisica*," pp. 38–9).

evidence that it provides for the intimate connection in Venturi's thinking between selective absorption and the chemistry of light.

"But how," Venturi asks, "does this extinction of certain determinate species of rays take place in the internal parts of bodies?" He grants that the cause of absorption may be explained by the wave theory, but he is truly enthusiastic about the explanation according to the emission theory:

> To grant that there exists an attraction between the ponderable matter of bodies and the multicolored particles of light such that when these particles pass into these bodies they gradually lose their progressive motion and their power to excite the retina and enter into new combinations with these bodies, one species of particle more quickly and easily than the others in proportion to their respective degree of reciprocal affinity with the given body: To grant all this is to assume the system of the *emission* of light, and to support one of the more illustrious theories of modern chemistry.

Thus Venturi has passed from his announced title, "*Physical* research on colors," to the "modern *chemistry*" of light. He does not insist on the certainty of the cause of absorption, but hopes that it will perhaps be discovered in the new century (which was just sixteen days away). He assures us, however, that we should "be satisfied for now if we have in the meanwhile established with certainty that this destruction does occur, and that we have clearly and distinctly distinguished four ways of coloring in bodies."[11]

In his concluding fourth chapter, Venturi applies his new understanding of selective absorption to an impressive variety of optical problems, such as the construction of colored filters to produce monochromatic light, and an explanation of dichroism, the phenomenon in which a substance is one color at one thickness and a different one at another thickness.[12] Venturi's explanation of color mixing and his distinction between additive and subtractive mixing are of greater interest. In his *Optical Lectures*, Newton introduced the concept of subtractive color mixing and distinguished it from additive or true mixing. Previously all methods of mixing colors were assumed to be physically equivalent. He explained subtractive mixing with a few simple illustrations, such as reflecting colored light from colored bod-

[11] Venturi, "Indagine fisica," pp. 737, 738. In passing, he judges Euler's theory (though he does not mention him by name) that the colors of bodies are due to a vibration of the corpuscles of the body, rather than to the incident light, to be "beyond the bounds of possibility" (p. 738).

[12] Ibid., pp. 739–40, 742–3.

ies or passing light through superposed colored glass plates.[13] When, to take another example, green light, which also contains some adjacent yellow rays, is passed through a red plate, which also transmits some of the neighboring yellow, then mostly yellow rays will emerge from the plate. The green will have been removed, or subtracted, and not mixed with the red to produce the yellow. For a variety of reasons, though, Newton held (incorrectly) that pigment mixing is an additive process; for example, each of the particles of a yellow and blue pigment reflects rays of its own color to the eye where they are then mixed or added to compose a green.[14] Though Venturi does not mention Newton here, it seems as if one of his aims is to clarify and correct Newton's account in the *Optical Lectures*.

When two dyes of different color are mixed to produce a third color, Venturi explained, the ingredients can either combine chemically (which he will not consider) or mechanically, where each substance preserves its own coloring properties. In the latter case "the two ingredients together exclude those sorts of colors that they excluded separately before mixing"; we call such mixtures "subtractive," and Venturi "superposition." To demonstrate this rule he transmitted light through two superposed vials, each containing a dye of a different color, and also through one vial containing the two dyes mixed together: The transmitted colors were the same in each case. Venturi contrasts this mode of mixing with that of a "*mosaic*" (our "additive") mixture in which yellow and blue spots evenly cover a surface, and the two sorts of reflected rays mix into green in the eye. According to Venturi, "many" consider this to be the way in which pigment mixing occurs. He concedes that this does occur sometimes, especially in painting, but generally his experiments show that "the rays transmitted to the eye in these mixtures, which I call that of *superposition*, are not the aggregate and sum of all the rays which were obtained from the two colors separately; rather they are the remainder and residue of those which are extinguished in each of the two combined liquids."[15] Consequently, a "mixture" by superposition, such as a

[13] Newton, *Optical Papers*, 1:510–11, 514–19. Venturi had read the *Optical Lectures*; see note 1.

[14] Ibid., pp. 108–13.

[15] Venturi, "Indagine fisica," pp. 749, 750. In one example he used Saxony blue of degree *D* (case XII) and cochineal in water of degree *D* (case XV); the former transmits some red, green, blue, indigo, and violet, and the latter red, orange, yellow, blue, indigo, and violet. Because the Saxony blue excludes (subtracts) orange and yellow and the cochineal green, the resultant transmitted color will be a mixture of the remaining red, blue, indigo, and violet, which is perceived as purple.

yellow and a blue dye, can result in a pure green color without any mixture of the components, yellow and blue.[16]

In midcentury, the recognition that additive and subtractive color mixing are different physical processes, and in particular that pigment mixing is subtractive, became an essential element in the development of the trichromatic theory of color vision by Herman von Helmholtz, Hermann Günther Grassmann, and James Clerk Maxwell.[17] The only aspect of Venturi's explanation of color mixing that remained untested was his assumption, then nearly universally held, that an additive mixture of yellow and blue produces green; and this was independent of the concept of absorption. A proper understanding of subtractive mixing and its fundamentally different nature from true, additive mixing follows directly from the concept of selective absorption, and we shall see that Herschel, too, fully grasped it (§9.1).

The most puzzling and difficult question concerning Venturi's "Indagine fisica" is to explain why it was virtually without influence. Because Hassenfratz's and Prieur's subsequent work was likewise not pursued, I can pose two alternative questions. Why was Venturi's work ignored in France, where he had just been an active participant in the scientific community, especially by Berthollet, Hassenfratz, and Prieur? And more generally, why was there such widespread lack of interest in studying selective absorption? I can offer nothing but conjectures to the first question, so that it remains a puzzle. I will address the second question after I treat the situation in France itself, but Brougham's review can provide some preliminary insights.

Although he welcomed Venturi's work as "a very valuable accession to optical science" and approved of his quantitative style of mathematical natural philosophy, which he hoped would counterbalance the more popular pursuit of "pneumatic chemistry,"[18] Brougham did not really appreciate the concept of selective absorption or the significance of absorption spectra. Venturi's failure to provide some sort of pictorial or graphical representation of absorption spectra was partly responsible for Brougham's failure to recognize that some colors are actually missing from such spectra. Brougham himself, however, did not grasp that the colors of bodies are compound and seemed to think that compound spectra represent impure substances: "Our author, by telling us that one kind of liquid sent through green

[16] Ibid., p. 753.

[17] See Hargreave, "Thomas Young's theory of color vision," pp. 271–302; and Paul D. Sherman, *Colour Vision in the Nineteenth Century*, Ch. 5; and also Ch. 9, note 9.

[18] [Brougham], "[Review of] *Indagine* fisica," p. 21.

as well as red rays, only says, that his coloured solution had a certain imperfection, [or] that it was not purely red or purely green."[19] Brougham was a competent student of optics, and not the buffoon frequently depicted. Rather, we should accept his difficulties as indicating that selective absorption was not as self-evident a concept as it now seems, a factor that probably contributed to the general lack of interest in the early investigations of absorption. At this point, however, we may note that all those historians who have attributed so much influence to Brougham's vitriolic reviews of Young's papers on the wave theory would be hard pressed to explain why his generally favorable account of Venturi's "Indagine fisica" in the same journal should have fallen on deaf ears just two years later.[20]

8.2. THE PHYSICISTS' DEFENSE OF LAW AND ORDER

The series of papers on dyeing, bleaching, and coloring parts that Berthollet had published in the 1780s culminated in 1791 in his epoch-making treatise on dyeing, *Eléments de l'art de la teinture.* Berthollet devoted part of his introductory chapter, "On coloring parts and their affinities," to arguing against Newton's theory, but still no one rose to defend it against the chemists' attack. Since Opoix had launched the chemists' offensive in 1776, physicists chose not to respond in print. Whether they judged the chemists' arguments to be contemptible or simply did not wish to draw attention to them, they did continue to hold and teach Newton's theory of colored bodies. In 1803, Haüy had expounded it in his *Traité élémentaire de physique,* which was an elementary textbook commissioned by Napoleon for the newly instituted national lycées. Only after Berthollet attacked Newton's theory by means of quotations from Haüy's *Traité* in the second edition of his *Eléments* in 1804 was Haüy provoked into replying to Berthollet in the new edition of his *Traité* (1806). Berthollet had no need to counterattack, for in the following year Hassenfratz and Prieur entered into the fray with their investigations of absorption spectra, which they argued refuted Newton's theory. The matter remained stalemated for nearly a decade, when Biot reopened the debate in his *Traité de physique* with a new defense of Newton's theory and a critique

[19] Ibid., p. 32.
[20] For example, Ernst Mach, *The Principles of Physical Optics,* pp. 275–6; Edgar W. Morse, "Young," p. 565; and Steffens, *Development of Newtonian Optics,* pp. 133–6. For a contrary view of the influence of Brougham's reviews on the acceptance of the wave theory, see Kipnis, *History of the Principle of Interference,* p. 162.

of the alternative affinity theory of coloration. For the physicists, Haüy and Biot, the defense of Newton's theory ultimately resolved itself into a defense of "law and order."

Berthollet's attack on Newton's theory (and on Delaval's first two papers in support of it) in the first edition of his *Eléments* was not as extended, clear, or cogent as Bancroft's three years later. In the second edition of his *Eléments* Berthollet acknowledged as much: "Some doubts had been offered about . . . the theory of Newton, in the first edition of this work; since which time Bancroft has opposed to it a great number of facts. We shall avail ourselves of these several observations in the discussion which we are about to undertake. . . ."[21] Berthollet made two strategic changes concerning theories of coloration in his revision of this chapter. First, he no longer put so much stress so soon on oxygen as the principal element affecting the affinities for light.[22] Second, he presented a more fully developed case against Newton's theory by means of large excerpts from Haüy's *Traité élémentaire* ("where physical knowledge is concentrated with equal elegance, clearness, and precision"), which he then countered with a variety of evidence.[23] In a third modification, which did not directly concern the theory of colored bodies, he no longer held that all aspects of dyeing could be explained by elective affinities. Berthollet's critical analysis of affinities in his *Essai de statique chimique* (which appeared the preceding year) was responsible for this change, and his analysis would seriously restrict the chemists' program to develop a theory of coloration by affinities.[24]

Much of Berthollet's critique of Newton covers the familiar ground of showing the implausibility and inconsistency of his physical theory with analyses of the actions of acids and alkalies, the grinding of pigments, and the oxidation of metals. As in his earlier writings, his

[21] C.-L. and A.-M. Berthollet, *Elements of the Art of Dyeing* (1824), 1:33. The first edition of *Eléments de l'art de la teinture* was published in 1791, and two English translations were quickly published, at London in 1791 and Edinburgh in 1792. Berthollet prepared the second edition of 1804 with the assistance of his son, Amédée-Barthelemy. This was belatedly translated as the *Elements of the Art of Dyeing* in 1824 and a second edition appeared in 1841. German translations of each edition were published in 1792 and 1806. Berthollet's attack on Newton's theory was further disseminated through his own abstract of the book; "*Elémens de l'art de la teinture*. . . . Extrait par l'auteur" (1791), pp. 138–9.

[22] Berthollet, *Eléments* (1791), 1:6–8, which was eliminated in the second edition. For the antiphlogistonists' early excessive emphasis on oxygen as the cause of color, see §6.2; and for Bancroft's criticism of their position, Ch. 7, note 15.

[23] C.-L. and A.-M. Berthollet, *Elements* (1824), 1:32.

[24] Keyser, "Between science and craft," pp. 233–8.

evidence is chemical rather than optical, except that he now introduces Delaval's experimental demonstration that colored bodies do not reflect light.[25] Berthollet probably learned of Delaval's third paper through Bancroft's book, though he more fully grasped its significance than Bancroft. Interestingly enough, this is one of the points to which Haüy did not reply, perhaps because Berthollet did not explain its significance sufficiently clearly and forcefully, or perhaps because he simply had no adequate response.

Though the rancorous tone of the earlier generation of chemists is gone, Berthollet's criticism of Newton's theory has the same epistemological foundation as Opoix's (§6.2). An intrinsic quality of a body such as its color must depend on particular substances, or its chemical composition, and not simply on hypothetical corpuscles with only mechanical properties, size, and density:

> We ought not to confound the fugitive colours produced by the reflection of plates, which follow the laws described by Newton, with the colours which continue, notwithstanding the changes of density and thickness. These appear to depend on properties in which the peculiar affinity for the different rays of light has an influence that withstands that of the dimensions and the density. On examining the facts, we perceive that condensed oxygen exercises a great power in this species of affinity.[26]

Perhaps more than any other chemist of his day, Berthollet desired to bring physics and chemistry closer together, and in this respect he differed greatly from Opoix. Although he regretted his break with Newton's theory of colored bodies, it was necessary "because we feel how important it is to connect all the effects due to the reciprocal actions of bodies, and we hope that future experiments will fill the gap which here seems still to separate physics [*physique*] and chemistry."[27] Berthollet was Laplace's ally in propagating the Newtonian program of short-range forces and corpuscles, and it served as one of the foundations for research at the Society of Arcueil. One of the central concepts uniting the two sciences for Berthollet was that of force or affinity, and when physics dealt with the corpuscular level

[25] C.-L. and A.-M. Berthollet, *Elements* (1824), 1:41–3.

[26] Ibid., p. 47; I have slightly altered the translation.

[27] Ibid., p. 49; following contemporary usage Ure translated "physique" as "natural philosophy." By "reciprocal actions" Berthollet alludes to affinities, as is evident from his remark about "the chemical properties of bodies, or those affinities by means of which they exert a reciprocal action"; Berthollet, *An Essay on Chemical Statics*, 1:viii; a translation of *Essai de statique chimique* (1803), 1:2.

– the realm of chemistry – it had to be based on affinities.[28] To him, Hellot's mechanical theory of dyeing and Newton's theory of colored bodies suffered from the same flaw: "We depart from *true theory*, which is solely the result of observation, when we ascribe to *laws purely mechanical* the adherence of the colouring particles to the substance which they dye . . . and the difference between permanent and fugitive colours."[29]

Haüy taught physics and mineralogy at the Ecole des Mines and since 1802 was professor of mineralogy at the Muséum d'Histoire Naturelle. In 1784, he had laid the foundation of the mathematical theory of crystal structure in his *Essai d'une théorie sur la structure des cristaux*, and in 1801 he published his major work on crystallography, *Traité de minéralogie*. In his theory of crystals, Haüy had derived the secondary forms of crystal structures from the geometrical form of the "integrant" or elementary constituent molecules. His method was essentially geometrical, as he constructed crystals from identically shaped molecules, or building blocks, according to certain laws, for example, a rhombic dodecahedron from cubes (Figure 8.2). He then used his crystal theory as a basis for a system of mineralogical classification in which species of minerals were defined by the geometrical form and chemical composition of the integrant molecules. In practice, however, Haüy classified minerals on the basis of their form and for the most part ignored their chemical composition. It should cause no surprise that chemists objected to his approach. Just one year before Berthollet criticized Haüy's exposition of Newton's theory of colored bodies, he likewise devoted a sixteen-page "note" in his *Essay on Chemical Statics* to criticism of his failure to take chemical analysis into account.[30]

Although Haüy came to science from natural history (botany and mineralogy) and not physics, his approach to a theory of crystals was in essence mathematical, being based on geometrical form and symmetry. We can therefore recognize why Newton's theory of colored

[28] Berthollet, *Essay on Chemical Statics*, 1:ix–xi. See Crosland, *Society of Arcueil*, pp. 233, 295.

[29] C.-L. and A.-M. Berthollet, *Elements* (1824), 1:56; italics added. For Hellot's theory, see Ch. 6, notes 53, 57.

[30] Berthollet, *Essay on Chemical Statics*, Note XIV, 1:432–48. On Haüy's crystallography, see Seymour H. Mauskopf, *Crystals and Compounds;* and John G. Burke, *Origins of the Science of Crystals*, Chs. 4, 5. Burke observes: "It should be noted, however, in fairness to Haüy, that his structural classification, rather than the chemical classification of his critics, conforms closely to modern practice" (p. 113). See also Mauskopf, "Minerals, molecules, and species," pp. 195–6, 199–202; and Melhado, "Chemistry and physics," p. 210.

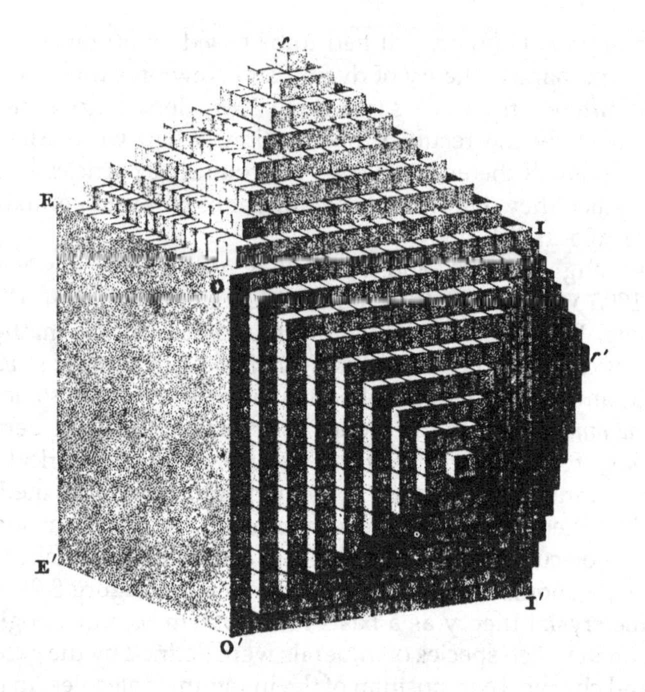

Figure 8.2. Haüy constructed a rhombic dodecahedron by placing around a cubic nucleus layers of cubic molecules that decrease by one row in successive layers. (From *Traité de minéralogie* [1801], vol. 5, Plate II, Figure 13.)

bodies would have naturally and profoundly appealed to him. The two theories share fundamental features: Both depend on the size, shape, and iterated arrangement of identical corpuscles and both are independent of chemical composition. In his exposition of Newton's theory, Haüy appreciated its dependence on Newton's hierarchical matter theory; and he explained how the "smallest molecules, or those which we call *integrant molecules*" are separated by pores and compose molecules of the second order, which in turn compose successively higher order of molecules separated by pores.[31] After quoting Haüy's account of Newton's theory, Berthollet disdainfully observed that "this manner of supposing mechanical arrangements [*dispositions mécaniques*], in order to employ them in the explanations of physical

[31] Haüy, *Traité élémentaire de physique* (1806), 2:245; the third edition of 1821 carries over the treatment of colored bodies unchanged from the second. On the English translation of the first edition of 1803 see Ch. 7, note 46.

properties, without directly proving them, may easily serve to prop hypotheses otherwise entirely destitute of probability."[32]

If Haüy at first considered it inappropriate to defend Newton's theory from Berthollet's criticisms in a high school textbook, he could not fail to respond in the second edition after his own exposition was directly attacked by Berthollet in the new edition of his *Eléments*. In a section added to his *Traité élémentaire*, Haüy defended Newton's conception of a physical theory in which chemical properties can have only a "remote influence" on the colors of bodies insofar as they alter the size and density of the corpuscles. He observed that although chemists may complain about Newton's theory, Newton himself had admitted chemistry into the physics of light by using short-range forces to explain reflection and refraction and in discovering that refraction is greater in inflammable bodies.[33] They should be content with this much chemistry. We can pass over the usual thrust and parry over such irresolvable issues as color changes caused by acids and alkalies and turn directly to Haüy's most important counterthrust, which would directly set the scene for Hassenfratz's introduction of absorption spectroscopy in France.

In his *Eléments*, Berthollet had correctly distinguished between simple and compound colors. Then "notre savant chimiste" asserted that the green oxide of copper, the green of plants, and "the greater number of hues which exist in nature" must be simple colors because they are caused by chemically simple or homogeneous coloring principles: "Were the green of plants due to two substances, one of which is yellow and the other blue, it would be extraordinary if we could not separate them."[34] Haüy responded by showing that these colors are compound; both greens, for example, are composed of yellow and blue. This observation "agrees perfectly" with Newton's theory, for in Newton's rings "the different species of rays, in mixing at all places of the air film intercepted between the two glasses, give rise to more or less compound colors." Newton's theory, which Haüy concedes is based on the "force of analogy," provides "simple and precise rules" for all coloration. "In adopting the contrary opinion" not only are two theories of color required – one for thin films and one for

[32] C.-L. and A.-M. Berthollet, *Elements* (1824), 1:36; *Eléments* (1804), 1:36.

[33] Haüy, *Traité élémentaire* (1806), 2:260–1.

[34] Ibid., p. 263; C.-L. and A.-M. Berthollet, *Elements* (1824), 1:53. Lamarck had in fact held that the green of plants arises from yellow and blue coloring molecules; *Recherches*, 2:158. Bancroft's ideas on the simple nature of the colors of bodies were similar to Berthollet's; see Ch. 7, note 19.

bodies – "but one is reduced to indicating in a vague way affinity as the cause of the coloration of opaque bodies, without being able to assign any law to its action nor establishing the mutual connection and dependence of the effects that are attributed to it."[35]

Haüy has articulated the appeal of Newton's theory to the quantitative physicist: It can account for all colors by one simple and precise law without arbitrarily having to postulate for every substance different affinities for different colors. Moreover, in contrast to Berthollet's theory it successfully predicts the compound colors of natural bodies. There is one very strange feature to Haüy's demonstration of the compound colors. Rather than viewing the bodies through a prism – and Berthollet had correctly defined compound colors as those resolved by a prism into a number of simple colors – he chose the odd method of using colored afterimages.[36] It was less odd that Berthollet was so confident of the chemical simplicity of coloring matter that he never put it to a test by viewing colored bodies through a prism, for chemists rarely employed that instrument.

Even before the second edition of his *Traité élémentaire* was published in 1806, Haüy's defense of Newton's theory appears to have stimulated further research in the Parisian scientific community. His demonstration that the colors of natural bodies are compound, as Newton's theory predicted, and his contrast of the unpredictable action of affinities with the precise rules for the appearance of colors in Newton's theory suggested an experimental test of the two theories and an investigation of absorption. By early 1805, Hassenfratz and Prieur took the obvious step and viewed colored bodies through a prism, though it was not until 1807 that they declared that their observations refuted Newton's theory.

Hassenfratz started his working life as a master carpenter, and after entering the mining service, he became an inspector of mines. He began his scientific career by serving as Lavoisier's laboratory assistant, and later, with P. A. Adet, he developed the system of chemical symbols that was part of the new nomenclature of Lavoisian chemistry that was published in 1787. From 1798 to 1815, he was professor of physics at the Ecole Polytechnique where his course was much criticized, though the part on optics was considered to be very good.[37]

[35] Haüy, *Traité élémentaire* (1806), 2:264.
[36] Ibid., pp. 263–4; C.-L. and A.-M. Berthollet, *Elements* (1824), 1:52–3.
[37] Fox, *Caloric Theory of Gases*, p. 231. On the new symbols see Crosland, *Historical Studies in the Language of Chemistry*, pp. 245–55.

His research was primarily in chemistry but extended to experimental optics. Prieur was trained as a military engineer, and played an important role in the scientific community as an administrator, helping to introduce the metric system and establish the Ecole Polytechnique. Together with Lazare Carnot, the other engineer on the Committee of Public Safety, he was responsible for successfully mobilizing the French scientific community (especially, its chemists, Berthollet, Chaptal, Guyton de Morveau, Fourcroy, and Hassenfratz) to solve the nation's severe matériel problems during the war with the allied European powers. As a scientist, however, he was an amateur, though his work on physiological colors was often cited.[38]

Both Prieur and Hassenfratz served together on commissions and editorial boards (in particular, for *Annales de chimie*) and collaborated with each other and many of the participants in our story – Berthollet, Biot, Fourcroy, Haüy, and Laplace. In the stimulating, highly charged Parisian scientific community people came together and clashed not only over scientific issues but also over politics, professional positions, and personalities.[39] Without access to unpublished papers, I have been unable to study seriously the influence of various alliances and rivalries on the debate on coloration, though I will indicate their presence when I can. In papers read to the Institut National (the renamed and reorganized Académie des Sciences) within a week of each other, Prieur and Hassenfratz introduced absorption spectra to study coloration. Whatever the interdependence of their work – and I am sure that there is some – their aims were different. Prieur's work on coloration was only one part of his attempt to establish a new theory of light and color. Hassenfratz, on the contrary, devoted his papers solely to resolving the debate on coloration between Berthollet and Haüy; and though not of the stature of the academicians, he was very much in the mainstream of contemporary science.

Hassenfratz read a memoir on transparent and opaque colored bodies to the institute on 11 March 1805, and as it was on the same subject as Prieur's memoir read a week earlier, he was assigned the same referees, Berthollet and Haüy – the principal antagonists – together with Laplace and Monge. This paper was neither reported upon nor published. Hassenfratz apparently withdrew it, for two years later, in 1807, he read at four sessions of the institute the

[38] Georges Bouchard, *Un organisateur de la victoire: Prieur de la Côte-d'Or*.

[39] See, for example, Fox, "Rise and fall of Laplacian physics"; and Crosland, *Society of Arcueil*.

very long paper published in the *Annales de chimie* in 1808 and was assigned new referees, J.A.C. Charles and Monge.[40] It was in this paper (insofar as can be judged from the published evidence) that he announced his refutation of Newton's theory. Hassenfratz devoted the first part of his paper to presenting the rival theories of coloration as advocated by Berthollet and Haüy. In his introduction he stresses that white light consists of an infinite number of colors, and not three.[41] This, I suspect, was an attempt on his part to distinguish his views from Prieur's, who argued for the disreputable three colors. In the second part of his paper Hassenfratz put Newton's theory to an experimental test by observing the absorption spectra of a variety of transparent colored glasses and liquids. To observe the spectrum of a colored glass plate C (Hassenfratz's Figure 1 in my Figure 8.3), he inserted the plate in a narrow beam of sunlight that was admitted through the hole *A* and then passed through the prism *B*.[42] He observed the spectra of fluids by putting them into a hollow glass prism.

According to the affinity theory, Hassenfratz explains, any arbitrary set of colors may be absorbed in passing through a substance, for it "assigns neither law nor order in the reflections or refractions of colors." Newton's theory, on the contrary, predicts that only certain sequences of colors can appear at a given thickness and density of any substance. Hassenfratz presents some simple and striking examples. Only two colors, red and violet, can be produced by a single species of rays, so that if a substance transmitted pure blue, green, or yellow rays it would contradict Newton's theory. Similarly, just four colors – orange red, pale green ("le vert-blanc" probably consisting of green and blue), violet indigo, and purple – can be composed of only two colors. However, "in the hypothesis of coloration by

[40] According to the institute's minutes, Hassenfratz read his "Mémoire sur les corps colorés, transparens et opaques" on 20 ventôse an 13, which is 11 March 1805; and Prieur read his "Considérations sur les couleurs et sur plusieurs de leurs apparances singulières" on 13 ventôse; *Procès-verbaux des séances de l'académie*, 3(1804–7):188. Hassenfratz's published memoir erroneously states that it was read on 27 January 1805, or before Prieur's. Hassenfratz later read his "Mémoire sur la coloration des corps" on 26 January, 23 February, 20 April, and 18 May 1807; ibid., pp. 495, 502, 517, 532. The title and division of this memoir correspond almost exactly with that published, "Mémoire sur la colorisation des corps." Because the paper read in 1805 had a different title, I assume that it is not the same as the published memoir.

[41] Hassenfratz, "Mémoire sur la colorisation," pp. 55–6. See also his review, "Versuche und Beobachtungen uber die Farben des Lichtes . . . par Chrétien-Ernest Wunsch" (1807).

[42] Hassenfratz, "Mémoire sur la colorisation," pp. 310–11.

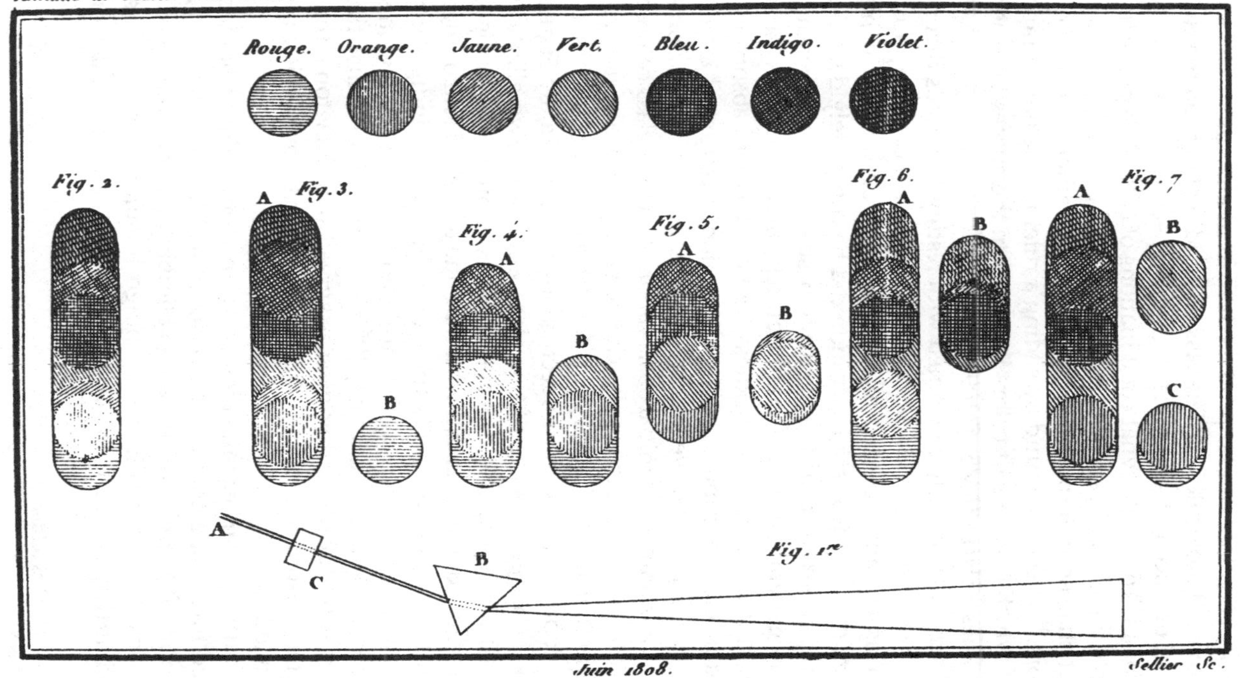

Figure 8.3. Hassenfratz's representation of absorption spectra. Figures 3 to 7 illustrate the spectra at two thicknesses of a red, yellow-orange, green, blue, and violet colored glass. Figure 2 shows the solar spectrum projected by the prism *B* (Figure 1) when no colored glass *C* is inserted in the incident beam *AC*. (From *Annales de chimie* 66[1808].)

affinity alone, *all* colors can be produced . . . from a single or several species of colored molecules."[43] The task of deciding between the two theories is in practice not so simple, for most spectra consist of more than two colors. To specify what combinations of colors are predicted by Newton's theory, Hassenfratz used Newton's nomograph (Figure 2.7), except that he determined the transmitted rather than the reflected colors.[44] These are found by sliding a ruler up the nomograph and noting which columns of colors it crosses in the spaces between the dotted lines, *AHI1*, *3LM5*, *7OP9*. . . .

Hassenfratz was primarily concerned with testing Newton's theory and not, as was Venturi, with describing the process of absorption. Although Hassenfratz recorded all the stages of absorption in his experiments, only in his first observation did he completely describe the changing appearance of the absorption spectra with increasing thickness. In general he gave a graphical depiction of the spectra at a very thin and a substantial thickness; the color where absorption begins; and a brief description of the absorption spectrum at a substantial thickness and how well it can be accounted for by Newton's theory. Thus it is only in his initial observation of a violet glass colored by manganese that he describes the characteristic feature of absorption: the different rates of absorption for different colors and its irregular attack on the parts of the spectrum (his Figure 7 in my Figure 8.3). When six or seven plates of this glass were used, two elliptical spectra (*B*, *C*) remained, one orange-red and the other gray-indigo. If the violet were not missing, this would be due to the red and orange of the second order of Newton's rings and the third of indigo and violet. There is, however, no way indigo alone can appear in Newton's rings, so this must be an "anomaly."[45] Hassenfratz found that the absorption spectra for four other glasses agreed with Newton's theory. For some reason, though, he did not henceforth count this case against Newton.

He then turned to an examination of twenty-one colored liquids and declared fifteen spectra to agree perfectly with Newton's theory, four to be doubtful, and two to be inexplicable (a green solution of alkalized scabious, and a solution of indigo in sulfuric acid).[46] The green solution (Figure 8.4a), for example, consisted of a circular or

[43] Ibid., pp. 293–4; italics added.
[44] Ibid., pp. 300–9.
[45] Ibid., pp. 313–14, 316–17.
[46] Ibid., pp. 16–17. Venturi (see note 6) also found that indigo dissolved in sulfuric acid, or Saxony blue, violated Newton's theory.

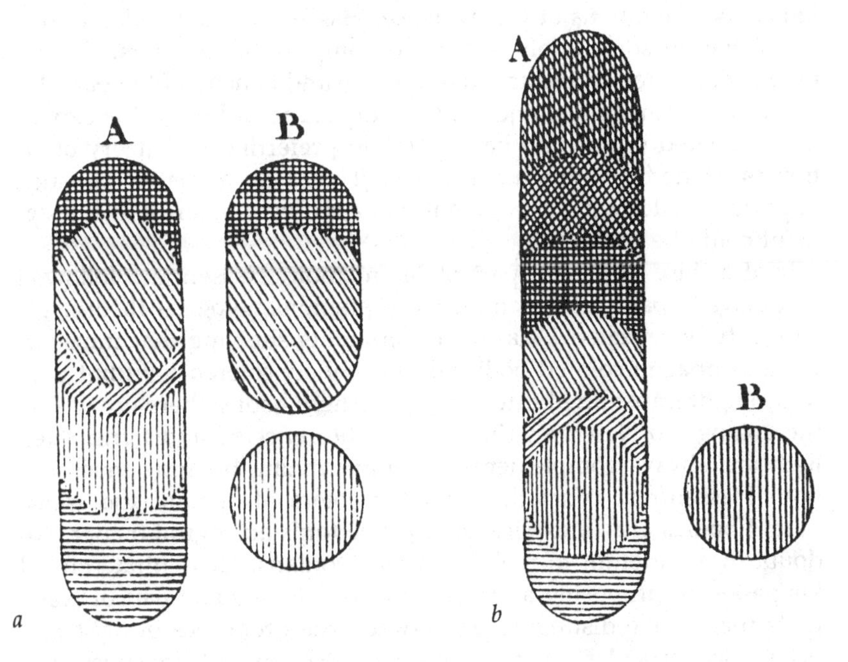

Figure 8.4. The absorption spectra for (*a*) a green solution of alkalized scabious and (*b*) a red infusion of dyer's moss according to Hassenfratz. (From *Annales de chimie* 67[1808].)

pure orange spectrum and an elliptical one of green and blue; but it is simply impossible to get a pure orange with green and blue in Newton's rings. The red infusion of dyer's moss (Figure 8.4b), one of the doubtful spectra, consisted of a pure orange spectrum, which would agree with Newton only "if it were red and if it started from the origin" or end of the spectrum. All of the doubtful spectra involved those which Hassenfratz granted might possibly contain a very feeble red extending to the end of the spectrum.[47] For this reason he did not count them against Newton's theory. Hassenfratz was then (unsuccessfully) attempting to become a member of the institute and was clearly in a difficult position.[48] He had to walk a very fine line in trying not to offend Laplace's powerful Newtonian group while making

[47] Ibid., pp. 9, 17.

[48] In just the period when he was working on colored bodies, Hassenfratz was nominated three times for various categories of membership in the institute, on 1 December 1806, 9 February 1807, and 2 March 1807; *Procès-verbaux*, 3(1804–7):462, 500, 504. In the December and February elections Prieur was a rival candidate.

judicious evaluations of the evidence. Hassenfratz concluded at this point that "if some anomalies in the composition of colors, if some unexplained facts in the action of the size and density of the particles ... suffice to make one reject this theory ... according to the experiments related one would be justified in preferring the theory of affinities alone."[49] Nonetheless, he felt that in a matter of such importance additional experiments were required before a final judgment could be passed.

In the third and final part of his memoir, Hassenfratz followed in Delaval's path and examined the phenomenological theory. He successfully repeated Delaval's experiments showing that transparent and opaque colored bodies do not reflect colored light but only return light transmitted through them. These bodies therefore violate the analogy to Newton's rings where the reflected and transmitted light are always complements. In summarizing his long investigation, Hassenfratz concluded that the colors of opaque and transparent bodies are "inexplicable" by Newton's analogical theory. No doubt in an attempt to mollify the Newtonians, Hassenfratz ended his paper by proposing a way to rescue at least some part of Newton's theory. He distinguished between the properties of light particles – in particular, their fits of easy reflection and transmission – and those of bodies in acting upon light. The fits, he argued, are necessary to explain not only the colors of thin films but also those of bodies, because the theory of affinities alone cannot account for the reflection of light, only its absorption. Therefore Newton's theory of fits (if not the refuted theory of colored bodies) must be combined with affinities to explain coloration fully.[50] The largely uncritical referees' report written by Charles for himself and Monge happily picked up Hassenfratz's face-saving compromise and put little emphasis on his rejection of Newton's analogical theory: "In the impossibility of explaining the coloration of all bodies by either of the two theories apart, Mr. Hassenfratz shows that in combining them everything is explained freely and with sufficient clarity."[51]

The week before Hassenfratz read his first paper on colored bodies to the institute, Prieur read the first of five memoirs on coloration

[49] Hassenfratz, "Mémoire sur la colorisation," pp. 24–5.
[50] Ibid., pp. 139–40, 144–50.
[51] 14 March 1808, *Procès-verbaux*, 4(1808–11):34.

that he would deliver to the institute during the next two years.[52] Prieur's memoirs were extracts from a larger, more comprehensive work on light and color, "Considerations on colors and several of their singular appearances," which, in addition to absorption, covered such subjects as the composition of white light, complementary colors, and the cause of diffraction. Only the first of these memoirs was published, but in 1806 and 1807, he published two additional papers that he described as extracts and continuations of the same "Considerations." Because of the controversial positions that he took in these papers and his outdated qualitative style of natural philosophy, it is unlikely that most members of the institute took very seriously his investigation of absorption and refutation of Newton's theory of colored bodies. No report was issued on his memoirs, and consequently they were not recommended for publication by the institute. Despite Prieur's importance as an administrator of science, as a scientist he was an amateur; and it is no doubt for these reasons that he was both put forward and rejected for membership in the institute.[53]

In his first paper Prieur described his observations of the absorption spectra of both opaque and transparent bodies. Opaque bodies were cut into a rectangular shape and placed on a black background or covered with a black cardboard with a rectangular opening. When the light reflected from them was viewed through a prism, colored fringes were seen on each edge. Transparent bodies were viewed directly through the cardboard slit. Prieur's method of observation was very much like Dutour's, but because he used such a broad slit, he observed compound boundary colors whose composition is far more difficult to determine than those of the simple spectral colors produced by the usual narrow slit. He therefore did not observe a single absorption spectrum for each substance, like all investigators since Dutour. This technique, together with the white light reflected from the opaque bodies (which is the reason why Venturi abandoned this method), is probably responsible for a number of Prieur's anomalous results. He found that absorption always begins with the complementary color to that of the body

[52] The memoirs were read on 4 March 1805 (see note 40), 28 July, 18 and 25 August, and 22 December 1806; ibid., 3(1804–7):188, 399, 414, 415, 472. The three published memoirs appeared in *Annales de chimie* and translations quickly appeared in *Philosophical Magazine*.

[53] Prieur was proposed for membership in the institute four times between 1796 and 1807; see note 48, and also Bouchard, *Un organisateur de la victoire*, p. 391.

and then proceeds "gradually, and never by jumps." The absorption pattern was irregular: "Sometimes it extends from one side only of the first rays absorbed; sometimes on both sides at the same time; and it there proceeds either by an equal progress from the right and left, or by advancing more rapidly on one of the sides."[54] In fact, Hassenfratz (as Venturi) found that the sequence of absorption is still more irregular, and that it can simultaneously occur at non-contiguous parts of the spectrum.

In his first paper Prieur simply presented a description of the phenomena of absorption and did not touch on the cause of coloration. His second published memoir must have received a very cool reception by the members of the institute. He argued that sunlight consists of only three colors, red, green, and violet, and that all the other spectral colors are composed of these. Through the course of the eighteenth century this idea was repeatedly proposed based on confusion with pigment mixing, where all colors can be produced from a mixture of three primaries, red, yellow, and blue. By 1800, however, all serious physicists, and even most chemists, understood that Newton's analysis of light by refraction showed that it consisted of innumerable colors in a continuous gradation. Prieur based his case on a different ground. He argued that analysis by refraction is incomplete and that absorption can further decompose the spectral colors. He found that yellow light from the solar spectrum becomes green when passed through a green filter and red through a red filter, so that yellow light must be composed of these two colors. Red and green therefore have the same degree of refrangibility and are incapable of being decomposed by the refraction of prisms. Laplace and Berthollet were present at some of these experiments.[55]

Prieur's observations of transparent bodies of all colors showed that they "transmit ultimately only the red, green, or violet rays. The progressive absorption never finishes by any other colours." By analyzing the solar spectrum by means of these colored filters, he concluded that it actually consists of three differently colored spectra that partly overlap each other.[56] Six years earlier Matthew Young had proposed a similar idea with less substantial evidence.[57]

[54] Prieur, "Extract from a memoir entitled 'Considerations on colours' " (1805), p. 295.
[55] Prieur, "On the decomposition of light" (1807), p. 214; original publication 1806. On the eighteenth-century history of the trichromatic spectrum, see Hargreave, "Thomas Young's theory of color vision," Ch. 5, and for Prieur, pp. 225–34.
[56] Prieur, "On the decomposition of light," pp. 167–8.
[57] M. Young, "On the number of the primitive colorific rays in solar light" (1800).

But the idea of a trichromatic solar spectrum based on analysis by absorption became respectable only in the 1830s, when Brewster, a scientist of greater stature, independently reintroduced it and absorption was more widely understood (§9.1). At the conclusion of this paper Prieur explained absorption and coloration in terms of affinities:

If the *affinity* of the body for the rays of light, be such as to absorb some into its substance, it will be coloured and will exert a preferable or stronger action on certain sorts of rays. In a small mass, the body will first absorb these rays to which it has a *preferable affinity*. The mass of the body being gradually increased, the destruction of the rays will proceed by new mixtures, still progressively; the kind least acted upon will remain the last.[58]

In 1807, Prieur finally took up Newton's theory of colored bodies in his third published memoir, but not before reinterpreting the physical cause of diffraction, the colors of thin films, and reflection and refraction. He proposed that all these phenomena have a single cause: Molecules of matter have a double "sphere of activity" that attracts light particles at very small distance and repels them at greater ones. Although Prieur suggested this only as a "simple probability" and recognized that calculations are required to see if it accurately predicts the phenomena, this style of qualitative natural philosophy was no longer practiced by French physicists. Nonetheless, his fundamental insight that diffraction and periodic colors are caused by strictly physical actions of bodies – as they are "determined only by the number or distance of their molecules, without any dependence upon their proper nature" – was crucial for distinguishing Newton's analogical theory of colored bodies from absorption, which is "chemical" or dependent on the nature of the bodies.[59]

Turning now to a direct refutation of Newton's theory, Prieur argues that, unlike thin films, which are one color by reflection and another by transmission, colored transparent bodies are seen by one set of rays in all directions. He then compares observed absorption spectra with those predicted by Newton's theory and concludes, like Hassenfratz (and Venturi), that "the colours resulting from absorption are *sometimes* owing to groups of rays very different from those which thin films can furnish. For instance, the latter never furnish colors

[58] Prieur, "On the decomposition of light," p. 218; "De la décomposition de la lumière," p. 259; I have here and below slightly altered the translation and added italics.

[59] Prieur, "Summary considerations upon variegated colours of bodies" (1807), pp. 335–7.

composed like those of bodies tinged violet by the oxide of manganese, nor like the blue of cobalt or indigo."[60]

Prieur's mode of representing the colors produced by absorption remains unknown because his "table" giving eight particular series or types of coloration to encompass all cases was never published. Without some tabular or graphical representation of absorption, together with a detailed comparison with the predicted color sequences of Newton's theory, it would be extremely difficult for anyone not already familiar with absorption to grasp the significance of Prieur's evidence, and at this time that would still include virtually everyone. In this respect Hassenfratz's memoirs were altogether more important. Prieur also notes that the course or rate of coloration with increasing thickness is entirely different with absorption and thin films and says that this is evident by comparing his (unpublished) tables with Newton's nomograph.[61]

Despite Prieur's advocacy of a three-color theory and his qualitative speculations, and Hassenfratz's superficial attempt to reach a compromise between the two theories at the conclusion of his investigation, the two had, it seems, presented sufficient evidence to show inadequacies in Newton's theory and prompt further research into absorption, if not to persuade its advocates to abandon it. Yet for fifteen years the problem lay fallow. Because the French physicists did not directly respond to these investigations for nearly a decade, we must attempt to uncover their attitude toward the alternative explanation by affinities at this time.

The experimental evidence that Prieur and Hassenfratz brought against Newton's theory was, for various technical reasons, inconclusive and differed on a number of important points.[62] Newton's theory did, however, give an adequate qualitative account of the spectra's appearance, and for most substances (over 75%) it even agreed in detail. It was quite possible, a Laplace or Biot could have reasoned, that only some additional assumption was required to bring the theory into perfect agreement with the phenomenon. Their objection certainly would not have been to affinities as such, for short-

[60] Ibid., p. 337; italics added. The results for manganese oxide and indigo agree with Hassenfratz's; see notes 45, 46.

[61] Ibid., pp. 337–8.

[62] Their spectra were not very well resolved because of the quality of the glass of their prisms and the breadth of the beams of light. Indeed, Prieur's beam was so broad that he observed only boundary colors. On the significance of improved optical glass, see Ch. 9, note 35.

range forces were at the center of Laplace's program to restore Newton's physics. As early as 1783, Laplace expressed his conviction that refraction and diffraction, capillary action, cohesion, crystalline structure, and chemical reactions result from short-range intermolecular forces, and that "one day the laws of affinities would be as perfectly known as those of gravity."[63] It was, however, only in 1805 that he began to execute this program in specific investigations. In Book X of his *Mécanique céleste,* which appeared that year, he produced the most accurate calculation of atmospheric refraction to date by adopting Newton's assumption of short-range forces between the corpuscles of air and light. In further pursuit of his Newtonian program, the next year he published in supplements to the *Mécanique céleste* and in several papers a comprehensive treatment of capillary action, which was similarly based on short-range intermolecular forces. During the following decade Malus, Biot, and Laplace himself applied this approach to double refraction and polarization and carried the emission theory of light to its zenith.[64]

Biot and Arago read a memoir "On the affinities of bodies for light, and particularly on the refractive forces of different gases" in 1806. They carefully measured the mean index of refraction n of various gases and confirmed Newton's physical theory of refraction according to which refractive power $(n^2 - 1)$ is proportional to density. The remainder of their paper was devoted to showing how measurements of the mean index of refraction could be used to determine chemical affinities and chemical composition. They justified their turn to chemistry by reasoning that

> since the action of bodies on light is exercised in a sensible manner only at very small distances, the intensity of this action is necessarily linked to the nature and arrangement of the particles of bodies, that is to say, to their most intimate properties. Thus, the physicist who observes the refractive powers of substances to compare them one to another acts exactly like the chemist who presents successively all the acids to the same base or all the alkalies to the same acid to determine their respective forces and degree of saturation. In our experiments the substance we present to all the bodies is light, and we evaluate their actions upon it by their refractive power, that is, by the increase in *vis viva* that the action of their particles tends to impress upon it.[65]

[63] Laplace, *Théorie du mouvement et da la figure elliptique des planètes,* p. xiii; compare Lavoisier's remark in 1782, quoted at Ch. 6, note 11.
[64] Fox, in Gillispie et al., "Laplace," pp. 358–60; Frankel, "Jean-Baptiste Biot," pp. 173–84.
[65] Biot and Arago, "Mémoire sur les affinités des corps pour la lumière," pp. 326–

This research was suggested by Laplace and Berthollet and carried out under their tutelage.[66] Thus far would French physicists follow Berthollet and the chemists on affinities, but they would not adopt the same approach for chromatic dispersion. In Laplace's treatment of refraction in the *Mécanique céleste,* where he laid down the analytical foundation for the mechanics of corpuscular optics, he devoted one brief paragraph to dispersion.[67] It was unnecessary for atmospheric refraction, so why raise a problem that was not amenable to mathematical physics? Biot and Arago devoted a longer paragraph to dispersion, noting that it is not proportional to the refracting force and varies according to the nature of the body. They indicated their belief that the refractive and dispersive forces have a common origin, as they generally increase and decrease together, and all rays of whatever color in every body obey the same law of refraction. Rather than assuming that the behavior of dispersion demonstrates the chemical nature of light, as the British did, they simply observed that "these phenomena seem to indicate that the molecules of bodies do not act equally on all the molecules of light, and that they attract some with a greater intensity, and others with a lesser intensity."[68]

Because chromatic dispersion resisted both theoretical and experimental efforts to describe it mathematically, French physicists consistently avoided it. Their objection to affinities for selective absorption, as Haüy had already made clear and Biot would reaffirm a decade later, was to their capricious, selective nature. They had to be arbitrarily assumed for each substance and color and were subject to no mathematical laws. Chromatic dispersion behaved the same way, and in the debate over the wave theory of light in the 1830s explaining these two apparently unlawful phenomena became test cases for the wave theory. The affinities responsible for refraction, double refraction, and capillarity, in contrast, were all shown to be subject to the laws of Newtonian mechanics and had been reduced to "rigorous calculations" and "analysis" by "very precise experiment" by Laplace and his followers.[69]

7. The slightly modified translation is from Frankel, "J. B. Biot," p. 62, who discusses this important paper; see also Crosland, *Society of Arceuil*, pp. 258–9; and Fox, *Caloric Theory of Gases*, pp. 197–202. For Newton's theory, see Ch. 6, note 13.

[66] Biot and Arago, "Mémoire sur les affinités," p. 304.

[67] Laplace, *Traité de mécanique céleste* (1805), Bk. X, Ch. I, *Oeuvres complètes*, 4:239.

[68] Biot and Arago, "Mémoire sur les affinités," p. 345. Compare this, for example, with Robison's interpretation; Ch. 7, note 30.

[69] Laplace, *Exposition du système du monde* (1835), Bk. IV, Ch. 18, "De l'attraction mo-

During the time when Hassenfratz and Prieur were delivering their papers, Biot articulated the requirements of the new mathematical and experimental physics. In dedicating his translation of Ernst Gottfried Fischer's *Physique mécanique* to Berthollet, Biot wrote: "People have been content to offer the public a certain series of brilliant experiments rather than try to fix exactly the laws of the phenomena and determine their relationships, which can only be done by mathematical reasonings."[70] French physicists' severe requirement for exact, mathematical description and their commitment to Newton's theory of colored bodies (which did predict so many of the phenomena) hindered them from further investigating selective absorption.

Laplace himself implicitly rejected the affinity theory of absorption by choosing to ignore it. In the greatly expanded chapter on molecular attractions in the third edition of his *Exposition du système du monde* (1808), he did not include selective absorptive affinities with the other known short-range forces.[71] It would have meant introducing a new force whose laws of operation were totally unknown while there was already an explanation for the colors of bodies that was perfectly adequate to explain the phenomenon. Laplace had been present at the readings of all of Prieur's and Hassenfratz's papers between 1805 and 1807, and even at some of Prieur's experiments. In the next edition of the *Système du monde,* in 1813, he still chose not to include selective absorptive affinities. Another factor probably contributed to the lack

léculaire," pp. 405, 418; in this sixth edition the text of this chapter was restored to that of the fourth edition (1813); see Gillispie et al., "Laplace," pp. 388–9.

[70] Fischer, *Physique mécanique* (1806), p. iii; a translation of *Lehrbuch der mecanischen Naturlehre* (Berlin, 1805); the English translation is from Frankel, "J. B. Biot," p. 45. The views of Fischer, who was a professor of mathematics and physics in various institutions in Berlin, reflect the widespread dissemination of the chemistry of light in Germany; see Ch. 6, note 47. In the section on light he briefly set forth Newton's theory of colored bodies and then noted: "However, it presents many obstacles when one wants to generalize the applications of this hypothesis. According to all appearances, there exists a sort of chemical attraction by means of which each body attracts certain constituent principles of light and combines them with it, so that the others can only be reflected according to the laws of the mechanics of light" (*Physique mécanique,* pp. 430–1). Biot's translation of Fischer appeared when he was carrying out his own investigation with Arago of the affinities of light in refraction, and he let Fischer's claim pass without comment. However, in the second edition in 1813 he added a footnote: "This manner of considering the phenomena indeed appears simpler at first glance, but when it is studied thoroughly, it is found to be infinitely less probable than that of Newton, which I hope to prove elsewhere" (p. 447). While he does not explain his reasoning, Biot had now apparently adopted Haüy's position.

[71] Ch. 17 in the third edition; compare note 69. The second edition was published in 1799.

of interest in pursuing the experimental investigation of absorption: In the years immediately following Prieur's and Hassenfratz's papers a series of exciting investigations and discoveries – on polarization, double refraction, and diffraction – that ultimately led to a revolution in optical theory must have seemed far more significant than absorption.[72]

Although French physicists drew a line – mathematization – beyond which they would not follow the chemists, we should not fail to see how much they did adhere to the chemistry of light. Light was one of the new, Lavoisian chemical elements (probably identical to caloric) and, like all other elements, possessed affinities for other substances. The younger generation of Arago, Biot, Fresnel, and Malus were all taught the chemistry of light at the Ecole Polytechnique by such professors of chemistry as Berthollet, Guyton de Morveau, and Fourcroy. It is likely that it was also imbibed in Hassenfratz's physics course. Because of the Laplacian program, the new quantitative physics, and the experimental and quantitative emphasis of Lavoisian chemistry, the French did not reach the speculative heights of British and German natural philosophers on the chemical activity of light. Fresnel alone of this generation rejected the chemistry of light, and this seems to have been responsible for his initial rejection of the emission theory and search for an alternative.[73]

Affinities served as the conceptual foundation for the chemistry of light and one of the bases for the alliance of chemistry and physics. But just as Venturi, Hassenfratz, and Prieur were carrying out their studies of absorption spectra and proposing affinities as an alter-

[72] The views of Antoine Libes, who observed the dispute on colored bodies from the sidelines, show how it had reached a new stalemate in the 1810s. Libes was professor of physics at the College Charlemagne in Paris; he had not adopted the new style of mathematical physics. In 1806, he was won over by Berthollet's arguments against Newton's theory in the second edition of his *Eléments*. Libes considered Berthollet to have countered the theory with "decisive [*décisives*]" experiments; *Nouveau dictionnaire de physique*, "Couleur," 1:275. He incorporated the identical material in his *Traité complet et élémentaire de physique* (1813), 1:139, but now he declared Berthollet's experiments to be only "seductive [*séduisantes*]." In the interim he was brought back to the physicists' camp by Haüy's counterattack, though he does not cite him (pp. 139–40). It is characteristic of the era that he ignores observations of absorption spectra.

[73] Before Malus undertook his investigations of light rays, double refraction, and polarization, he wrote on the chemistry of light, but these speculative elements never appeared in those analytical and experimental works. They did, however, resurface in research on the new "chemical" and "calorific" rays that he began shortly before his death with the chemist Jacques-Etienne Bérard. See Buchwald, *Rise of the Wave Theory*, pp. 25–6, 112–13; and André Chappert, *Etienne Louis Malus (1775–1812)*, pp. 50–6, 237–41.

native to Newton's theory of colored bodies, Berthollet delivered a serious blow to the received doctrine of affinities. In a series of papers in 1801, and then in his *Statique chimique* two years later, he demonstrated the erroneous nature of the assumption made throughout the eighteenth century that the strength of affinities between any two substances was a constant characteristic of the combination.[74] He showed that chemical reactions depended on many factors besides affinities, such as volatility, cohesion, and especially the relative masses of the combining substances. Berthollet did not abandon his belief in the universal action of affinities or short-range forces, as many then thought. Rather, he criticized the idea that they were invariant forces, depending only on the nature of the pairs of substances involved, as well as the methods by which they were determined.

After 1801, chemists no longer unquestioningly accepted affinities, and as they realized the obstacles to determining their absolute values, they gradually abandoned affinity tables. At virtually the same time, Volta discovered the electric battery as a way to generate continuous electric currents, and very quickly chemists thought of affinities as electrical rather than gravitational forces. The entire foundation of Newtonian chemistry was collapsing. The final blow to classical affinity theory came a few years later with Dalton's atomic theory and law of definite proportions.[75] By 1810, the center of chemists' interests was shifting away from the study of mechanisms of reaction, or affinities, to measuring atomic weights and chemical proportions. Consequently, in the period immediately following the observations of Hassenfratz and Prieur, chemists would no longer have been inclined to pursue selective absorption as a means to study elective affinities.

In his vigorous defense of Newton's theory of colored bodies in his *Traité de physique* in 1816, Biot had to confront the alternative interpretation of absorption by elective affinities, however they were understood. What is perhaps most striking about Biot's analysis is the way in which he, as it were, selectively adopted absorption. He introduced selective absorption phenomenologically without affinities, but rejected the investigation of absorption spectra. Newton, too, treated selective absorption phenomenologically, but unlike

[74] See Frederic L. Holmes, "From elective affinities to chemical equilibria."
[75] Trevor H. Levere, *Affinity and Matter;* Thackray, *Atoms and Powers,* pp. 230–3, 252–82.

Newton, who carefully kept the phenomenological and analogical theories apart in separate books of the *Opticks,* Biot treated them together. Consequently, the lack of complete congruence of the two theories became more apparent.

After sketching the elements of Newton's analogical theory, Biot points out that although the theory implies that the reflected and transmitted colors should be complements (as in Newton's archetypal phenomena), this is never strictly observed. To account for the missing colors he introduces the concept of selective absorption. He gives the example of gold foil, which is yellow by reflection and green by transmission, whereas in thin films the complement of a reflected yellow is always blue (the color that Newton in fact saw in gold foil). It must therefore be assumed that in the transmitted blue light, which is a compound of blue, violet, and green, some of the blue and violet rays are absorbed in the substance of the gold foil to produce the observed green.[76] Thus Biot found it necessary to supplement Newton's analogical theory with the additional assumption of selective absorption. Absorption occurs after the colors are separated by reflection and transmission in the corpuscles. Using many of Newton's own observations, Biot then describes the principal feature of absorption, namely, that bodies absorb different colors at different rates, so that as light passes through greater thicknesses successive colors gradually vanish from the transmitted beam. He emphasizes that while the order of absorption is fixed for each substance, it varies for different substances. Like Haüy, Biot altogether ignores Delaval's experiment (cited by both Berthollet and Hassenfratz) that bodies have no distinct power of reflection, so that the reflected and transmitted colors are in most bodies identical.

In a clear but implicit criticism of Hassenfratz's (and Venturi's) method of investigating absorption, Biot describes the "most certain way of proceeding" to determine the reflected and transmitted colors for any substance: Simple spectral colors should be passed successively through various thicknesses of a substance; the quantities of each color reflected and absorbed should be measured "comparatively"; and then by Newton's color-mixing circle the color composed from these simple colors should be calculated. Monochromatic colors must be used, Biot argues, because if the light is even slightly compound, the unequal absorption of different colors will alter the result.[77] Without mentioning any names, Biot has

[76] Biot, *Traité,* 4:127.
[77] Ibid., p. 130.

explicitly rejected the method of absorption spectroscopy, or passing compound light through a substance and afterward decomposing it by a prism. His concern for the purity of the colors was misguided, for his proposed method, which he never actually carried out, would by means of a complex (and inaccurate) calculation ultimately yield the compound color transmitted or reflected by the substance. It is, however, the components or spectra rather than the compounds that are of interest. The spectra, as we have seen with Venturi and Hassenfratz, can then be used to compare the absorption of different substances directly and also to test Newton's analogical theory according to which the components of every order of each color differ, even though the colors compounded of them may be virtually identical.

In preparation for his argument in support of Newton's theory of colored bodies, Biot resolves the theory into three fundamental principles: (1) Matter is hierarchically structured, (2) the refractive power of the highest order of corpuscles is much greater than that of the surrounding medium, and (3) the reflection and transmission of light in these corpuscles follow the same laws as in thin films. Because these three principles are so tied to one another, he declares that if the last one is confirmed, the entire theory will follow. Biot has conveniently ignored the principle of selective absorption that he had just set forth as essential to the theory to account for the failure of the reflected and transmitted rays to be complementary. Having admitted selective absorption as a supplementary principle, he had by Ockham's principle already come perilously close to admitting the rival theory, and there was no need to emphasize that point. Indeed, in the next paragraph he concedes:

> Undoubtedly one could explain, as certain celebrated scholars have attempted, the proper colors of bodies by chemical attractions and repulsions that preferentially determine the absorption or reflection of certain hues. It would then, however, be necessary to attribute to these forces all the variety of effects which occur by fits with such simplicity; that is, it would be necessary to attribute in certain orders of colors an action which extends only to certain luminous particles, and even to a certain definite proportion of these particles; because the colors of natural bodies are never simple, and Newton's theory alone shows why they cannot be otherwise.[78]

Thus, like Haüy before him, Biot argues that to explain the complex mixture of colors for each body the affinities would have to be arbi-

[78] Ibid., p. 131.

trarily assumed: "Nothing proves that the affinity is really capable of producing these choices of luminous particles." Newton's theory, on the contrary, predicts the mixture of colors by the analogy to thin films, which is confirmed by experiment. Moreover, he cleverly argues that whatever be the nature of the substance – water, oil, glass, or air – when it is reduced to a thin film, "the order and succession of colors are always found alike, independently of the affinity."[79] Finally, Biot objects that not only must the affinity be assumed to vary in such a way that it can account for any given color, but the sequence of color changes brought about by any cause "must also faithfully follow the periods assigned by Newton's table; because when a body gradually changes color as a result of any chemical action whatever... the hues through which it passes always follow exactly the order entered in this table and derived from those of [colored] rings."[80] Pursuing the approach that Delaval adopted in his first two papers ("although he did not always properly interpret the examples that he had chosen"),[81] Biot then places the thrust of his case in support of Newton's theory upon showing that color changes always follow the sequence predicted by Newton's table of colors. To consider just his opening instance he not only supports Newton's assignment of the green of plants to the third order but generalizes it. He presents evidence that the succession of colors that the leaves and flowers of plants pass through in the various phases of their growth always obeys the order of Newton's table.

Biot had presented perhaps the strongest case on behalf of Newton's theory of colored bodies that was possible in 1816, but he had accomplished this by totally ignoring the experimental evidence of Delaval (confirmed by Hassenfratz) and by rejecting the method of absorption spectroscopy. Biot's argument against the rival affinity theory was now principally a methodological one. Affinities had to be arbitrarily assumed in order to yield the complex sequences of observed colors, which, Biot still insisted, followed the pattern of Newton's rings. Nonetheless, his charge that affinities are arbitrary is not a frivolous one. When they are invoked as the physical cause of selective absorption, one has not really provided a physical explanation, for whatever the pattern of absorption that is observed in a particular substance, it is explained simply by asserting that that is how affinities behave. Affinities are merely a verbal solution, for there

[79] Ibid. To be sure, absorption is insignificant in such a thin *colorless* film.
[80] Ibid., p. 132.
[81] Ibid., p. 134.

were no theoretical principles governing their behavior. As a chemical explanation, though, affinities (no matter how vaguely they were then envisioned) were more satisfying, insofar as they reduced the interaction of light and matter to ordinary chemistry. Chemists were accustomed to the arbitrary or unique nature of chemical reactions and properties.

Having reduced the choice between the two theories to a methodological one, Biot was not on very firm ground. He had found it necessary to supplement Newton's theory with absorption, though without appealing to affinities; and he had even conceded that other than for its arbitrary nature, the affinity theory could explain the colors of bodies equally well. By the principle of parsimony, why invoke two separate processes to explain coloration, when one alone – selective absorption – could account for the phenomenon? A stalemate had been reached in France, or at least in the institute, with each side adhering to its own theory. Chemists were really little affected by the situation, for their main business with color was to analyze and produce in the laboratory and factory dyes, pigments, and other coloring matters, and they were increasingly successful at it. All that they then required was some heuristic theory, and they had one. Physicists, however, were in a more embarrassing situation, for their explanation of one of the most evident properties of bodies, their color, was in serious doubt.

If Biot expected his arguments to stem the tide of defections from Newton's theory of colored bodies, he was to be disappointed. After about 1815 the French physics community no longer presented a united front behind it. This change is related to the decline of the Laplacian style of physics as well as to their gradual loss of control over the powerful, centralized scientific institutions. In 1815, François-Sulpice Beudant's elementary physics book restricted Newton's theory to a single case of coloration, namely, the colors of thin films, and proposed that all the others could be explained (following "several distinguished physicists") by a "chemical action" of absorption.[82] Beudant was a mineralogist who had attended the Ecole Polytechnique and studied under Haüy. At this time he was professor of physics at Marseilles and would later become professor of mineralogy and physics at the Sorbonne and a member of the academy. His treatment of colored bodies appears to be based upon Hassenfratz's,

[82] Beudant, *Essai d'un cours élémentaire et général des sciences physiques*, pp. 529–31; this went through six editions by 1838. He includes the classic experiment with colored fluids to show that colored bodies do not reflect colored light.

who was his physics teacher at the Ecole Polytechnique, but he altogether ignores absorption spectra (which is probably justified in an elementary text) and focuses on the phenomenological problem of reflected and transmitted colors.

In 1817, Isaac-Bénédict Prevost, a Genevan natural philosopher and naturalist, published a paper on colored bodies in the *Annales de chimie et de physique* in which he similarly rejected Newton's theory based on an experimental study of reflected colors, adopted affinities, and ignored absorption spectra. In some "additions" to his paper, he explained that when he wrote the first part, he had not yet seen Biot's *Traité,* though he was familiar with Haüy's and Beudant's books.[83] Arago, as the new coeditor of the renamed *Annales de chimie,* was responsible for physics contributions and the publication of Prévost's paper. He may have viewed this paper as a way to strike at his archrival, Biot. Arago would later publish papers by Fresnel, Ampère, and Fourier that probably would not have been printed in the Laplacian-dominated *Mémoires de l'académie des sciences.*[84]

By 1823, the anti-Laplacians, led by Arago, were in control of the academy and other leading institutions. "Laplace was reduced to silence," Guglielmo Libri wrote, "M. Biot absented himself from the Institut for several years and M. Arago remained master of the battlefield."[85] Arago was able to weaken the position of Laplace and Biot by utilizing his support of Fresnel and the wave theory of light, which gathered adherents only toward the latter part of the decade. Thus fundamental changes occurred both in physical theory and in the power structure. Since about 1815 Dalton's atomic theory was making inroads in France against Berthollet's affinity theory and becoming widely adopted.[86] In 1822, Fresnel himself vigorously attacked the chemistry of light and its assumption that light was subject to elective affinities (§9.2). These changes were reflected in explanations of coloration in the 1820s. Newton's theory gradually faded from French

[83] Prevost, "Sur le mode d'émission de la lumière qui part des corps colorés"; and "Nouvelles additions au mémoire sur le mode d'émission de la lumière," p. 437. Many years later both Xavier de Maistre and l'abbé Moigno rejected Newton's theory of colored bodies and cited Hassenfratz's papers while still ignoring the evidence of absorption spectra; de Maistre, "Sur la cause des couleurs dans les corps naturels" (1831), p. 18; and Moigno, *Répertoire d'optique moderne* (1847), 2:479–553, on p. 491.

[84] Crosland, *Society of Arcueil,* pp. 404–6; Fox, "Rise and fall of Laplacian physics," pp. 112, 125; and Frankel, "Corpuscular optics and the wave theory of light," p. 172.

[85] Quoted in Frankel, "Corpuscular optics," p. 173, from Libri, "Les sciences en France," *Revue des deux mondes* (1840), p. 797.

[86] Fox, "Rise and fall of Laplacian physics," pp. 115–16.

physics texts, though in 1825, a sketchy account of it could still be found in César-Mansuète Despretz's textbook.[87] No alternative explanation had yet been adopted in France. To reach the denouement of our story, we must return to England.

[87] Despretz, *Traité élémentaire de physique,* p. 620. A third edition of Haüy's textbook, with its defense of Newton's theory, had been published in 1821. Neither Eugene Péclet in his *Cours de physique* (1823) nor Claude Pouillet in his *Elemens de physique expérimentale* (1827–30) treats the colors of bodies. Pouillet was a protégé of Biot and worked on diffraction experiments with him in 1816, but by 1830 he supported Fresnel's wave theory; Frankel, "Corpuscular optics," p. 173. Jacques Babinet and Charles Bailly, *Résumé complet de la physique des corps imponderable* (1825), 1:200–3, give a phenomenological account of coloration, but do not mention Newton's theory. By 1832, Péclet, apparently adopting Herschel's approach, explained coloration by absorption in his *Traité élémentaire de physique* (1832), 2:346–9.

9

ABSORPTION SPECTROSCOPY IN BRITAIN: 1822–1833

One year after Biot's *Traité* appeared, Joseph von Fraunhofer published a paper whose disarming title, "On the refractive and dispersive power of different species of glass, in reference to the improvement of achromatic telescopes," scarcely indicated that it was destined to alter the debate on the nature of coloration – indeed, to alter physics itself. In this paper Fraunhofer described more than six hundred dark lines running across the solar spectrum that since have been named after him, and also the corresponding bright emission lines in the spectra of such luminous sources as flames and electric sparks. Although Fraunhofer himself did not investigate selective absorption, once his remarkable discovery became widely known – in about 1823, when it was translated into French and English – the study of emission and absorption spectra, which were assumed from the first to be related, became a central aspect of physics.[1] This was particularly true in Britain, for in 1822 two of its leading physicists, John Herschel and David Brewster, published studies of the spectra produced by selective absorption. Like Fraunhofer, who made his discovery while searching for a source of monochromatic light to determine indices of refraction with great precision, Herschel and Brewster were led to study selective absorption in their quest for monochromatic illumination. Initially, their concern was applied, fashioning monochromatic sources for experimental use, and descriptive, characterizing the phenomenon. Only at the end of the twenties

[1] Fraunhofer, "On the refractive and dispersive power of different species of glass." Brewster translated the paper from a French translation in *Annales de chimie* and not directly from the original German paper of 1817. On the diffusion of Fraunhofer's discovery, see the fine paper by Frank A.J.L. James, "The discovery of line spectra," pp. 59, 69, n. 39.

and the beginning of the thirties did they publicly turn to examining the causes of absorption, to relating it to the cause of the colors of bodies, and to refuting Newton's theory.

As the wave theory of light gained increasing acceptance at the end of the 1820s, so the concept of chemical affinities between light particles and corpuscles of matter lost its explanatory power, except to Brewster, who never abandoned the emission theory. The need to develop new explanatory models of absorption to replace the long-prevailing affinities became a central issue in the British debate over the wave theory. The level and scale of this debate reflect the maturity of the new, quantitative physics in Britain. In this same period Brewster's use of selective absorption to carry out a further analysis of the solar spectrum into just three colors attracted substantial interest. By the early thirties, then, absorption and coloration were an integral part of optics, and physicists no longer ignored absorption spectra as they had thirty years earlier. One of the more striking aspects of this story is the sharp discontinuity of the work of the 1820s and 1830s with that at the turn of the century. The only notable link between the two periods was Biot's *Traité*, which vigorously defended Newton's theory and denied the significance of absorption spectra for the study of coloration.

9.1. BREWSTER AND HERSCHEL

In July 1822, Herschel saw an abstract of a paper delivered the preceding April by Brewster on the development of a monochromatic lamp, the spectral composition of flames, and absorption in colored media.[2] Herschel had worked on these same problems in 1819. He immediately wrote up his earlier work and sent the paper to Brewster, who recognized that Herschel's approach was different and generously offered to publish it together with his own paper.[3] The initial purpose of each was to create monochromatic light sources, Brewster

[2] "Proceedings of the Royal Society of Edinburgh," *Edinburgh Philosophical Journal* 7(1822):163. The paper was published as "Description of a monochromatic lamp for microscopical purposes" (1823); in 1824, it was translated in *Annalen der Physik*.

[3] Brewster to Herschel, 12 October 1822, Royal Society MS HS 4.256. Herschel's letter to Brewster was published as the paper "On the absorption of light by coloured media" (1823). Herschel anonymously reviewed his own and Brewster's papers in "Analysis of scientific books and memoirs" (1825); Herschel is identified as the author in the *Royal Society Catalogue of Scientific Papers*, vol. 3 (London, 1869), p. 324, no. 28. See M. A. Sutton, "Sir John Herschel and the development of spectroscopy in Britain."

for use with microscopes to avoid chromatic aberration and Herschel for the study of polarization in crystals, and they succeeded in this aim. Struck by the irregularly compound nature of the light transmitted by their colored filters, they both began to examine the process of selective absorption itself. They utilized experimental methods quite similar to those of their Continental predecessors. This similarity should come as no surprise, for the use of prismatic spectra to study the composition of colors was an obvious technique to anyone schooled in Newton's theory of color. The only prior observation of absorption spectra they cite is the 1803 paper "Experiments and calculations relative to physical optics" by Thomas Young, whose works were now being reread in the wake of Fresnel's revival of the wave theory.[4] Herschel even believed that Young was the first to have observed absorption spectra. It is evident, though, that Brewster and Herschel discovered absorption spectra as a consequence of their research programs, and not from prior knowledge of Young's or any other predecessor's work. Their observations at this time, like their experimental techniques, did not go beyond their predecessors', although by the end of the decade their spectra were better resolved because of the higher quality glass of their prisms. The most significant way in which their work differed from earlier research was in its representation.

Brewster, the experimentalist, adopted the simplest, clearest, and most extravagant mode of representation (Plate 1, p.335): colored illustrations of spectra with absorbed colors blackened out. This undoubtedly would have made the total absence of light and color immediately evident to those like Brougham, who simply could not imagine the appearance of absorption spectra.[5] Herschel's innovation in presenting his observations was more significant and the way of the mathematical physicist: He constructed spectral transmittance curves (Figure 9.1). In his graphs the horizontal axis represents the colors of the spectrum and the vertical one the quantity of light transmitted at a given thickness. Figures 2, 3, and 4, for example, represent a blue glass whose thickness is first doubled and then increased many times, whereas Figure 5 combines three successive thicknesses of a glass that passes from a "rich ruddy purple" at curve 1 to a "fine

[4] Brewster, "Description of a monochromatic lamp," p. 439; and Herschel, "On the absorption of light" (1823), p. 446; for Young's observation see Ch. 7, note 37.

[5] See Ch. 8, note 4. Young, in his *Course of Lectures,* had published a colored illustration of the spectrum of a flame that showed dark bands, but this was an emission spectrum; see Ch. 7, note 36.

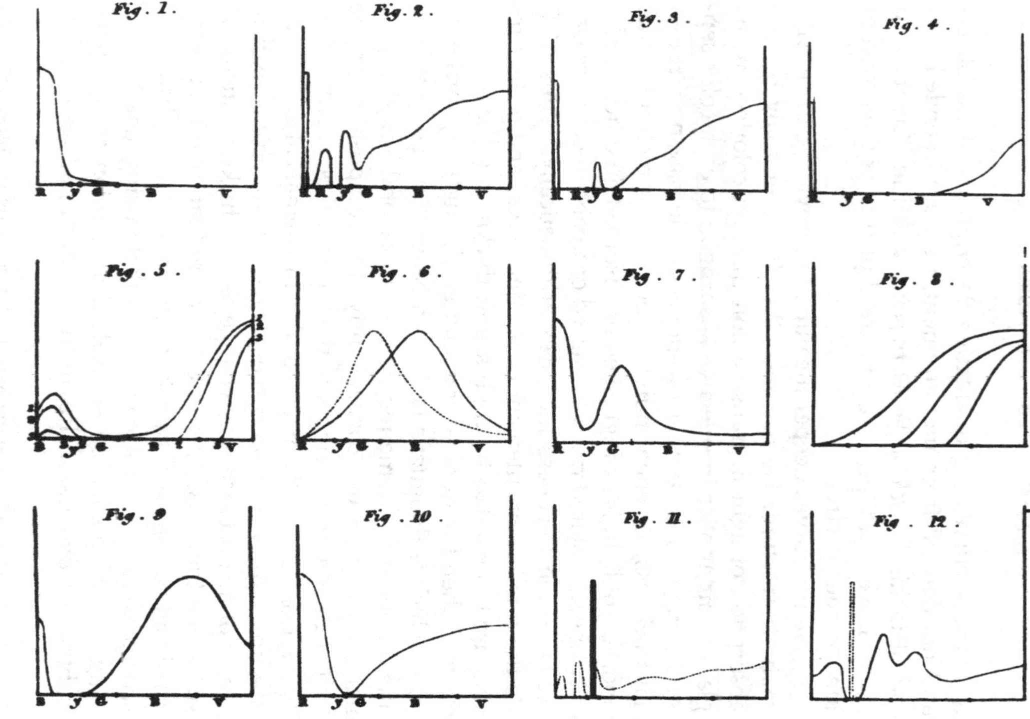

Figure 9.1. Herschel's transmission/absorption curves. Figure 1 represents a ruby red glass; Figures 2 to 4 increasing thicknesses of a blue glass; Figure 5 three thicknesses of a purple glass; Figures 6 and 7 two different types of green media, one with a single maximum and another with two; Figures 8 and 9 another type of blue with only one maximum, in contrast to those in Figures 2 to 4; and Figure 10 a type of purple different from that in Figure 5. Figures 11 and 12 are flame spectra. (From *Transactions of the Royal Society of Edinburgh* 9[1823], Plate XXVIII.)

violet" at 3. These curves are qualitative descriptions of his observations and not derived from measurement; they are intended – as he later put it in his *Encyclopaedia Metropolitana* article "Light" – to represent, "as it were, a type, or geometrical picture of the action of the medium on the spectrum."[6]

Herschel gave the curves deeper meaning by providing them with a mathematical structure. He assumed that Bouguer's law for the general absorption of light in transmission through colorless media such as glass or air applies in selective absorption to each color separately. According to this "simplest hypothesis," in passing through each equal thickness, or equal number, of colorific molecules, the same proportion of light is extinguished or absorbed; in other words, the quantity of each colored ray transmitted or unabsorbed decreases geometrically as the thickness increases arithmetically.[7] Thus, if C represents the number or intensity of incident rays, and y the proportion that are transmitted through a unit thickness ("the index of transparency"), then the number transmitted through any thickness t will be Cy^t, which is Bouguer's law.[8] In a beam of white light let C now be the number of extreme red rays, C' those of the next degree of refrangibility, and so on; consequently, the composition of the transmitted beam after traversing any thickness will be expressed by $Cy^t + C'y'^t + C''y''^t + \dots$. By means of this mathematical expression and his absorption/transmission curves, Herschel was able to explain that the color of a substance may change as the thickness increases (dichroism) if the transmission curve has more than one maximum, for eventually all but one will vanish. In the treatment of absorption in his *Encyclopaedia* article he also explained that mixing together and superposing colored media are equivalent absorption phenomena, that is, subtractive processes.[9] Recall that Venturi had also applied

[6] Herschel, "Light," §490, p. 431. The mathematical description of absorption in the *Encyclopaedia* (§§488–91) is fundamentally the same as that in the paper ("On the absorption of light," pp. 447–8).

[7] Herschel's recognition that altering the thickness of the absorbing medium or the concentration of its colorific molecules are equivalent is known as Beer's law; "On the absorption of light" (1823), p. 447. Venturi also recognized this law in carrying out his experiments on colored fluids. See August Beer, "Bestimmung der Absorption des rothen Lichts in farbigen Flussigkeiten" (1852), p. 83.

[8] Pierre Bouguer first formulated this fundamental law of photometry in his *Essai d'optique sur la gradation de la lumière* (Paris, 1729), and later in his *Traité d'optique sur la gradation de la lumière* (Paris, 1760), Bk. III, Sect. I, Art. II; *Optical Treatise on the Gradation of Light*, p. 155. Herschel cites Bouguer earlier, "Light," §40, p. 347.

[9] "In combinations of media, the ray finally transmitted is the residuum of the action

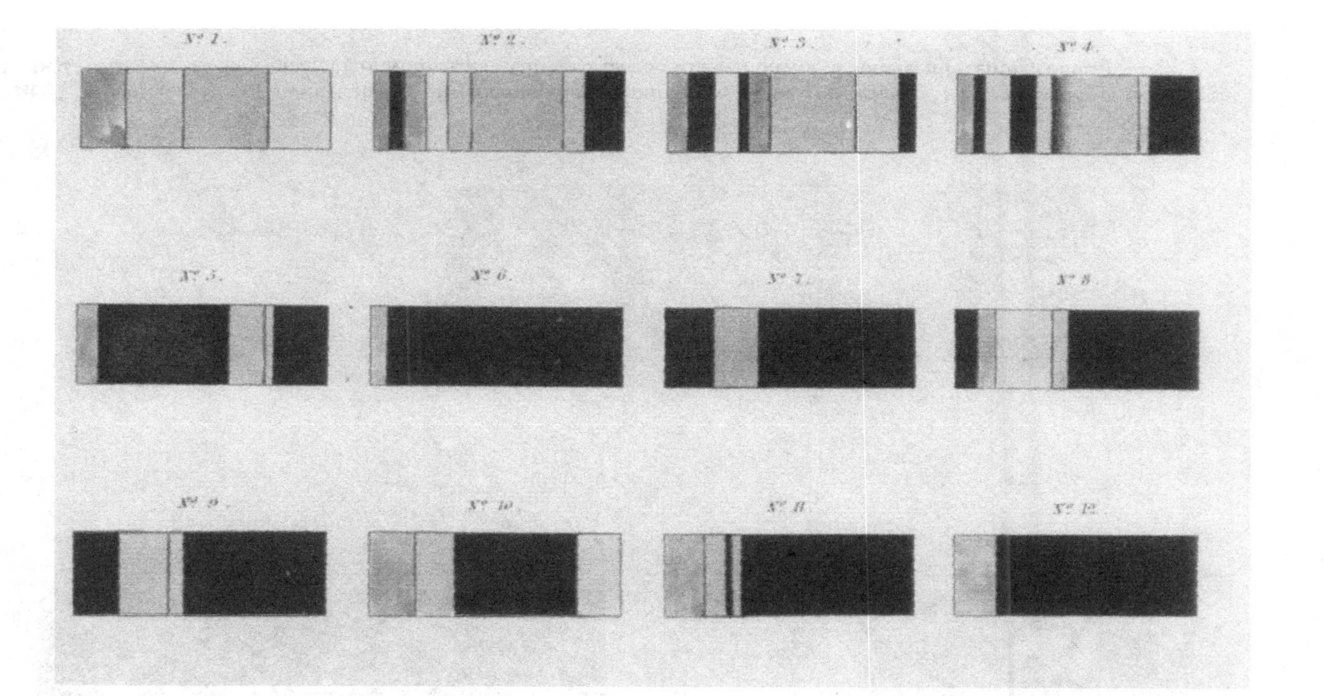

Plate 1. Brewster's rendering of absorption spectra. Figures 2 and 3 represent spectra of different kinds of blue glass, and 4 and 5 the two combined at increasing thicknesses. Figures 6 and 7 represent spectra of a sky-blue paste and a very thick green glass. Figures 8 and 9 are spectra at increasing thicknesses of copper sulfate, which is bluish green. Figures 10, 11, and 12 represent the spectra of a thick yellowish red glass, a solution of lake (which is red), and a thick red glass. (From *Transactions of the Royal Society of Edinburgh* 9[1823], Plate XXVII.)

These plates are available in colour as a download from www.cambridge.org/9780521117593

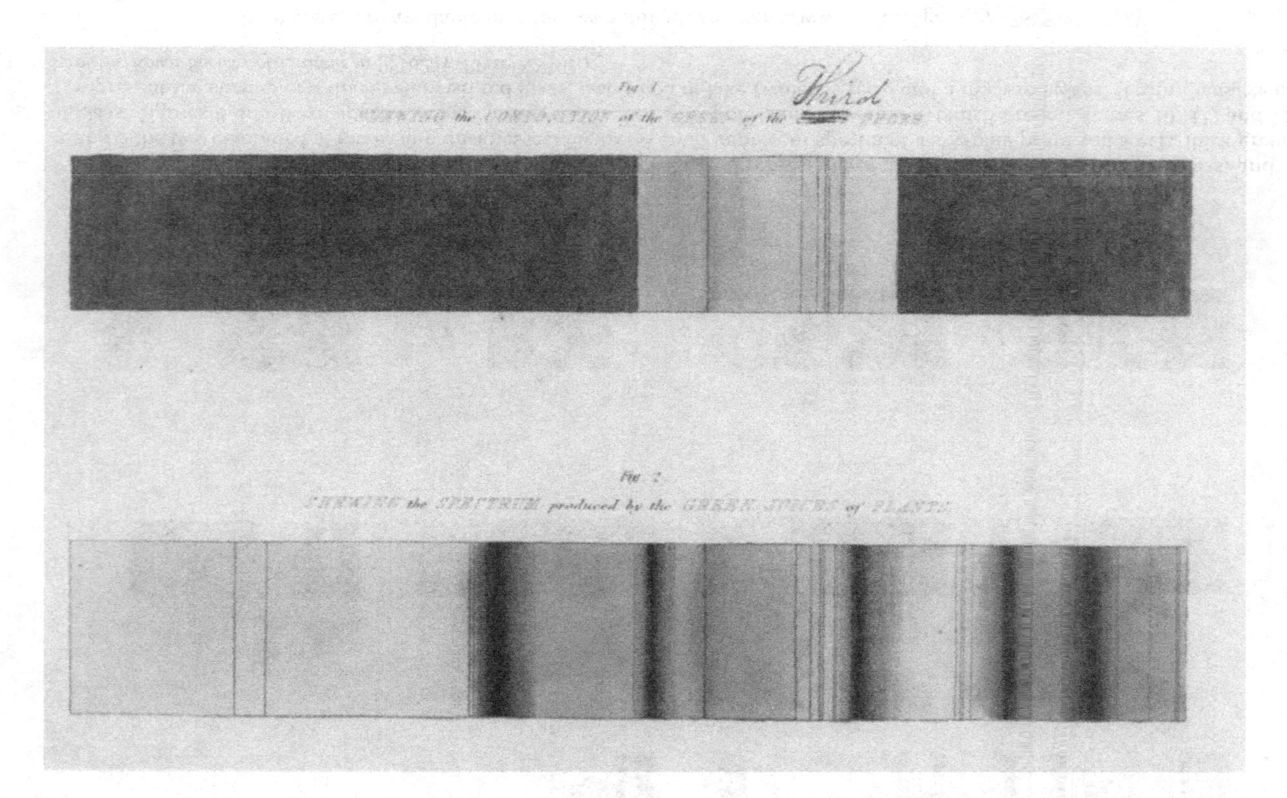

Plate 2. Brewster's test of Newton's theory of colored bodies. (From a copy of the *Transactions of the Royal Society of Edinburgh* 12[1834], Plate XII, at the University of Minnesota, with a correction in an anonymous nineteenth-century hand.)

his understanding of absorption to the same problem. At least as important as these explanations was Herschel's success in bringing some semblance of physical law and order to a phenomenon that was so capricious and irregular. This waywardness had been the French physicists' principal objection to the theory of affinities, and the issue would erupt once again in Britain in the early thirties.

Why did Brewster and Herschel stop to investigate absorption spectra, when no one else had taken notice of them since the efflorescence at the turn of the century? Inasmuch as they learned of Fraunhofer's work only after their papers appeared, it was not yet because of an awareness of the broader significance of spectral studies. They were motivated by their belief that the optical investigation of matter – the absorption spectra – would yield knowledge about its chemical composition. Since the work of Haüy at the end of the eighteenth century, crystallography and mineralogy were making extraordinary progress.[10] In the first decades of the nineteenth century chemists such as Wollaston and Eilhard Mitscherlich investigated the relation of chemical composition to crystal structure. The study of the optical properties of crystals (double refraction and polarization and its chromatic effects) was at the center of the optical revolution of the early nineteenth century, and new discoveries on these properties were pouring forth at a steady rate from Malus, Biot, Fresnel, Arago, and Brewster himself.

Brewster was actively involved in the 1810s in attempting to use measurements of such optical properties as refractive and dispersive

of each. . . . it is indifferent in what order the media are placed. They may therefore be mixed, unless a chemical action take place" ("Light," §502, p. 433). Helmholtz never claimed that his famous explanation of the subtractive nature of pigment mixing by selective absorption was original; see Ch. 8, note 17. Indeed, in 1852, he stated: "The theory of pigmentary colours here presented is simply derived from the generally recognized laws of physics" ("On the theory of compound colors," p. 529; "Ueber die Theorie der zusammengesetzten Farben," p. 60). His source for this "generally recognized" knowledge was most likely Herschel's article "Light," for in his *Treatise on Physiological Optics* he refers to Herschel's article a number of times. In particular, in the discussion of compound colors (2:164) he cites §507, p. 434, on color blindness, which is in the same section as absorption. This is not to deny that Helmholtz recognized a very significant physical principle, namely, that the addition of spectral colors is not equivalent to the mixing of pigments and, in particular, that mixing yellow and blue spectral colors yields white rather than green, as with pigments. Neither Herschel nor Venturi drew these conclusions; see "Light," §516, p. 436. The *Treatise on Physiological Optics* is a translation of Helmholtz's *Handbuch der physiologischen Optik*, 3rd ed. (1909); the first edition of vol. 2 appeared in 1860 and already contained the reference to Herschel.

[10] Burke, *Origin of the Science of Crystals*, Ch. 5.

power, birefringence, and polarization to determine the chemical composition of substances. In 1814, he declared:

> In examining the changes which light undergoes during its passage through transparent bodies, we not only receive information respecting the properties of that mysterious agent; but we are in some measure made acquainted with the composition of the substances themselves, and with the manner in which their ingredients are combined. The optical phenomena, therefore, which bodies exhibit in their action upon light, are so many tests, to which the philosopher may have recourse, either in supplying the place of chemical analysis, or in correcting and modifying its results. A difference in the optical properties of two bodies, is generally an infallible indication of a difference in their elementary principles. . . .
>
> It is highly desirable, therefore, that the Chemical Philosopher would avail himself more frequently of the agencies of light in the prosecution of his inquiries. . . . An opportunity is thus lost . . . of contributing most essentially to the progress of optical knowledge. It is by the alliance, indeed, of Chemistry with Optics, that great revolutions are yet to be effected in Physics; and the time is probably not very distant, when, by their united efforts, we shall be able to develope those mysterious relations among the elementary principles of matter.[11]

Just the year before his first publication on absorption, Brewster published a paper whose title, "On the connexion between the optical structure and chemical composition of minerals," clearly stated the nature of his program. Given his interests at that time, there is every reason to accept his claim eleven years later that "the first and the principal object of my inquiries" on absorption was "the discovery of a general principle of chemical analysis, in which simple and compound bodies might be characterised by their action on definite parts of the spectrum."[12]

Herschel, as Gregory Good has shown, was similarly concerned to apply optical properties to mineralogical and chemical analysis.[13] Indeed, Brewster was a major influence in forming his interests. After his graduation from the University of Cambridge in 1813, Herschel attended to the study of chemistry and mineralogy and soon became expert in them. Herschel was of course familiar with the British tradition in philosophical chemistry and the chemistry of light, which he first learned from his father, but from the first he rejected that tradition and its belief that light was a chemical substance. In 1814,

[11] Brewster, "On the optical properties of sulphuret of carbon," pp. 285–6.
[12] Brewster, "Observations on the lines of the solar spectrum" (1834), p. 519.
[13] Good, "J.F.W. Herschel's optical researches," pp. 142–62.

he wrote his close friend Charles Babbage that "the world is run mad. . . . And ancient chaos will resume his sway and the chemical dance will be danced no longer to the music of the spheres, but to the wreck of matter and the crush of glass."[14] In 1818, he started to investigate chromatic polarization in crystals in order to understand crystal structure. It was his efforts to produce monochromatic light for this research that led him to his study of absorption.

Neither Brewster nor Herschel in 1822 raised the issue of the general process of coloration or Newton's theory. Both knew Newton's theory, were aware of Biot's defense of it some six years earlier, and undoubtedly accepted it, for they still did five years later.[15] They also knew that selective absorption (as an affinity) had been proposed as an explanation of coloration from Biot's discussion and rejection of it. After developing monochromatic sources, their sole aim in their 1822 papers was to characterize selective absorption and emission spectra from flames. Rather than comparing the observed spectra with those predicted by Newton's theory, as their predecessors had done, they compared the spectra to one another in order to discover types or patterns of absorption in different media. In finding some rough general patterns for absorption in media of each of the principal colors, they again succeeded in bringing a semblance of order to this irregular phenomenon.

In 1833, Brewster reflected on the direction of his research since the publication of his initial experiments on selective absorption: "These experiments were continued at irregular intervals, with the view [i] of obtaining distinguishing characters of coloured media, [ii] of investigating the cause of the colours of natural bodies, and [iii] of examining more correctly the phenomena of the overlapping colours of equal refrangibility . . . ," or decomposing the spectrum by means of absorption in addition to refraction; but "[iv] the discovery of a

[14] Ibid., p. 56; quoted from Royal Society MS HS.20.17.

[15] Herschel cited the optical portion of Biot's *Traité* (though not on colored bodies) in a paper read in December 1819, "On the action of crystallized bodies on homogeneous light," p. 86. In 1822, Brewster approvingly cites Biot on colored bodies in his *Edinburgh Encyclopaedia* article "Optics," p. 623. As will be shown shortly, Herschel was still not prepared to abandon Newton's theory altogether in his article "Light," and as late as 1829, Brewster still enthusiastically supported it. That Herschel and Brewster were not the only physicists who still supported Newton's theory of colored bodies is apparent from Leopoldo Nobili's "Memoir on colors in general" (1837), which was originally published in the *Bibliothèque universelle* (1830). Working within a framework of Newton's theory, Nobili, professor of physics at the University of Florence, proposed some modifications and new interpretations of it based on his experiments with thin films deposited by electroplating.

general principle of chemical analysis . . . still remained to be investigated."[16] Three more areas of research should be added to Brewster's, (v) the bearing of absorption on the truth of the rival wave and emission theories of light, (vi) the physical cause of absorption and emission, and (vii) the production of monochromatic light by colored filters. Herschel, too, was engaged in all these pursuits, as well as (viii) explaining color mixing. I will, of course, focus on the second area, but will also touch on the others insofar as they illuminate it and show the changing conception of absorption. The new physical orientation of research on absorption should already be apparent. The principal item on the old agenda was (ii) and its interpretation by chemical affinities, with Prieur including (iii) and Venturi (vii) and (viii). When chemistry did impinge on the new research program, as in (iv) and (vi), it was treated from a physical perspective.

In his article "Light," which Herschel completed in December 1827 for the *Encyclopaedia Metropolitana* (1828), he switched his allegiance to the wave theory of light. The "Essay on Light," as Herschel referred to it, was actually a major treatise on optics. As the earliest comprehensive account of the wave theory, it was quite influential and soon was translated into French (1829–33) and German (1831).[17] The section on absorption built on his earlier paper, and though it was more sophisticated and comprehensive, it was not fundamentally different. What concerns us, however, are the two pages on the colors of natural bodies at the conclusion of the "Essay." Herschel opens with a marvelously apt description of Newton's theory as one "of extraordinary boldness and subtilty, in which great difficulties are eluded by elegant refinements, and the appeal to our ignorance on some points is so dexterously backed by the weight of our knowledge on others, as to silence, if not refute, objections which at first sight appear conclusive against it." In his account of Newton's theory Herschel focused on two points that had troubled Newton himself; namely, that an independent power of absorption must be attributed to the primordials, for otherwise all bodies would glow like luminous ones, and that in transparent colored bodies the reflected and transparent beams are in general not complementary. In fact, Herschel produced an experiment to show that "Transparent coloured media . . . have *no* reflected colour." He filled a vessel, which was blackened on the interior to stifle the reflection

[16] Brewster, "Observations on the lines of the solar spectrum," p. 519.

[17] On the publication date of the article, see Cantor, *Optics after Newton*, p. 162. The French translation by P. F. Werhulst and A. Quetelet, *Traité de la lumière*, was particularly influential because of the large *Supplément* by Quetelet in Vol. 2.

of transmitted light, with various colored liquids and reflected light from them. Like all those who carried out the experiment since Kepler, Herschel found that no colored light is reflected from the fluids. Completely unaware of Delaval's work, he declared:

> We are not aware that the objection so put has been sufficiently considered, or even propounded. To us its weight appears considerable, and we cannot but believe that some other cause besides mere internal reflexions must interfere to prevent the complementary colour from reaching the eye; and that absorption, with its kindred phenomenon, or rather its extreme case, opacity, is not satisfactorily accounted for in this theory, but must rather be admitted as (at present,) an ultimate fact, of which the cause is yet to seek.[18]

Earlier, in his account of absorption, Herschel had argued that all colored bodies are seen by rays that have actually penetrated their substance; if the rays were reflected from the surface, all bodies in sunlight would appear white, and their hue could not be altered by varying their thickness.[19] He now considers the colors of bodies to fall into two classes: "true" colors, which arise from the absorption of rays in the interior of bodies, and which include such common colors as dyes, pigments, and flowers; and "false" or "superficial," which arise from interference and satisfy Newton's theory, and which include such variable colors as feathers, scratched surfaces, and oxidized metals. That Herschel should think that any of this was original shows the complete discontinuity with the earlier tradition of Newton's critics. The continuity, rather, is with the quantitative physicists, with Biot's *Traité* serving as the link. Herschel, though, went farther than Biot in his criticism of Newton's theory. Whereas Biot was content to have two causes of coloration operating simultaneously, thin films and absorption, Herschel rejected such a multiplicity of causes and eliminated the former. Yet he was still not prepared to abandon Newton's theory altogether. He held that there were "a variety of cases to which the Newtonian doctrine strictly applies, for there is no denying that cases of colour, not *merely* superficial, do occur in which the Newtonian doctrine, to say the least, is highly probable." To support this contention he played the old game of invoking sequences of color changes that follow Newton's scale of colors for thin films; the sequences were drawn from Biot and Brewster. He also maintained that the blue of the sky "is, no doubt, a blue of the first order," just as Newton had claimed.[20]

[18] Herschel, "Light," §§1134, 1142, pp. 580–1.
[19] Ibid., §485, p. 430.
[20] Ibid., §1143, p. 581.

Herschel's criticism of Newton's analogical theory and substitution of absorption as the principal cause of coloration was carried out solely on phenomenological grounds; he did not utilize absorption spectra to put the predictions of Newton's theory to a detailed test. Only in 1833, when he delivered a paper on absorption and the wave theory at a meeting of the British Association for the Advancement of Science, did he fully reject Newton's theory. The earlier playful admiration of Newton's theory is gone and replaced by a sense of irritation at his own earlier credulity:

The speculations of Newton on the colours of natural bodies, however ingenious and elegant, can hardly, in the present state of our knowledge, be regarded as more than a premature generalization; and they have had the natural effect of such generalizations, when specious in themselves and supported by a weight of authority admitting for the time of no appeal, in repressing curiosity, by rendering further inquiry apparently superfluous, and turning attention into unproductive channels. I have shown, I think satisfactorily, however, in my Essay on Light, that the applicability of the analogy of the colours of thin plates to those of natural bodies is limited to a comparatively narrow range, while the phaenomena of absorption, to which I consider the great majority of natural colours to be referrible, have always appeared to me to constitute a branch of photology *sui generis* to be studied in itself by the way of inductive inquiry.[21]

No new evidence is presented to justify this more assertive rejection, but there can be little doubt that it is based on the case against Newton's theory that Brewster had been presenting for the preceding two years. In turn, I am confident that it was Herschel's earlier critique of Newton's theory in his "Essay on Light" that prompted Brewster to undertake his more comprehensive attack, for he still supported the theory in a work published in 1829. As in 1822, their work must be considered in tandem, even though for various reasons they were no longer on friendly terms and mentioned each other as little as possible in their publications.

In the article "Optics" that Brewster wrote for the *Edinburgh Encyclopaedia* in 1822, he showed himself to be an admiring supporter of Newton's theory. He declared it to be "one of the finest results" of Newton's investigations of thin films and "a striking specimen of the

[21] Herschel, "On the absorption of light by coloured media, viewed in connexion with the undulatory theory," p. 401; the following year translations appeared in *Annalen der Physik* and *Correspondance mathématique et physique*. See also the abstract with the same title in *Report of the British Association* (1833) [1834]:373–4.

sagacity and address of that extraordinary man."[22] His presentation of the theory is rather pedestrian other than for a number of sequences of color changes from recent chemistry (later cited by Herschel) that he adduces in its behalf. It is to be expected that his brief concluding bibliography should include Biot's *Traité*, but it is surprising to find Delaval's third paper cited without any discussion (and presumably recognition) of its subversive conclusions. As late as 1829, in his essay "Optics," in the *Library of Useful Knowledge*, Brewster was still espousing Newton's theory, and he added still more sequences of color changes in its support.[23]

In 1831, Brewster announced in his *Treatise of Optics*, a popular work forming part of the *Cabinet Cyclopaedia*, that Newton's theory "will not stand a rigorous examination under the lights of modern science" and referred to his forthcoming *Life of Sir Isaac Newton* (1831) for a more detailed critique.[24] Apparently without any sense of irony, Brewster chose to present one of his earliest refutations of Newton's theory of colored bodies in his biography of the great man. Likewise, he chose to include in the *Life of Newton* an early account of his "New analysis of solar light" by which he rejected Newton's analysis of sunlight by refraction – the foundation of Newton's theory of color – as incomplete. According to Brewster, sunlight consists of only three primary colors – red, yellow, and blue – and rays of each of these colors possess the entire range of degrees of refrangibility (Figure 9.2).[25] Therefore, when they are refracted by a prism, the three sorts of rays are dispersed through the entire spectrum and by their mixture

[22] Brewster, "Optics" (1830), p. 621. Although the complete set was published in 1830, beginning in 1808 individual volumes appeared as they were completed. Geoffrey Cantor has kindly informed me that the Leeds University Library copy of Vol. 15, Pt. 2, which contains Brewster's article, bears the date 1822 on its cover; personal communication, 20 December 1988.

[23] [Brewster], "Optics" (1829), Pt. IX, pp. 59–65; each part is separately paginated. This was probably composed a year or so before publication, that is, before or contemporaneous with Herschel's *Encyclopaedia* article. Without a salaried professional position, Brewster was compelled to support himself by writing and editing, whence his popular works and texts are so numerous; see A. D. Morrison-Low and J.R.R. Christie, "*Martyr of Science*," which also contains a useful bibliography of Brewster's enormous output that identifies him (nos. 457–8) as the author of this contribution.

[24] Brewster, *Treatise on Optics*, pp. 280, 283. This elementary treatise went through at least eighteen editions and printings and was especially popular in the United States. It was also translated into French (1833) and German (1835).

[25] Brewster, *The Life of Sir Isaac Newton*, pp. 70–3; and also in his *Treatise on Optics*, pp. 71–4. *The Life of Newton* went through one revision and fifteen editions by 1875 and was translated into German in 1833. The greatly expanded *Memoirs of the Life, Writings, and Discoveries of Sir Isaac Newton* (1855), Ch. 8, also contains his refutation of Newton's theory.

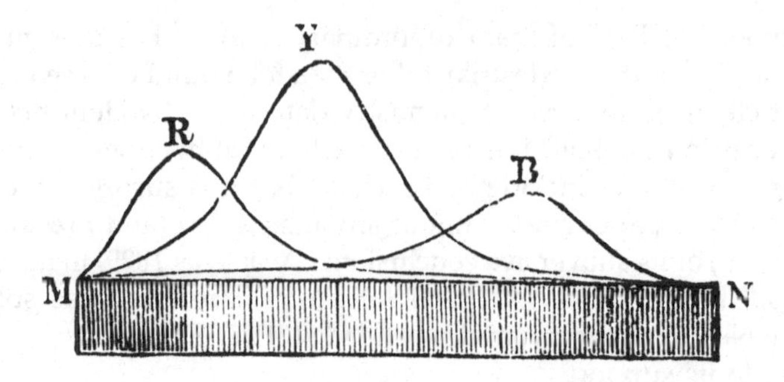

Figure 9.2. According to Brewster's analysis of the solar spectrum by absorption, the prismatic spectrum *MN* is composed of overlapping red, yellow, and blue spectra, each of which consists of rays distributed through all degrees of refrangibility from *M* to *N* but with a maximum of intensity at *R*, *Y*, and *B*, respectively. (From *A Treatise on Optics* [1831], p. 74.)

produce all the colors of the Newtonian spectrum. Because at each point of the spectrum there are rays of each color with the same degree of refrangibility, they cannot be further separated by a prism: Whence Newton was mistaken to conclude that to each degree of refrangibility there corresponds a unique color and that sunlight consists of innumerable colors. The spectrum can be further decomposed by absorption, "which science had hitherto scarcely recognized."[26] To consider a single example of Brewster's method, if green light becomes yellow when passed through a yellow filter (which absorbs blue and transmits yellow), then analysis by absorption shows that green light consists of a mixture of yellow and blue.[27]

Brewster's theory enjoyed a mode of popularity through the mid-1840s, and Herschel tentatively endorsed it in his "Essay on Light" (though not afterward). After sustaining attacks by Airy, Whewell, Macedonio Melloni, and Draper in 1847, the theory was decisively refuted by Helmholtz in 1852, when he demonstrated numerous sources of observational error, especially from extraneous scattered and reflected light.[28] I am not concerned with Brewster's "new anal-

[26] Brewster, "On a new analysis of solar light," p. 124; read 21 March 1831; a slightly abridged version with the same title was published in *Edinburgh Journal of Science*, new ser., 5(1831):197–206.

[27] Brewster, *Life of Sir Isaac Newton*, pp. 71–2; *Treatise on Optics*, p. 72.

[28] Herschel, "Light," §506, p. 434. See Hargreave, "Thomas Young's theory of color vision," Ch. 8; Sherman, *Colour Vision in the Nineteenth Century*, Chs. 2, 3; and Ch. 8, note 55.

ysis" in itself, but only to show the central place that absorption had assumed in physics. His "new analysis" also once again indicates the virtually complete break with the earlier experimental tradition, for Brewster was totally unaware of the earlier versions of analysis by absorption proposed by Matthew Young and Prieur (§8.2).

Returning to Brewster's refutation of Newton's theory of colored bodies, I will focus on his scientific paper, "On the colours of natural bodies," which was read to the Royal Society of Edinburgh in December 1832, rather than on the popularizations.[29] In the *Life of Newton,* Brewster does present some historical analysis of Newton's theory, which is omitted from the paper, but at this stage of our story that need not concern us. We should note, though, that in the *Life of Newton* he recounts that he repeated and confirmed Delaval's experiment.[30] I suspect that he omitted this from the paper, because Herschel had effectively refuted the phenomenological theory in his "Essay on Light" even though he was unaware of Delaval's prior work. Brewster, therefore, felt free to confine himself to the analogical theory.

"On the colours of natural bodies" is devoted exclusively to an experimental refutation of the heart of Newton's theory, that the colors of bodies are caused by their component corpuscles in exactly the same way as the colors of thin films. Brewster takes the green of plants as his "*experimentum crucis,*" as it is the most prevalent color in nature and Newton gave the most detailed description of it.[31] His method of comparing observed absorption spectra of transparent colored substances with the spectra predicted by Newton's theory is essentially the same as that of Venturi, Prieur, and especially Hassenfratz, but his experimental results are dramatically more decisive. Newton held that the green of plants is that of the third order, which consists principally of green together with some yellow and blue. Whether the spectrum is deduced from Newton's nomograph or ac-

[29] Brewster, "On the colours of natural bodies" (1834); and more widely disseminated in 1836 in *Philosophical Magazine* and *Annalen der Physik.* The date "Read 22d December 1833" in the published paper is evidently an error, for the summary of the proceedings of the Royal Society of Edinburgh that was sent by the society's secretary to *L'Institut: journal des sociétés savantes,* 2(1834):223, is given as 3 December 1832. This is consonant with Brewster having read a version of the paper at the British Association meeting in June 1832. Inasmuch as he cites conflicts between observed absorption spectra and those predicted by Newton's theory in the *Life of Sir Isaac Newton,* pp. 86–7, these experiments were underway in 1831 or earlier.

[30] Brewster, *Life of Sir Isaac Newton,* p. 86.

[31] Brewster, "On the colours of natural bodies" (1834), p. 539. Newton's descriptions are quoted at Ch. 2, note 111, and Ch. 3, note 64.

tually observed with a prism – a step not carried out by Brewster's predecessors – the third order green of Newton's rings is found to consist of all of the green space with the contiguous least refrangible part of the blue space and the most refrangible yellow part (Plate 2, Figure 1, p. 336). To observe the spectrum of the green of plants, he used an alcoholic solution of "green colouring matter" (chlorophyll) extracted from the leaves of common laurel and nineteen other plants, all of which gave identical results. Absorption begins with three red bands; at a greater thickness the blue and violet are attacked; and then absorption begins at the middle of the green. "The general effect" of absorption is "rudely" illustrated by his figure:

> The simple inspection of Figure 2 [Plate 2] affords the most unquestionable evidence that the green colour under consideration is neither the green of the third order nor of any other order of periodical colours, and that in its general character, as well as in the character of its component bands, it has no resemblance whatever to any colour produced by the action of thin plates.[32]

Brewster then turned to the sequences of color changes that had been a major bulwark of the theory. They had even been invoked earlier by Brewster himself in his articles in the *Edinburgh Encyclopaedia* and *Library of Useful Knowledge*. In particular, Newton claimed that when leaves wither they pass to yellows, oranges, and reds of the third order. Biot not only had confirmed this result but extended it by showing that all color changes in vegetation follow the order predicted by Newton's theory. Brewster observed solutions extracted from withered yellow leaves of common laurel and from deep black-violet leaves of privet:

> Now, in both these experiments, the action of the colouring matter of the decayed leaves is decidedly different from that of the green juice, and there is no appearance whatever of the tints having any such relation as that which subsists between adjacent colours of the same order.
>
> From facts like these, which it is impossible to misinterpret, we are entitled to conclude, that the green colour of plants, whether we examine it in its original verdure, or in its decaying tints, has no relation to the colours of thin plates.[33]

After examining (but not reporting) the spectra of nearly 150 additional colored substances, including dyes, colored glasses and minerals, and the blue of the sky (which was invoked by Herschel),

[32] Ibid., p. 541.
[33] Ibid., p. 544. For Biot, see Ch. 8, note 81.

Brewster confirmed his conclusion: The colors of these bodies are not the same as those of thin films. And he ended the paper by explaining the "true cause" of the colors of natural bodies:

When light enters any body, and is either reflected or transmitted to the eye, a certain portion of it, of various refrangibilities, is lost within the body; and the colour of the body, which evidently arises from the loss of part of the intromitted light, is that which is composed of all the rays which are not lost; or, what is the same thing, the colour of the body is that which, when combined with that of all the rays which are lost, compose the original light.[34]

At this point we can consider Newton's theory finally to have been refuted and replaced.

In contrast to his predecessors, Brewster devoted his paper to a single task: refuting Newton's theory of colored bodies, which he dispatched decisively and dramatically. He did not burden his reader with testing rival physical and chemical explanations or with dozens upon dozens of trials, many of which were inconclusive. Nor was there any ambiguity to his spectral comparisons. In prior tests about three quarters of the spectra agreed with Newton's theory. Now all spectra were found to be in disagreement. The principal reason for this striking difference was, I believe, the quality of Brewster's glass prisms. In the late 1820s, Brewster and Herschel went to the effort of procuring prisms from Fraunhofer and his associates in Munich. Fraunhofer had succeeded in greatly improving the quality of optical glass, which for centuries had burdened opticians with striae, bubbles, and inhomogeneities. To appreciate the improvement, we need only note that Brewster's spectra (Plate 2) revealed the Fraunhofer lines, unlike any earlier absorption spectra, including his own and Herschel's from 1822.[35]

[34] Ibid., p. 545. In 1837, Brewster once again turned to refuting Newton's theory, "On the connexion between the phenomena of the absorption of light, and the colors of thin plates"; reprinted in *Philosophical Magazine* in 1842. His aim in this paper was to disprove Newton's claim that if a thin film were shattered, the heap of fragments would have the same color as the single film; see Ch. 3, note 54. There is no need to devote attention to it, as devotees of Newton's theory may turn directly to the original paper.

[35] M. A. Sutton, "Spectroscopy and the chemists," pp. 19–20. In "On the colours of natural bodies" (1834), which was read in December 1832, Brewster states that he used a "*fine* prism" (p. 540; italics added). In his paper "Observations on the lines of the solar spectrum," read a few months later on 15 April 1833, he says he used "two *very fine* rock-salt prisms, executed by myself" and "a *fine* plate glass prism, executed by Fraunhofer, and which I owe to the kindness of Mr Talbot" (p. 527; italics added). It thus seems that in his refutation of Newton's theory he observed the absorption spectra with Fraunhofer's "fine" prism. In his 1833 paper, p. 523,

Brewster's and Herschel's investigations must be considered together: Brewster primarily for his experimental investigation of absorption and refutation of Newton's analogical theory, and Herschel for his more theoretical accounts of absorption and coloration and his refutation of the phenomenological theory. Their work was widely known beyond Great Britain through numerous translations.[36] Why was it not ignored, as had happened to the work at the turn of the century? Selective absorption had emerged out of their experimental research, together with Fraunhofer's crucial discoveries, as an independent *physical* phenomenon; and there was a sufficiently large community of quantitative physicists – no longer natural philosophers – to pursue it, especially in Britain and Germany. The study of absorption now demanded precision measurements and specially designed and manufactured apparatus. Their discovery fell on prepared ground, for the scientific community had already been exposed to the general concept of selective absorption through the earlier research tradition in color – even Biot had treated it. In their initial work neither confounded his experimental descriptions with speculation as to the physical (or chemical) cause of absorption or engaged in polemics. Chemically conceived affinities for light were gone, though Brewster used them as a heuristic in the early 1830s.[37] Moreover, they had no opposition other than themselves, that is, they were physicists, not chemists, attempting to convince physicists. Chemists were now not even involved in the study of absorption; it belonged to physical optics. Nor can we ignore the fact that the scientific stature of Brewster and Herschel, as well as Fraunhofer, who died in 1826, was greater than that of their predecessors. Finally, in 1832, Brewster precipitated a debate in Britain on the physical cause of absorption and whether it could be explained in the wave theory. This exchange, carried out largely at British Association meetings and in *Philosophical Magazine*, made absorption a subject of broad interest to the scientific community.[38]

Brewster still lamented "the difficulty of procuring out of the mass of glass to be employed, prisms sufficiently pure to show such narrow lines as E, or the two which constitute D."

[36] For example, Peclet's account of absorption in his textbook *Traité élémentaire de physique* (1832) appears to derive from Herschel's; see Ch. 8, note 87. In his *Répertoire d'optique* (1847), 2:490, Moigno summarizes Brewster's paper.

[37] Only in the penultimate sentence of his refutation of Newton's theory did Brewster mention a "specific affinity" as the cause of absorption, but he immediately added that it was not "rigorously demonstrated"; "On the colours of natural bodies" (1834), p. 545.

[38] See James, "The debate on the nature of the absorption of light."

This debate on absorption brings to the fore the fundamental change in physical models of the interaction of light and matter caused by the adoption of the wave theory. Thomas Young was unable to accept the idea of elective affinities for light, inasmuch as an attraction between corpuscles of matter and light waves was incomprehensible (§7.2). The rise of the wave theory entailed the rejection of the chemistry of light and its affinities. It also required the development of alternative conceptions of the interaction of light and matter based on the transfer of motion or energy rather than matter. These changes are exemplified by the views of Brewster, who never abandoned the emission theory and affinities, and Herschel, who adapted acoustic analogies in his reply to Brewster's challenge.

Though Brewster did not publicly state his ideas on the cause of absorption until 1831, they are such a common feature of the Newtonian traditions we have followed that it is highly probable that he long held them. In the popular *Life of Newton* he argued that the heating and chemical effects of light could not be due to its "mere mechanical action," or to vibrations of the aether, but must arise from its "combination" with bodies: "That the light detained within bodies has been stopped by the attractive force of the particles seems to be highly probable, and the mind will not feel any repugnance to admit that the particles of all bodies . . . have a *specific affinity* for the particles of light." Brewster applied this concept to coloration by selective absorption, where "the particles of the body have exercised a *specific attraction*" on particular colors in the incident light. "In compound bodies, like some of the artificial glasses," he explained, "the particles will attract and detain rays of light of different colours, as may be seen by analyzing the transmitted light with a prism, which will exhibit a spectrum deprived of all the rays which have been detained."[39] Thus Brewster was willing to adopt what Biot and Laplace had rejected when confronted with the evidence and arguments of Hassenfratz and Prieur. He introduced specific absorptive affinities, which were exercised selectively on different colored rays by different chemical elements. This was a natural extension of the approach that Biot and Arago had adopted in 1806, when they attempted to use (nonspecific) refractive affinities to analyze the chemical composition of gases.[40] Specific affinities served Brewster only as a heuristic in his physical reasoning and experiments, and they were soon abandoned when they conflicted with observation.

[39] Brewster, *Life of Sir Isaac Newton*, pp. 93, 94; italics added.
[40] See Ch. 8, note 65.

In April 1833, Brewster read a paper, "Observations on the lines of the solar spectrum," in which he gave a historical account of his research on absorption since his 1822 paper. "The first and principal object of my inquiries," he related, was "the discovery of a general principal of chemical analysis, in which simple and compound bodies might be characterised by their action on definite parts of the spectrum."[41] In 1826, W. H. Fox Talbot had proposed such a program of spectroscopic chemical analysis with emission spectra, and two years later Herschel made the same suggestion in his "Essay on Light."[42] There is, however, no reason to doubt Brewster's independence for the related absorption spectra, for the idea of specific attractions between particular substances and particular colors was the very essence of the Newtonian affinity tradition.[43] Continuing the narration of his investigations, Brewster related that from his observations of the spectra transmitted through numerous colored media he conjectured that

> as some of these bodies attacked the spectrum at *two, three, four,* and even *five* or more points at once, it became probable that the number and intensity of such actions depended on the number and nature of the elements which entered into the composition of the body, or, what is nearly the same thing, that it was the sum of all the separate actions of such elements; and hence the next step in the inquiry was, to determine the action of elementary bodies on the solar spectrum.[44]

This hypothesis was "completely destroyed" when he observed the spectrum of nitrous acid gas (NO_2) and found *thousands* of lines and not the few, corresponding to the number of elements, that he expected. As James has shown, all attempts at chemical spectroscopic analysis before the work of Gustav Kirchoff and Robert Bunsen in 1860 similarly failed, but my concern is, rather, with Brewster's use of affinity models.[45]

Brewster's final effort to apply this model was in seeking a common cause for the line spectra of gases and the band spectra of colored liquids and solids:

> From the various experiments which I had made on the absorptive action of coloured media, I was led to a general principle, which, in that stage of

[41] Brewster, "Observations on the lines of the solar spectrum," p. 519.

[42] Talbot, "Some experiments on coloured flames"; Herschel, "Light," §524, p. 438.

[43] James, "Debate on the nature of absorption," p. 340, implicitly rejects Brewster's claim that he developed the idea of chemical analysis in the 1820s, but he does not recognize how Brewster did arrive at this idea. It was unquestionably by affinities, and the only open issue is whether he had this concept in the 1820s.

[44] Brewster, "Observations on the lines of the solar spectrum," p. 520.

[45] James, "Debate on the nature of absorption"; and "Creation of a Victorian myth."

the inquiry, appeared to possess considerable importance. The points of maximum absorption exhibited a distinct coincidence with some of the principal dark lines in the solar spectrum, and thus indicated that these lines marked, as it were, weak points of the spectrum, on which the elements of material bodies, *whether they existed in the solar atmosphere or in coloured solids and fluids*, exercised a particular influence.

This idea that the corpuscles of matter possess a greater specific affinity for rays of certain degrees of refrangibility also went up in smoke with his observation: "The phenomena of ordinary absorption could not be identified with those of the definite actions by which the solar lines are produced."[46] Brewster's unsuccessful attempts to explain absorption by affinities represents the end of a long optical tradition.[47]

At the second meeting of the British Association, in June 1832, Brewster delivered a "Report on the recent progress of science" and was able to draw on some of these observations, which were already underway. At the end of his "Report" he turned to absorption ("One of the finest fields of optical inquiry, and one almost untrodden") and laid down a challenge to the wave theory presented by his observation of upward of a thousand lines in nitrous acid gas:

On the Newtonian hypothesis of emission. . . . When a beam of white light is transmitted through a certain thickness of a particular gas, *a thousand* different portions of that beam are stopped in their passage, in consequence of a specific action exerted upon them by the material atoms of the gas. . . . Such a specific affinity between definite atoms and definite rays, though we do not understand its nature, is yet perfectly conceivable.

Of course, this is "perfectly conceivable" only to a supporter of the emission theory. In the wave theory, Brewster continued, the same phenomenon would have to be interpreted differently: "A thousand different waves or rays of light of different velocities or refrangibilities, are incapable of propagating undulations through the aether of a transparent gas, while all waves or rays of intermediate velocities and refrangibilities are freely transmitted through the same medium." Thus red light with wavelengths of 250 and 252 millionth of an inch freely pass through this gas, whereas another red ray 251 millionth of an inch is entirely stopped. "There is no fact," Brewster insisted,

[46] Brewster, "Observations on the lines of the solar spectrum," pp. 524–5; italics added.
[47] In his paper "Sur la cause des couleurs" (1831), de Maistre also still used elective affinities to explain coloration. He rejected Newton's theory by studying reflection; see Ch. 8, note 83.

"analogous to this in the phaenomena of sound, and I can form no conception of a simple elastic medium so modified by the particles of the body which contains it, as to make such an extraordinary selection of the undulations which it stops or transmits."[48]

Within two years, Brewster's challenge drew replies from Airy, Herschel, Humphrey Lloyd, and Whewell (twice), and absorption became a central element in the debate over the acceptance of the wave theory.[49] Because I am concerned only with the new physical conception of absorption that replaced affinities, Herschel's reply, which was delivered at the British Association meeting the following year, will suffice to illustrate this change. He first rejected the explanation in the emission theory – without mentioning Brewster by name here or anywhere else in the paper – as one that simply "appeals to our ignorance," for we have no knowledge of what happens to the matter of light extinguished in bodies. In the wave theory, on the contrary, one only has to account for a loss of motion or decay of vibrations, which is well understood. Herschel's objective was not to offer a particular explanation of absorption but only to show that one is possible in the wave theory:

> Now it is sufficient for our present purpose if, without pretending to analyse the actual structure of any optical medium, we can indicate structures and combinations in which air, in lieu of the aether, is the undulating medium. ... For that which experiment, or theory ... shows to be possible in the case of musical sounds, will hardly be denied to have its analogue or representative among the phaenomena of colour, when referred to the vibrations of an aether.[50]

Herschel provided an example of a compound vibrating system that is incapable of transmitting a given frequency: Imagine a tube (Figure 9.3) having two branches *BCD* and *bcd* that differ in length by half of a given wavelength. If a note of that particular wavelength is sounded at *A*, the wave will divide at *B* and be reunited at *D*, where the two waves will destructively interfere. If a medium were composed of such tubes, it would be incapable of transmitting certain

[48] Brewster, "Report on the recent progress of optics," pp. 319, 321. He published similar objections in his "Observations on the absorption of specific rays" (1833).
[49] Airy, "Remarks on Sir David Brewster's paper" (1833); Herschel, "On the absorption of light"; Lloyd, "Report on the progress and present state of physical optics" (1834), pp. 317–23; and Whewell, "Presidential address" (1833), pp. xvi–xvii, and "Suggestions respecting Sir John Herschel's remarks" (1834). See also Talbot, "On the nature of light" (1835).
[50] Herschel, "On the absorption of light" (1833), pp. 402, 405.

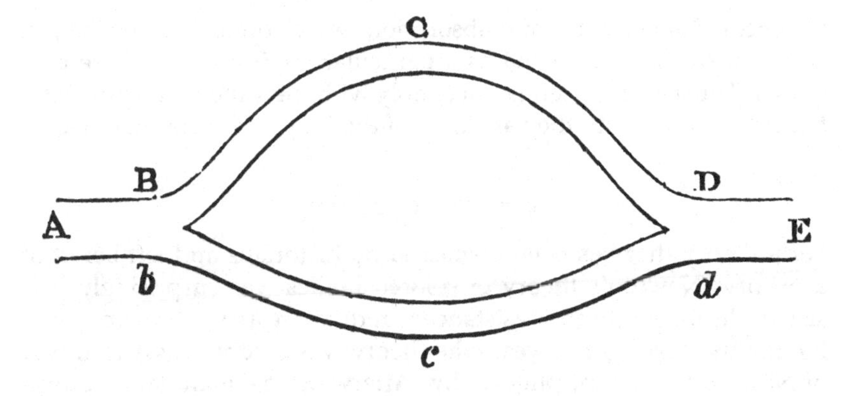

Figure 9.3. Herschel's acoustic analog to line absorption. (From *Philosophical Magazine*, ser. 3, 3[1833], p. 405.)

frequencies and consequently is an analog to line absorption.[51] To account for band and emission spectra, he imagined tuning forks with circular disks fixed to one prong. Herschel's application of interference and forced vibrations to the propagation (or lack of propagation) of motion or energy is characteristic of the new models for absorption and, indeed, of the new optics. The following year Fabian Jakob von Wrede proposed a model in which the incident rays of light interfered with rays partially reflected from the atoms of matter, and this even allowed the successful calculation of absorption lines for certain media.[52]

The study of absorption was now an integral part of physical optics, and its later history became largely that of spectroscopy. The nature of selective absorption and its application to the colors of natural bodies became, as it were, absorbed into optics.[53] The colors of natural bodies, though, was no longer a separate subject with its own unified theory as it was for Newton and his predecessors. As the numerous physical processes involved in coloration, such as scattering (opa-

[51] Ibid., pp. 405–6. On Herschel's models see David B. Wilson, "The reception of the wave theory of light by Cambridge physicists," pp. 174–81; and James, "Debate on the nature of absorption," pp. 350–2.

[52] Wrede, "Attempt to explain the absorption of light according to the undulatory theory" (1837); published originally in Swedish in 1834. See James, "Debate on the nature of absorption," pp. 354–60.

[53] See Baden Powell, *A Short Elementary Treatise on Experimental Optics* (1833), pp. 116–19; F.W.G. Radicke, *Handbuch der Optik* (1839), 2:188; and Emil Verdet, *Leçons d'optique physique*, 5:149–51, lectures delivered at the Sorbonne in 1865–6.

lescence), fluorescence, and absorption, were sorted out, the subject was broken up and treated as a particular application of those processes. Thus our story ends not simply with the vanquishing of Newton's theory but with the vanishing of the very subject of that theory.

9.2. CONCLUSION

For a theory that has been dismissed by historians and left bereft of a history, Newton's theory of colored bodies was surprisingly persistent. Its longevity and resistance to refutation are readily explained. For its first seventy-five years the theory was accepted as a vital part of Newtonian natural philosophy. Afterward, to quantitative physicists, then its sole supporters, it was valued for its comprehensiveness, coherence with optical theory, and above all for its specific predictions of the composition of the colors of bodies. A truly impressive number – to us, a surprising number – of observed sequences of color changes in nature and chemical reactions did in fact appear to follow those predicted by Newton's table of colors. Moreover, the general appearance of absorption spectra, once these were observed, was found to be very much like that derived from Newton's nomograph. Consequently, even when Hassenfratz, for example, claimed to have refuted Newton's theory by comparison with absorption spectra, his results could rather have been interpreted as a confirmation in twenty out of twenty-six cases. Agreement in over three quarters of the cases could reasonably be seen as an indication only that some modification of the theory was required. Delaval's refutation of the phenomenological theory by demonstrating that bodies do not reflect colored light might have been expected to be more damaging, for it had none of the observational ambiguities that absorption spectra initially possessed. Though his experiment was periodically cited or independently repeated, until Herschel forcefully indicated its implications for the analogical theory, the intimate relation of the two theories must be judged to have been not at all obvious. It eluded even such a discerning scientist as Young.

A more complex explanation is required to understand the almost total lack of continuity between spectroscopic investigations of selective absorption (and refutations of Newton's theory of colored bodies) at the turn of the century and similar activity twenty-five years later in Britain. So complete is the break with the earlier Continental tradition of Dutour, Venturi, Hassenfratz, and Prieur that there is no trace of it in histories of spectroscopy from Heinrich Kayser's substantial historical introduction to his *Handbuch der Spectroscopie* at the beginning of this century through many more recent works.[54] Wollaston and Young alone are mentioned because Brewster and Herschel had cited them.

To understand this discontinuity, we must recognize that research before and after Fraunhofer was carried out within two distinct traditions. The earlier tradition of absorption spectroscopy was inextricably connected with the problem of colored bodies and associated with the chemistry of light and its goal of wresting the theory of coloration from physics. Venturi, Hassenfratz, and Prieur based their spectroscopic investigations upon the physico-chemical model of selective affinities. Dutour and Young, who did not utilize affinities, were not investigating absorption at all and did not use that concept, although we may recognize their work as absorption spectroscopy. Dutour aimed to demonstrate that the colors of bodies are compound; and Young was simply attempting to repeat Wollaston's observations when he first viewed absorption spectra, and he interpreted them in terms of Newton's theory of colored bodies, even when he grew disenchanted with it. The discontinuity with the earlier research tradition is undoubtedly partly due to the simple vagaries of history. Had Venturi's research been known, or had Hassenfratz's series of papers been translated into English and German (as much of his other work was), or had Biot added citations to his discussion of absorption in his *Traité*, the transition between the two traditions would probably not have been so abrupt. This problem was exacerbated by the vast amount of fundamental optical research that was produced between 1800 and 1825 and had engulfed the earlier literature on colored bodies, which had lost its urgency. When absorption spectra were rediscovered as a physical phenomenon in a different cultural and disciplinary context in the 1820s, there was a vast gap with the earlier literature. But to gain a deeper understanding of the discontinuity, let us look more carefully at the various communities involved in our history.

Quantitative physicists resisted the spectroscopic investigations of absorption by Venturi, Hassenfratz, and Prieur. Brougham only partly understood the phenomenon of selective absorption, but he grasped enough of it to object to the chemical interpretation, in particular, that so much light could be retained in bodies. Haüy and Biot likewise

[54] Kayser, *Handbuch der Spectroscopie*, Vol. 1, Ch. 1; William McGucken, *Nineteenth-Century Spectroscopy;* and the papers by James and Sutton.

objected to the chemistry of absorption, but their complaint was with the arbitrary nature of affinities, which obeyed no laws. The selective affinities between light and bodies, in their judgment, were like chemical reactions in which each substance behaved in its own way, and not the physicists' orderly, predictable phenomenon. In the later research tradition in absorption and spectroscopy, quantitative physics successfully recovered the study of absorption from the conceptual framework of the chemistry of light, which had actually yielded it by default, for after 1808, it stimulated no further spectroscopic studies. This tradition began with the investigations of Fraunhofer, Brewster, and Herschel, which were almost completely unrelated to the earlier one. Affinities (which were now a dying concept) played little role in the later tradition, and coloration only entered after nearly a decade of research on absorption. This trio succeeded in convincing the community of physicists that absorption was an independent physical process by treating it phenomenologically, without the chemists' selective affinities. Because the problem of the highly irregular and unlawlike nature of absorption was a persistent one for physicists, the phenomenological approach was essential to convince them that absorption with all its vagaries was a real physical process.[55] An explanation of the colors of bodies emerged as an application of the concept of absorption and was not central to the tradition, as it had been earlier.

To explain why those physicists and chemists who adopted the chemistry of light did not pursue coloration and selective absorption by spectroscopic means at the turn of the century a number of distinct factors must be taken into account. Physicists who accepted a chemical interpretation of light, such as John Leslie, had effectively relinquished the study of coloration to chemists. Berthollet and Bancroft had turned the study of coloration into a branch of chemistry, concerned, for example, with determining the role of oxygen in coloration and with isolating and analyzing coloring principles. Despite the al-

[55] This was not such a simple matter as it may seem. When Brewster stirred up the debate on absorption in 1832, leading members of the physics community – Powell, Lloyd, Talbot, and Whewell, in addition to Herschel and Brewster – agreed that it was a significant phenomenon. Airy, however, obstinately insisted that absorption "is a sort of extraneous interruption which either leaves the ordinary laws in full vigour, or wholly destroys, not the laws, but that which is the subject of the laws. Reflexion, refraction, interference, double refraction, polarization, go on with absorption just as if there were no such thing in nature. . . . the suppositions (whatever they are,) that are to account for absorption are necessary only now and then" ("Remarks on Sir David Brewster's paper," p. 423).

liance of chemistry and physics in this period, the two fields were still distinct, and this sort of research was a task for chemists, not physicists. Physicists – again, consider Leslie – were concerned with the optical rather than the strictly chemical aspect of light and turned to the more exciting calorific and chemical radiation (infrared and ultraviolet), which had just been discovered by William Herschel and Johann Wilhelm Ritter in 1800 and 1801.[56] With absorption spectroscopy then so closely tied to the problem of coloration, it would not have appeared to have broader implications. These physicists (and chemists) accepted these experiments as a refutation of Newton's theory of colored bodies and selective affinities as an alternative explanation of coloration, but what more was there to investigate – patterns of absorption or its relation to chemical composition? The goal of systematically using optical techniques to study the properties of matter did not get seriously underway until the following decade, and, in any case, this approach was primarily developed by the very quantitative physicists who supported Newton's theory.

If physicists in the chemical tradition had effectively abandoned the study of absorption spectra and coloration, why did chemists not pursue it? In the first place, after Berthollet's critique of the received concept of absolute affinities in the opening years of the nineteenth century, studying elective affinities by means of light would not have seemed very fruitful. Even if a chemist wished to pursue such a line of research, a more sophisticated theoretical underpinning than physics could yet (or long after) provide was required. No one, for example, could explain the relation (if any) between absorptive and refractive affinities. In the second and more important place, just as physicists did not for the most part study chemical reactions at the bench, so chemists did not wield prisms and examine spectra. Except for Prieur, those who observed absorption spectra before the 1820s were either physicists or closely connected to physics: Dutour was an experimental physicist in the style of Nollet; Young and Venturi were physicists; and though Hassenfratz was primarily a chemist, he was also the professor of physics at the Ecole Polytechnique. As an amateur scientist and engineer, Prieur was an anomaly and would best be classified as a natural philosopher, which then had little place in French science.

A recent debate over the history of spectroscopy between 1820 and

[56] Leslie, in fact, rejected Herschel's claim to have discovered a new sort of radiation and carried out experiments to disprove it; see E. S. Cornell, "The radiant heat spectrum from Herschel to Melloni," pp. 126–7.

1860 serves to show how foreign an instrument the prism was to the chemist and also how much the context of absorption spectroscopy had changed in the twenty years from 1805 to 1825. The debate centers about the creation of spectrochemical analysis by Bunsen and Kirchoff in 1860 (yet a third spectroscopic tradition) and the nature of spectroscopy in the preceding four decades. Recognizing the existence of a continuous tradition of absorption spectroscopy since Fraunhofer's discovery, historians, most recently Sutton, have asked why spectrochemical analysis was not established earlier. James replied that spectroscopy and the study of absorption in that period were part of physics, not chemistry; consequently, to pose such a question is misguided, for it interprets the past in terms of later developments. The resolution of their differences need not concern us but, rather, their points of agreement.

Sutton opens his paper "Spectroscopy and the chemists" by quoting Stokes's observation in 1885 that "it is remarkable for how long chemists neglected the precious means of discrimination lying at their very hands in the use of the prism."[57] He then proceeds to show that before 1859 only two chemists, William Allen Miller in 1845 and John Hall Gladstone in 1857, utilized a prism to provide spectra for chemical analysis. James accepts this point and goes even farther: "There is no puzzle in the fact that chemists did not use spectra if it is remembered that it was really only physicists who were interested in spectra." Those who did (unsuccessfully) attempt chemical analysis by spectroscopy, Brewster, Herschel, and Talbot, he argues, were not chemists. He also argues that "the absorption of light was a problem in physics, not a problem in chemistry."[58] To appreciate how radically the study of absorption had altered within twenty years, we need only observe that to describe the period prior to 1820 properly we have to invert James's assessment – namely, the absorption of light was then a problem in chemistry or, rather, the chemistry of light, and not physics. Yet his claim that the use of spectra belonged to physics rather than chemistry may be confidently extended to optical instrumentation in general and to the period before 1820. We must, of course, recognize that disciplinary borders were never so impermeable that no chemist ever used a prism (as Scheele, Senebier, and Wollaston, for example, did) or a polarimeter (as Pasteur), or that no

[57] Sutton, "Spectroscopy and the chemists," p. 16; from Stokes, *On Light as a Means of Investigation* (London, 1885), p. 35.

[58] James, "The creation of a Victorian myth," pp. 4, 6. See in reply Sutton, "Spectroscopy, historiography, and myth"; and James "Spectrochemistry and myth."

physicist (as Biot and Herschel) ever carried out a chemical experiment.

The use of optical phenomena to study the properties of matter did not become common among chemists until the latter nineteenth century. Spurred on by discoveries in crystal optics at the beginning of the century, physicists were utilizing optics with greater frequency to study the internal structure of matter. At the beginning of this period Brewster urged the "Chemical Philosopher" to "avail himself more frequently of the agencies of light" and predicted that this alliance would cause "great revolutions."[59] By midcentury physicists were frustrated with chemists' failure to adopt optical techniques for chemical analysis. In 1860, at the end of his long career, Biot lamented the fate of his polarimeter: "Chemists are nothing but cooks, they do not know how to take advantage of the admirable instrument that I have put in their hands."[60] We have already encountered Stokes's criticism of chemists in 1885, but twenty years earlier he had admonished them for failing to observe absorption spectra. Berzelius had identified biliverdin (the green substance obtained from bile) and chlorophyll, but Stokes found that only chlorophyll fluoresces and that the absorption bands of the two were altogether different:

> In fact, no one who is in the habit of using a prism could suppose for a moment that the two were identical; for an observation which can be made in a few seconds, which requires no apparatus beyond a small prism, to be used with the naked eye, and which as a matter of course *would* be made by any chemist working at the subject, *had the use of the prism made its way into the chemical world*, is sufficient to show that chlorophyll and biliverdin are quite distinct.[61]

Thus, independent of the other factors already adduced, the fact that chemists throughout this period rarely used prisms explains why they did not pursue absorption spectroscopy after the series of pioneering investigations at the turn of the century.

Fresnel, whose work sounded the death knell for the chemistry of light, had difficulty with selective absorption and coloration, though

[59] Quoted in full at note 11.

[60] Charles-Emile Picard, "La vie et l'oeuvre de Jean-Baptiste Biot," p. xxxi. See also Biot, "Introduction aux recherches de mécanique chimique" (1860); and for his later career, Frankel, "Jean-Baptiste Biot," pp. 366–74.

[61] Stokes, "On the supposed identity of biliverdin with chlorophyll, with remarks on the constitution of chlorophyll," *Mathematical and Physical Papers*, 4:236–7; first italics Stokes's; originally published in *Proceedings of the Royal Society*, 1864.

it was that especially French problem of insisting on a mathematical description. In 1822, in the supplement to the French translation of Thomas Thomson's *Chemistry*, he published a general account of his wave theory of light that synthesized his research papers. Thomson's treatise, we should recall, contained one of the clearer formulations of the chemistry of light and explanations of the colors of bodies by selective affinities.[62]

In the penultimate paragraph of his long essay, Fresnel allowed: "Without doubt there are still many obscurities to be elucidated, especially such as relate to the absorption of light, for instance, in ... the proper colours of bodies." He said virtually nothing about coloration other than that it arises from absorption, for his principal concern was to explain that absorption must involve a transfer of energy and not matter. He then concluded the essay with a peroration against the chemistry of light:

> If light is only a certain mode of vibration in a universal fluid ... we must no longer suppose that its chemical action on material substances consists in a combination of its molecules with theirs; but in a mechanical action, which the particles of this fluid exercise on the ponderable particles, and which causes them to enter into new arrangements, into more stable forms of equilibrium, from the peculiar kind or energy of the vibrations to which they are exposed. It is obvious that the hypothesis adopted on the nature of light and heat may change our conception of their chemical actions, and that it is very important not to be mistaken on the true theory. . . . If anything can contribute very essentially to advance this great discovery and to reveal the secrets of the internal constitution of bodies, it must be the minute and indefatigable study of the phenomena of light.[63]

Fresnel was so concerned with refuting the chemistry of light – and this was apparently his initial motivation for adopting a wave theory[64] – that he added a "Postscript. On the chemical action of light" in which he related an experiment by Arago that "overthrows the sup-

[62] Thomson, *Système de chimie*, trans. Jean Riffault, 4 vols. (Paris, 1818). I have not seen this work, which is a translation of the fifth London edition (1817), but the American edition that is also based on that edition contains the explanation of coloration by selective affinities virtually unchanged from the first edition; *A System of Chemistry* (1818), Bk. I, Div. I, Ch. I, §13, 1:33; see Ch. 7, note 45. Fresnel's "De la lumière," which appeared in *Supplément à la traduction française . . . du Système de chimie* (Paris, 1822), which I also have not seen, is included in Fresnel, *Oeuvres complètes*, and translated into English as "Elementary view of the undulatory theory of light" (1827–9).

[63] Fresnel, "Elementary view," pp. 161, 162; *Oeuvres complètes*, 2:140, 141; I have altered the translation throughout.

[64] See Buchwald, *Rise of the Wave Theory*, pp. 112–13.

position adopted by several savants according to which the chemical effects of light depend on its combination with bodies."[65]

Fresnel returned to coloration in 1826 (just months before his death) in a referee's report on a paper that Opoix had submitted to the institute half a century after his earlier paper to the academy began the chemists' revolt against Newton's theory. Though Fresnel does not explain the full process of coloration by selective absorption, it is evident that he understands it. He could have easily acquired this knowledge from a number of sources: Biot's *Traité;* his collaborator and senior colleague, Arago; or the articles in the *Annales de chimie* by Hassenfratz, who was his physics teacher at the Ecole Polytechnique.[66] Of course, he would have had to purge the affinities from these earlier investigations. Fresnel began his report by dismissing another paper, by a Mr. Déal, with a single sentence, which it is evident he also intended to apply to Opoix: "I will limit myself to saying that his work is a new proof of the necessity of mathematical knowledge to explain the phenomena of light." He then turned to the ironic situation of contemporary understanding of the cause of the colors of bodies:

> It is, however, truly remarkable that while we can easily calculate the tints of soap bubbles and of the rings formed between two lenses pressed against one another, and of the exceedingly variable phenomena of coloration . . . in crystalline plates, *we still cannot explain in a satisfactory way and calculate the proper colors of bodies, the most common and oldest observed optical phenomenon.* I do not claim that we have not already explained it up to a certain point, as well as other phenomena in which a part of the incident light turns out to be absorbed; but a complete and rigorous theory, confirmed by agreement of calculation and observations, has not yet been provided on this subject.

Because Fresnel considered only mathematical theories to be worthy of "a science as advanced as optics," it is now evident why he never gave a qualitative description of coloration by selective absorption.[67] If Newton's own theory of the colors of natural bodies was now gone, the problem of coloration was once again ensconced within quantitative physics, and his ideal of a mathematical-physical theory of colored bodies still very much alive.

[65] Fresnel, "Elementary view," p. 163; *Oeuvres complètes,* 2:144.

[66] It is not clear whether Fresnel knew the pair of papers by Brewster and Herschel from 1822, though I strongly suspect that he did.

[67] Fresnel, "Rapport verbal sur la théorie des couleurs et des corps inflammables de M. Opoix" [30 octobre 1826], *Oeuvres complètes,* 2:724–33, on pp. 724–5; italics added.

APPENDIX 1

JORDAN'S CRITICISM OF TRANSDUCTION IN NEWTON'S THEORY OF COLORED BODIES

Jordan published three works on optics in 1799 and 1800 in which he tested and criticized most of Newton's explanations of optical phenomena. The full title of the book that will concern us tells much of the story: *New Observations concerning the Colours of Thin Transparent Bodies, Shewing Those Phaenomena to Be Inflections of Light, and That the Newtonian Fits of Easy Transmission and Reflection Derived from Them Have No Existence, but Fail Equally in Their Establishment and in Their Application by Newton to Account for the Colours of Natural Bodies* (1800).[1] Although Jordan identified himself in these works only by his initials, his identity should have been no secret after Thomas Young referred to him by name in his papers on the wave theory and *Lectures on Natural Philosophy* (1807).[2] Jordan's criticism of Newton, which was then not unusual, did not lead to his isolation, and his works were in addition cited and utilized by Brewster, Cavallo, and Arago.[3]

By 1800, Newton's theory of colored bodies had been under attack in Britain for fifteen years, largely by the group to which Jordan belonged, qualitative natural philosophers. I am not concerned with his scientific criticisms of Newton's explanations, for they were already made by others, but rather in his thoroughgoing critique of

[1] The two books published in 1799 are *The Observations of Newton concerning the Inflections of Light* and *An Account of the Irides or Coronae*.

[2] In his "Outlines of experiments and inquiries respecting sound and light," p. 128, Young refers to "'Mr. Jordan, the ingenious author of a late publication on the Inflection of Light." See also "An account of some cases of the production of colours," p. 390; and *Course of Lectures*, 2:317, 622.

[3] Cantor, *Optics after Newton*, pp. 82, 85, 89; Buchwald, *Rise of the Wave Theory*, p. 117; see also Ch. 7, note 43, and Ch. 8, note 20. Cantor has a brief account of Jordan's optical research, and identifies his profession as "colonial agent," pp. 80, 206; see also Kipnis, *History of the Principle of Interference*, pp. 71–3, 144–5.

Newton's method of transduction in establishing his theory. His first line of attack is against Newton's use of transduction in interpreting the experimental evidence. In Prop. 2, Newton argued that the least parts of all bodies are transparent from the observation that all opaque substances when made sufficiently thin become transparent. Jordan objected:

> But these Facts do not establish the Point. They only show that the Bodies themselves are transparent under these Circumstances, not the Parts of the Bodies, and much less the very least Parts themselves. When Light passes through a perfectly transparent Body or any of the Bodies of the Experiment, there must be Parts by which, and not through which the Light passes, and of these Parts it certainly cannot in these Cases be inferred that *they* are transparent, and they are either the least or composed of the least Parts.[4]

He applies a similar argument against other observations invoked by Newton.[5]

Jordan then presents a deeper analysis of Newton's conception of colored bodies and his indiscriminate shifting between consideration of the microscopic particles composing bodies and macroscopic portions of them. He perceptively observed that Newton simultaneously considers transparent bodies to consist of small portions similar to the body itself and also of constituent particles that in turn make up those portions. This is apparent in Newton's interpretations of his empirical evidence as well as in his conception of colored bodies, which, he argued, were composed of corpuscles as large as fragments of thin films.

In order to show the confusion inherent in transduction, Jordan first introduces some definitions. The corpuscles and pores that compose a transparent body such as glass he calls particles^1 and pores.^{1} (I have added superscripts to help us keep track of the various levels of structure.) This is the lowest level of structure that Jordan considers, and it corresponds to Newton's highest order particles. He then introduces the concept of distinct portions or parts^2 and interstices^2 of bodies like glass; these will contain many particles^1 and pores.^{1} "Now the direct and principal Consequence of Newton's considering all Bodies as Aggregates of this second Class," Jordan objects, "is an inextricable Confusion of Parts^2 and Particles,^{1} Interstices^2 and Pores,^{1} and the constant and uncontroulled Substitution of these for the Oth-

[4] *New Observations*, pp. 55–6.
[5] Ibid., pp. 82–3.

ers, as Occasion or the Argument required."[6] He illustrates Newton's confusion by considering his proof of Prop. 3, that between the parts of opaque bodies there are pores filled with media of other densities. To demonstrate this Newton appealed to such evidence as opaque bodies like paper and the oculus mundi stone becoming transparent when soaked in water or oil, and transparent bodies like glass when ground to a powder or water whipped into a froth becoming opaque. All of this evidence, Jordan charges, comes from bodies of the second class or aggregates,

> and by calling the Interstices[2] Pores,[1] and considering the Parts[2] as Particles,[1] [he] has applied to Pores[1] and Particles[1] what is only shewn of Parts[2] and Interstices,[2] and endeavoured thus to render the Proposition general of all Bodies. . . . but no Argument drawn from these Facts can be applied to solid continuous Bodies. Between these and those and their internal Constitutions there is no more Resemblance than between the Particles[1] and Pores[1] of solid Glass and the Intervals between gross Pieces of the same, or Pieces of the same Glass reduced to Powder. These Pieces or Parts[2] are distinct Bodies.[7]

Jordan has captured the way Newton subtly shifts from one level to another in using transduction. What was simply conceptual confusion to Jordan in 1800 was to a mechanical philosopher in the seventeenth century a legitimate mode of reasoning to gain knowledge of the invisible constitution of bodies.

We need not follow all of Jordan's analysis of Newton's reasoning. It should be noted that he comments on related problems in applying the theory of fits to the theory of colored bodies, something Newton himself ignored. The problem here also involves transduction, for the fits should behave at each corpuscle composing a body just as they do at a thin film, but this is not always consistent with the observed phenomena or the behavior of the fits within a uniform body. In particular, except at the central dark spot some light is always reflected at the surface of a thin film, just as at any thick body, but this cannot occur at each corpuscle composing a transparent colored body.[8] Finally, it should also be mentioned that he also rejects Newton's methodology for showing that the colors of natural bodies and those of thin films are phenomena of the same kind, or that there is an "affinity of their properties." His approach is to show that all of these phenomena can be explained other than by treating the corpuscles of

[6] Ibid., p. 69.
[7] Ibid., pp. 70–1.
[8] Ibid., pp. 84–6; see also Melvill, "Observations," pp. 82–3, for a related objection.

colored bodies as pieces of thin films. For instance, he rejects Newton's interpretation of color changes when different liquids are mixed by turning directly to Newton's invocation of his analogical theory: "to ascribe these Effects to the Formation of thin Plates in the Liquors by the Parts thereof, seems as unphilosophical as to suppose the Existence of a Bubble of Water or Fragments of a Bubble amidst the solid and unseparated Parts of the Liquid."[9]

[9] *New Observations*, pp. 79–80; see Ch. 3, note 57, for Newton's interpretation.

APPENDIX 2

A DERIVATION OF THE DIAMETERS OF THE COLORED RINGS IN THICK PLATES ACCORDING TO THE THEORY OF FITS

In calculating the diameters of the colored rings in thick plates, Newton did not derive an equation for them according to the theory of fits. In the following derivation, I make a number of simplifications and approximations without thorough justification. My aim is to assist the modern scholar in understanding how the rings are produced according to the theory of fits, and not to justify or develop that theory. That task was undertaken first by Benvenuti and then by Biot (who derived these equations) as serious advocates of the theory of fits.[1] This derivation parallels that sketched according to the wave theory at the beginning of §4.2. It explicitly invokes the concept of path difference that underlaid Newton's calculation and makes it akin to the solution by the wave theory, though the physical explanations according to the two theories are altogether different.

Following Newton, the equation is derived for a perpendicularly incident light ray *CP* (Figure A2.1), which is scattered at *N* at the second surface of the mirror *PP'VN* and refracted at *P'* on the first surface, where it proceeds to *Q* on the screen *LCQ*, which is perpendicular to the mirror's axis *CN* and passes through its center *C*. Let the mirror's thickness *NP* be d, the oblique path *NP'* be s, the interval of the fits at perpendicular incidence be I_0, and the index of refraction of the glass be n. For the perpendicular ray

$$d = \frac{mI_0}{2n},$$

where m is an integer that will be even if the ray is in a fit of easy transmission and odd for reflection. For the scattered ray

[1] Benvenuti, *De lumine*, Prob. III, pp. 69–71; Biot, *Traité*, 4: 168–74.

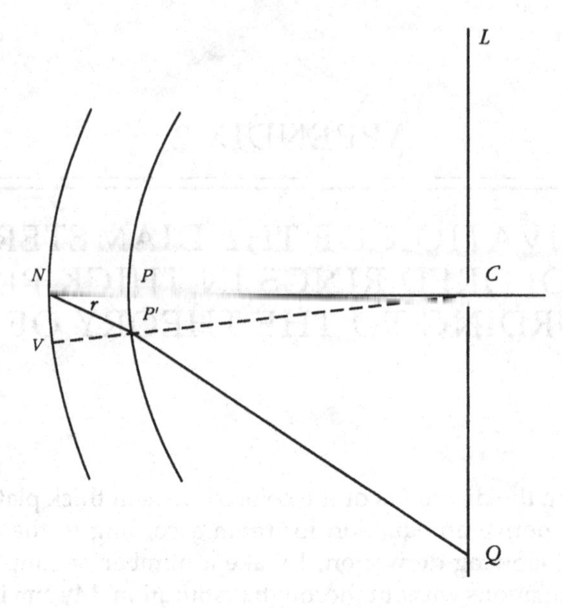

Figure A2.1. Explanation according to the theory of fits of the colored rings produced by the spherical mirror $PP'VN$ silvered on the back side NV. The incident rays CPN are scattered at point N.

$$s = \frac{(m - p)I}{2n},$$

where p is an integer and I the interval at an angle of diffuse reflection $PNP' = r$; if p is even, the ray will be in the same state at P' as at P. According to Prop. XV of the theory of fits (Eq. 4.4), for an oblique ray

$$I = \frac{I_0}{\cos u \cdot \cos r} \approx \frac{I_0}{\cos^2 r};$$

where the angle u, determined according to the law of 106-mean-proportionals (Eq. 4.5), is taken to be very nearly equal to r, because that angle is very small. Therefore

$$s\cos^2 r = d - \frac{pI_0}{2n},$$

but since $d = s\cos r$, then

$$\cos r = 1 - \frac{pl_0}{2nd}.$$

If the preceding is squared while terms in l_0^2 are ignored, then

$$\frac{pl_0}{nd} = 1 - \cos^2 r = \sin^2 r.$$

By taking the angle of refraction $NP'V \approx PNP' = r$, it follows from the sine law of refraction that

$$\sin i = n \sin r = \sqrt{\frac{npl_0}{d}},$$

where the angle $CP'Q$ is the angle of incidence i. Let the ring's radius CQ be $D/2$, and the mirror's radius CP be R. Then, because P and P' are very nearly coincident, $D = 2R\sin i$. Consequently, there results

$$D^2 = \frac{8nR^2}{d} \cdot \frac{pl_0}{2} = \frac{4nR^2}{d} \cdot \frac{p\lambda}{2},$$

for $l_0 = \lambda/2$, where λ is the true wavelength. When p is even, there will be a bright ring, and when odd a dark ring, because the central spot is bright. This derivation by the theory of fits thus gives the identical equation (4.1) as that according to the wave theory.

BIBLIOGRAPHY

Achinstein, Peter. "Newton's corpuscular query and experimental philosophy." In Phillip Bricker and R.I.G. Hughes, eds. *Philosophical Perspectives on Newtonian Science*. Cambridge, Mass.: MIT Press, 1990, pp. 135–73.

Airy, George Biddell. "Remarks on Sir David Brewster's paper 'On the absorption of specific rays, &c.' " *Philosophical Magazine*, ser. 3, 2(1833):419–24.

Albertus Magnus. *De anima. Opera omnia*. Vol. 7, pt. 1. Edited by Clemens Stroick. Münster: Monasterii Westfalorum, 1968.

Alembert, Jean le Rond d'. "Couleur." In *Encyclopédie*, 4(1754):327–32.

Algaroti, Francesco. *Sir Isaac Newton's Philosophy Explain'd for the Use of the Ladies. In Six Dialogues on Light and Colours*. Translated by Elizabeth Carter. 2 vols. London, 1739.

Anderson, Godfrey Tryggve, and Dennis Kent Anderson. "Edward Bancroft, M.D., F.R.S.: Aberrant 'practitioner of physick.' " *Medical History* 17(1973):356–67.

Averroes. *Epitome of "Parva naturalia."* Translated by Harry Blumberg. Cambridge, Mass.: Mediaeval Academy of America, 1961.

Babinet, Jacques, and Charles Bailly. *Résumé complet de la physique des corps impondérables*. 2 vols. in one. Paris: A. Boulland, 1825.

Bacon, Roger. *The Opus Majus of Roger Bacon*. Translated by Robert Belle Burke. 2 vols. Philadelphia: University of Pennsylvania Press, 1928.

Bancroft, Edward. *Experimental Researches concerning the Philosophy of Permanent Colours; and the Best Means of Producing Them, by Dyeing, Callico Printing, &c*. Vol. 1. London, 1794.

———. *Experimental Researches concerning the Philosophy of Permanent Colours; and the Best Means of Producing Them, by Dyeing, Calico Printing, &c*. 2 vols. 2nd ed. London: T. Cadell and W. Davies, 1813.

Barrow, Isaac. *The Mathematical Works of Isaac Barrow, D.D.* Edited by William Whewell. 2 vols. Cambridge University Press, 1860.

———. *The Usefulness of Mathematical Learning Explained and Demonstrated: Being Mathematical Lectures Read in the Publick Schools at the University of Cambridge*. Translated by John Kirkby. London, 1734; rpt. London: Frank Cass, 1970.

Bechler, Zev. " 'A less agreeable matter': The disagreeable case of Newton and achromatic refraction." *British Journal for the History of Science* 8(1975):101–26.

———. "Newton's search for a mechanistic model of colour dispersion: A suggested interpretation." *Archive for History of Exact Sciences* 11(1973):1–37.

———. "Newton's 1672 optical controversies: A study in the grammar of scientific dissent." In Yehuda Elkana, ed. *The Interactions between Philosophy and Science.* Atlantic Highlands, N.J.: Humanities Press, 1974, pp. 115–42.

———. ed. *Contemporary Newtonian Research.* Dordrecht: Reidel, 1982.

Beer, August. "Bestimmung der Absorption des rothen Lichts in farbigen Flüssigkeiten." *Annalen der Physik,* ser. 2, 86(1852):78–88.

Beer, John J. "Eighteenth-century theories on the process of dyeing." *Isis* 51(1960):21–30.

Bennett, J. A. *The Mathematical Science of Christopher Wren.* Cambridge University Press, 1982.

[Bentley, Thomas.] "[Review of] *An Experimental Inquiry into the Cause of the Changes of Colours in Opake and Coloured Bodies* . . . by Edward Hussey Delaval." *Monthly Review* 57(1777):22–31.

Benvenuti, Carlo. *Dissertatio physica de lumine. Ex editione Romana anni MDCCLIV recusa.* Vienna, 1761.

Bergman, Torbern. "Analyse et examen chimique de l'indigo, tel qu'il est dans le commerce, pour l'usage de la teinture." *Mémoires de mathématique et de physique, présentés à l'académie royale des sciences, par divers savans, & lûs dans ses assemblées* 9(1780):121–64.

Berthollet, Claude-Louis. "De l'influence de la lumière." *Journal de physique* 29(1786):81–5.

———. "Description du blanchîment des toiles et des fils par l'acide muriatique oxigéné, et de quelques autres propriétés de cette liqueur relatives aux arts." *Annales de chimie* 2(1789):151–90.

———. "Additions à la description du blanchîment, etc." *Annales de chimie* 6(1790):204–9.

———. *Eléments de l'art de la teinture.* 2 vols. Paris, 1791.

———. "*Elémens de l'art de la teinture*. . . . Extrait par l'auteur." *Annales de chimie* 9(1791):138–56.

———. *Essai de statique chimique.* 2 vols. Paris, 1803; rpt. New York: Johnson Reprint, 1972.

———. *An Essay on Chemical Statics.* Translated by B. Lambert. 2 vols. London: J. Mawman, 1804.

———. "Mémoire sur l'acide marin déphlogistique." *Mémoires de l'académie royale des sciences* 1785 [1788]:276–95.

———. "Mémoire sur l'acide Prussique." *Mémoires de l'académie royale des sciences* 1787 [1789]:148–62.

———. "Mémoire sur l'action que l'acide muriatique oxigéné exerce sur les parties colorantes." *Annales de chimie* 6(1790):210–40.

Berthollet, Claude-Louis, and Amédée-Barthelemy Berthollet. *Eléments de l'art de la teinture, avec une description du blanchiment par l'acide muriatique oxigéné.* 2 vols. 2nd ed. Paris: F. Didot, 1804.

———. *Elements of the Art of Dyeing; with a Description of the the Art of Bleaching by Oxymuriatic Acid.* Translated by Andrew Ure. 2 vols. 2nd ed. London: Thomas Tegg, 1824.

———. *Elements of the Art of Dyeing and Bleaching.* Translated by Andrew Ure. 2 vols. new ed. London: Thomas Tegg, 1841.

Beudant, François-Sulpice. *Essai d'un cours élémentaire et général des sciences physiques.* Paris: Tilliard Fréres, Croullebois, and Verdiäre, 1815.

[Bewley, William.] "[Review of] 'Experiments and observations...' by Edward Delaval." *Monthly Review* 35(1766):374–5.

Biot, Jean-Baptiste. "Introduction aux recherches de mécanique chimique, dans lesquelles la lumière polarisée est employée auxiliairement comme réactif." *Annales de chimie,* ser. 3, 59(1860):206–326.

———. *Traité de physique expérimentale et mathématique.* 4 vols. Paris: Deterville, 1816.

Biot, Jean-Baptiste, and Dominique-François-Jean Arago. "Mémoire sur les affinités des corps pour la lumière, et particulièrement sur les forces réfringentes des différens gaz." *Mémoires de la classe des sciences mathématiques et physiques de l'Institut national de France* 7(1806):301–87.

Birch, Thomas. *The History of the Royal Society of London, for Improving of Natural Knowledge, from its First Rise.* 4 vols. London, 1756–7; rpt. Brussels: Culture et Civilisation, 1968.

Blackburne, William. "Communication from Dr. Blackburne respecting caloric, light, and colours." *Philosophical Magazine* 6(1800):334–5.

Blay, Michel. *La conceptualisation Newtonienne des phénomènes de la couleur.* Paris: J. Vrin, 1983.

Boas, Marie. "Acid and alkali in seventeenth century chemistry." *Archives internationales d'histoire des sciences* 9(1956):13–28.

Boerhaave, Hermann. *A New Method of Chemistry; Including the Theory and Practice of that Art: Laid Down on Mechanical Principles, and Accommodated to the Uses of Life.* Translated by Peter Shaw and Ephraim Chambers. London, 1727.

———. *A New Method of Chemistry; Including the History, Theory, and Practice of the Art.* Translated by Peter Shaw. 2 vols. 2nd ed. London, 1741.

Bonnet, Charles. "Lettre... sur les moyens de conserver diverses especes d'insectes & des poissons dans les cabinets d'histoire naturelle; sur le bel azur dont les champignons se colorent à l'air; & sur les changemens de couleur de divers corps par l'action de l'air ou de la lumière." *Journal de physique* 3(1774):296–301.

Bošković, Rudjer Josip. *Dissertatio de lumine.* 2 parts. Rome, 1749.

———. *Theoria philosophiae naturalis rêdacta ad unicam legem virium in natura existentium.* Venice, 1763.

———. *A Theory of Natural Philosophy.* Translated from the Venice edition of 1763 by J. M. Child. Cambridge, Mass.: MIT Press, 1966.

Bouchard, Georges. *Un organisateur de la victoire: Prieur de la Côte-d'Or, membre du comité de salut public.* Paris: R. Clavreuil, 1946.

Bouguer, Pierre. *Optical Treatise on the Gradation of Light.* Translated by W. E. Knowles Middleton. University of Toronto Press, 1961.

Boyd, Julian P. "Silas Deane: Death by a kindly teacher of treason?" *William and Mary Quarterly,* ser. 3, 16(1959):165–87, 319–42, 515–50.

Boyer, Carl B. *The Rainbow, from Myth to Mathematics.* New York: Thomas Yoseloff, 1959.

Boyle, Robert. *Experiments and Considerations Touching Colours.* London, 1664; rpt. New York: Johnson Reprint, 1964.

———. *The Works of the Honourable Robert Boyle.* Edited by Thomas Birch. 6 vols. new ed. London, 1772; rpt. Hildesheim: G. Olms, 1975–6.

Brewster, David. "Description of a monochromatic lamp for microscopical purposes,

&c. with remarks on the absorption of the prismatic rays by coloured media." *Transactions of the Royal Society of Edinburgh* 9(1823):433–44.

———. *The Life of Sir Isaac Newton*. New York: J. & J. Harper, 1831.

———. *Memoirs of the Life, Writings, and Discoveries of Sir Isaac Newton*. 2 vols. Edinburgh, 1855; rpt. New York: Johnson Reprint, 1965.

———. "Observations on the absorption of specific rays, in reference to the undulatory theory of light." *Philosophical Magazine*, ser. 3, 2(1833):360–3.

———. "Observations on the lines of the solar spectrum, and on those produced by the earth's atmosphere, and by the action of nitrous acid gas." *Transactions of the Royal Society of Edinburgh* 12(1834):519–30.

———. "On a new analysis of solar light, indicating three primary colours, forming coincident spectra of equal length." *Edinburgh Journal of Science*, new ser., 5(1831):197–206.

———. "On a new analysis of solar light, indicating three primary colours, forming coincident spectra of equal length." *Transactions of the Royal Society of Edinburgh* 12(1834):123–36.

———. "On the colours of natural bodies." *Report of the British Association* (1832) [1833]:547–9.

———. "On the colours of natural bodies." *Transactions of the Royal Society of Edinburgh* 12(1834):538–45.

———. "On the connexion between the optical structure and chemical composition of minerals." *Edinburgh Philosophical Journal* 5(1821):1–8.

———. "On the connexion between the phenomena of the absorption of light, and the colours of thin plates." *Philosophical Transactions* 127(1837):245–52.

———. "On the optical properties of sulphuret of carbon, carbonate of barytes, nitrate of potash, with inferences respecting the structure of doubly refracting crystals." *Transactions of the Royal Society of Edinburgh* 7(1815):285–302.

———. "Optics." In *The Edinburgh Encyclopaedia*. 18 vols. Edinburgh: William Blackwood, 1830, 15:460–662 [First published in 1822].

[———.] "Optics." *Library of Useful Knowledge. Natural Philosophy*. Society for the Diffusion of Useful Knowledge. Vol. 1. London: Baldwin and Cradock, 1829.

———. "Report on the recent progress of optics." *Report of the British Association* (1832) [1833]:308–22.

———. *Treatise on Optics. The Cabinet Cyclopaedia*. London: Longman, Rees, Orme, Brown & Green, 1831.

Brougham, Henry. "Experiments and observations on the inflection, reflection, and colours of light." *Philosophical Transactions* 86(1796):227–77.

[———.] "[Review of] *Indagine fisica sui colori*. . . . Di Giambattista Venturi. . . . (1801)." *Edinburgh Review* 6(1805):20–43.

Browne, C. A. "A sketch of the life and chemical theories of Dr. Edward Bancroft." *Journal of Chemical Education* 14(1937):103–7.

Brunello, Franco. *The Art of Dyeing in the History of Mankind*. Translated by Bernard Hickey. Vicenza: Neri Pozza Editore, 1973.

Brunet, Pierre. *L'introduction des théories de Newton en France au XVIII*[e] *siècle, avant 1738*. Paris, 1931; rpt. Geneva: Slatkine Reprints, 1970.

Buchwald, Jed Z. "Experimental investigations of double refraction from Huygens to Malus." *Archive for History of Exact Sciences* 21(1980):311–73.

———. *The Rise of the Wave Theory of Light: Optical Theory and Experiment in the Early Nineteenth Century*. University of Chicago Press, 1989.

Burke, John G. *Origins of the Science of Crystals.* Berkeley: University of California Press, 1966.

Burkhardt, Richard W., Jr. *The Spirit of System: Lamarck and Evolutionary Biology.* Cambridge, Mass.: Harvard University Press, 1977.

Burlingame, Leslie J. "Lamarck's chemistry: The chemical revolution rejected." In Harry Woolf, ed. *The Analytic Spirit: Essays in the History of Science in Honor of Henry Guerlac.* Ithaca: Cornell University Press, 1981, pp. 64–81.

Burtt, Edwin Arthur. *The Metaphysical Foundations of Modern Physical Science.* rev. ed. Garden City, N.Y.: Doubleday, 1954.

Butts, Robert E., and Joseph C. Pitt, eds. *New Perspectives on Galileo.* Dordrecht: D. Reidel, 1978.

Cantor, Geoffrey. *Optics after Newton: Theories of Light in Britain and Ireland, 1704–1840.* Manchester University Press, 1983.

Carrier, Martin. "Die begriffliche Entwicklung der Affinitätstheorie im 18. Jahrhundert: Newtons Traum – und was daraus wurde." *Archive for History of Exact Sciences* 36(1986):327–89.

Casini, Paolo. "Newton: The classical scholia." *History of Science* 22(1984):1–58.

Cavallo, Tiberius. *The Elements of Natural or Experimental Philosophy.* 4 vols. London: T. Cadell and W. Davies, 1803.

Chappert, André. *Etienne Louis Malus (1775–1812) et la théorie corpusculaire de la lumière.* Paris: J. Vrin, 1977.

Chaptal, Jean-Antoine. *Chimie appliquée aux arts.* 4 vols. Paris: Deterville, 1807.

———. *Elements of Chemistry.* Translated by William Nicholson. 3 vols. London, 1791.

———. "Observations sur les caves et le fromage de Roquefort." *Annales de chimie* 4(1789) [1796]:31–61.

Chaulnes, Michel-Ferdinand d'Albert d'Ailly, duc de. "Observations sur quelques expériences de la quatrième partie du deuxième livre de l'Optique de M. Newton." *Mémoires de l'académie royale des sciences* 1755 [1761]:136–44.

Cheyne, George. *Philosophical Principles of Natural Religion: Containing the Elements of Natural Philosophy, and the Proofs for Natural Religion, Arising from Them.* London, 1705.

———. *Philosophical Principles of Religion: Natural and Reveal'd.* 2nd ed. London, 1715.

[Chisholm, Alexander.] "[Review of] *Memoirs of the Literary and Philosophical Society of Manchester,* Vol. II." *Monthly Review* 74(1786):350–62.

Clow, Archibald, and Nan L. Clow. *The Chemical Revolution: A Contribution to Social Technology.* London: Batchworth Press, 1952.

Cohen, I. Bernard. "The first English version of Newton's *hypotheses non fingo.*" *Isis* 53(1962):379–88.

———. *Franklin and Newton. An Inquiry into Speculative Newtonian Experimental Science and Franklin's Work in Electricity as an Example Thereof.* Memoirs of the American Philosophical Society, vol. 43. Philadelphia: American Philosophical Society, 1956.

———. "Hypotheses in Newton's philosophy." *Physis* 8(1966):163–84.

———. *Introduction to Newton's 'Principia.'* Cambridge, Mass.: Harvard University Press, 1971.

———. *The Newtonian Revolution: With Illustrations of the Transformation of Scientific Ideas.* Cambridge University Press, 1980.

———. *See also* Koyré and Cohen.

Coleby, L.J.M. *The Chemical Studies of P. J. Macquer.* London: George Allen & Unwin, 1938.

Cornell, E. S. "The radiant heat spectrum from Herschel to Melloni. I. The work of Herschel and his contemporaries." *Annals of Science* 3(1938):119–37.

Crombie, A. C. "Newton's conception of scientific method." *Bulletin of the Institute of Physics* 8(1957):350–62.

Crosland, Maurice P. "Chemistry and the chemical revolution." In G. S. Rousseau and Roy Porter, eds. *The Ferment of Knowledge: Studies in the Historiography of Eighteenth-Century Science.* Cambridge University Press, 1980, pp. 389–416.

———. "The development of chemistry in the eighteenth century." *Studies on Voltaire and the Eighteenth Century* 24(1963):369–441.

———. *Historical Studies in the Language of Chemistry.* London, 1962; rpt. New York: Dover Publications, 1978.

———. *The Society of Arcueil: A View of French Science at the Time of Napoleon I.* London: Heinemann, 1967.

Daumas, Maurice. *Scientific Instruments of the Seventeenth and Eighteenth Centuries and Their Makers.* Translated by Mary Holbrook. London: Portman Books, 1989.

Dear, Peter. "Jesuit mathematical science and the reconstitution of experience in the early seventeenth century." *Studies in History and Philosophy of Science* 18(1987):133–75.

Defoe, Daniel. *Robinson Crusoe.* Edited by Michael Shinagel. New York: W. W. Norton, 1975.

Delaval, Edward Hussey. *An Experimental Inquiry into the Cause of the Changes of Colours in Opake and Coloured Bodies. With an Historical Preface Relative to the Parts of Philosophy therein Examined, and to the Several Arts and Manufactures Dependent on Them.* London, 1777.

———. "An experimental inquiry into the cause of the permanent colours of opake bodies." *Memoirs of the Literary and Philosophical Society of Manchester* 2(1789):147–272.

———. *An Experimental Inquiry into the Cause of the Permanent Colours of Opake Bodies.* Warrington, 1785.

———. "Experiments and observations on the agreement between the specific gravities of the several metals, and their colours when united to glass, as well as those of their other proportions." *Philosophical Transactions* 55(1765):10–38.

———. "Recherches expérimentales sur la cause des changements de couleur dans les corps opaques naturellement colorés." *Nouveaux mémoires de l'académie royale des sciences et belles-lettres* (1774) [Berlin, 1776]:154–94.

Descartes, René. *Discourse on Method, Optics, Geometry, and Meteorology.* Translated by Paul J. Olscamp. Indianapolis: Bobbs-Merrill, 1965.

———. *Principles of Philosophy.* Translated by Valentine Rodger Miller and Reese P. Miller. Dordrecht: Reidel, 1984.

Despretz, César-Mansuète. *Traité élémentaire de physique.* Paris: Méquignon-Marvis, 1825.

Dewhurst, Kenneth. *Dr. Thomas Sydenham (1624–1689): His Life and Original Writings.* Berkeley: University of California Press, 1966.

Dictionary of Scientific Biography. Edited by Charles Coulston Gillispie and Frederic L. Holmes. 18 vols. New York: Charles Scribner's Sons, 1970–90.

Dobbs, Betty Jo Teeter. *Alchemical Death and Resurrection. The Significance of Alchemy in the Age of Newton.* Washington, D.C.: Smithsonian Institution Libraries, 1990.

———. *The Foundations of Newton's Alchemy, or "The Hunting of the Greene Lyon."* Cambridge University Press, 1975.

———. "Newton's alchemy and his theory of matter." *Isis* 73(1982):511–28.

Donovan, Arthur L. "Lavoisier and the origins of modern chemistry." In Donovan, ed. *Chemical Revolution*, pp. 214–31.

———. *Philosophical Chemistry in the Scottish Enlightenment: The Doctrines and Discoveries of William Cullen and Joseph Black*. Edinburgh University Press, 1975.

———. ed. *The Chemical Revolution: Essays in Reinterpretation*. *Osiris*, ser. 2, 4(1988).

Drake, Stillman, and I. E. Drabkin, trans. *Mechanics in Sixteenth-Century Italy: Selections from Tartaglia, Benedetti, Guido Ubaldo, and Galileo*. Madison: University of Wisconsin Press, 1969.

Drake, Stillman, and C. D. O'Malley, trans. *The Controversy on the Comets of 1618*. Philadelphia: University of Pennsylvania Press, 1960.

Draper, John William. "On some analogies between the phaenomena of the chemical rays, and those of radiant heat." *Philosophical Magazine*, ser. 3, 19(1841):195–210.

Duncan, A. M. "Some theoretical aspects of eighteenth-century tables of affinity." *Annals of Science* 18(1962):177–94, 217–32.

Dutour, Etienne François. "Considérations optiques. Mémoire sur la décomposition de la lumière dans le phénomene des anneaux colorés entre deux lames de verre." *Journal de physique* 1(1773):339–74.

———. "Considérations optiques. VIe mémoire. Examen des phénomenes sur lesquels on se fonde pour nier que la lumière soit réfléchie immédiatement par la surface des corps." *Journal de physique* 3(1774):116–26

———. "Considérations optiques. Septieme mémoire. Sur le caractere des athmospheres optiques." *Journal de physique* 5(1775):120–9.

———. "Considérations optiques. XIe mémoire. Expériences sur la décomposition et les combinaisons des rayons auxquels les corps, soit transparens, soit opaques, doivent leurs couleurs." *Journal de physique* 7(1776):230–9, 341–53.

———. "Recherches sur le phénomene des anneaux colorés." *Mémoires de mathématique et de physique, présentés à l'académie royale des sciences, par divers savans, et lûs dans ses assembleés* 4(1763):285–312.

Edelstein, Sidney M. "Historical notes on the wet-processing industry. VI. The dual life of Edward Bancroft." *American Dyestuff Reporter* 43(1954):712–13, 735.

Eder, Joseph Maria. *History of Photography*. Translated from the 4th German ed. (1932) by Edward Epstean. New York, 1945; rpt. New York: Dover Publications, 1978.

Encyclopaedia Britannica. 18 vols. 3rd ed. Edinburgh, 1797.

Encyclopédie, ou dictionnaire raisonné des sciences, des arts et des metiers. 28 vols. Paris, 1751–72; rpt. New York: Readex Microprint Corporation, 1969.

Euler, Leonhard. *Leonhardi Euleri opera omnia*. Sub auspiciis societatis scientiarum naturalium helveticae. Edited by Ferdinand Rudio et al. 73 vols. to date. Leipzig/Berlin/Zurich, 1911–

———. *Letters of Euler on Different Subjects in Natural Philosophy. Addressed to a German Princess*. Translated by Henry Hunter. Edited by David Brewster and John Griscom. 2 vols. New York: J. & J. Harper, 1833.

Farlie, Susan. "Dyestuffs in the eighteenth century." *Economic History Review*, ser. 2, 17(1964–5):488–510.

Fichman, Martin. "French Stahlism and chemical studies of air, 1750–1770." *Ambix* 18(1971):94–122.

Figala, Karin. " 'Die exakte Alchemie von Isaac Newton': Seine 'gesetzmässige' Interpretation der Alchemie – dargestellt am Beispiel einiger ihn beeinflussender Autoren." *Verhandlungen der Naturforschenden Gesellschaft in Basel* 94(1984):157–228.

———. "Newton as alchemist." *History of Science* 15(1977):102–37.

Fischer, Ernst Gottfried. *Physique mécanique*. Translated by Jean-Baptiste Biot. Paris: Bernard, 1806.

———. *Physique mécanique*. Translated by Jean-Baptiste Biot. 2nd ed. Paris: J. Klostermann Fils, 1813.

Forrester, Stanley D. "The history of the development of the light fastness testing of dyed fabrics up to 1902." *Textile History* 6(1975):52–88.

Fouchy, Jean-Paul Grandjean de. "Eloge de M. le duc de Chaulnes." *Histoire de l'académie royale des sciences* (1769):180–8.

Fourcroy, Antoine-François. "Mémoire sur la coloration des matières végétables par l'air vital, et sur une nouvelle préparation de couleurs solides pour la peinture." *Annales de chimie* 5(1790) [new ed., 1800]:80–91.

Fox, Robert. *The Caloric Theory of Gases from Lavoisier to Regnault*. Oxford: Clarendon Press, 1971.

———. "The rise and fall of Laplacian physics." *Historical Studies in the Physical Sciences* 4(1974):89–136.

Frankel, Eugene. "Corpuscular optics and the wave theory of light: The science and politics of a revolution in physics." *Social Studies of Science* 6(1976):141–84.

———. "J. B. Biot and the mathematization of experimental physics in Napoleonic France." *Historical Studies in the Physical Sciences* 8(1977):33–72.

———. "Jean-Baptiste Biot: The career of a physicist in nineteenth century France." Ph.D. diss. Princeton University, 1972.

Fraunhofer, Joseph von. "On the refractive and dispersive power of different species of glass, in reference to the improvement of achromatic telescopes, with an account of the lines or streaks which cross the spectrum." *Edinburgh Philosophical Journal* 9(1823):288–99, 10(1824):26–40.

Fresnel, Augustin. "Elementary view of the undulatory theory of light." *Quarterly Journal of Science, Literature and Art*, new ser., 1(1827):127–41, 441–54; 2(1827):113–35; 3(1828):198–215; 4(1828):168–91, 389–407; 5(1829):159–65.

———. *Oeuvres complètes d'Augustin Fresnel*. Edited by Henri de Senarmont, Emile Verdet, and Léonor Fresnel. 3 vols. Paris, 1866–70; rpt. New York: Johnson Reprint, 1965.

Gabbey, Alan. "Newton's *Mathematical Principles of Natural Philosophy:* A treatise on 'mechanics'?" In Harman and Shapiro, eds. *Investigation of Difficult Things*, pp. 305–22.

Galilei, Galileo. *Dialogue concerning the Two Chief World Systems – Ptolemaic and Copernican*. Translated by Stillman Drake. 2nd ed. Berkeley: University of California Press, 1967.

Gehler, Johann Samuel Traugott. *Physikalisches Wörterbuch oder Versuch einer Erklärung der vornehmsten Begriffe und Kunstwörter der Naturlehre*. 4 vols. Leipzig, 1787–91.

Gernsheim, Helmut. *The Origins of Photography*. New York: Thames and Hudson, 1982.

Gibbs, F. W. "William Lewis, M.B., F.R.S. (1708–1781)." *Annals of Science* 8(1957):122–51.

Gillispie, Charles Coulston. *Science and Polity in France at the End of the Old Regime*. Princeton University Press, 1980.

Gillispie, Charles Coulston, et al. "Laplace." In *Dictionary of Scientific Biography*, 15:273–403.

Glanvill, Joseph. *Essays on Several Important Subjects in Philosophy and Religion*. London, 1676.

Good, Gregory. "J. F. W. Herschel's optical researches: A study in method." Ph.D. diss. University of Toronto, 1982.

Gough, J. B. "Lavoisier and the fulfillment of the Stahlian revolution." In Donovan, ed. *Chemical Revolution*, pp. 15–33.

Gravesande, Willem Jacob 's. *Mathematical Elements of Natural Philosophy, Confirm'd by Experiments: Or, an Introduction to Sir Isaac Newton's Philosophy*. Translated by J. T. Desaguliers. 2 vols. 6th ed. London, 1747.

———. *Physices elementa mathematica, experimentis confirmata, sive introductio ad philosophiam Newtonianam*. 2 vols. Leyden, 1742.

Greenberg, John L. "Mathematical physics in eighteenth-century France." *Isis* 77(1980):59–78.

Gregory, David. *David Gregory, Isaac Newton and Their Circle. Extracts from David Gregory's Memoranda 1677–1708*. Edited by W. G. Hiscock. Oxford: Printed for the Editor, 1937.

Gregory, Joshua C. "The Newtonian hierarchic system of particles." *Archives internationales d'histoire des sciences* 7(1954):243–7.

Gren, Friedrich Albrecht Carl. *Grundriss der Naturlehre in seinem mathematischen und chemischen Theile neu bearbeitet*. Halle, 1793.

Grimaldi, Francesco Maria. *Physico-mathesis de lumine, coloribus, et iride*. Bologna, 1665; rpt. Bologna: Arnaldo Forni, 1963.

Grotthuss, Theodor von. "Auszug aus vier Abhandlungen physikalisch-chemischen Inhalts." *Annalen der Physik* 61(1819):50–64.

Guerlac, Henry. "Chemistry as a branch of physics: Laplace's collaboration with Lavoisier." *Historical Studies in the Physical Sciences* 7(1976):193–276.

———. "Newton's optical aether: His draft of a proposed addition to his *Opticks*." *Notes and Records of the Royal Society of London* 22(1967):45–57.

———. "Some French antecedents of the chemical revolution." *Chymia* 5(1959):73–112.

Hacking, Ian. *The Emergence of Probability: A Philosophical Study of Early Ideas about Probability, Induction and Statistical Inference*. Cambridge University Press, 1975.

Hakfoort, Casper. *Optica in de eeuw van Euler: Opvattingen over de natuur van het licht, 1700–1795. Optics in the Age of Euler: Conceptions of the Nature of Light, 1700–1795 (with a Summary in English)*. Amsterdam: Rodopi, 1986.

Hanson, Norwood Russell. "Hypotheses fingo." In Robert E. Butts and John W. Davis, eds. *The Methodological Heritage of Newton*. Oxford: Basil Blackwell, 1970, pp. 14–33.

———. "Waves, particles, and Newton's 'fits.'" *Journal of the History of Ideas* 21(1960):370–91.

Hargreave, David. "Thomas Young's theory of color vision: Its roots, development, and acceptance by the British scientific community." Ph.D. diss. University of Wisconsin, 1973.

Harman, P. M., and Alan E. Shapiro, eds. *The Investigation of Difficult Things. Essays on Newton and the History of the Exact Sciences in Honour of D. T. Whiteside*. Cambridge University Press, 1992.

Harrison, John. *The Library of Isaac Newton*. Cambridge University Press, 1978.

Hartley, David. *Observations on Man, His Frame, His Duty, and His Expectations*. 2 vols. London, 1749; rpt. Gainesville, Fla.: Scholars' Facsimiles and Reprints, 1966.

Hassenfratz, Jean-Henri. "Mémoire sur la colorisation des corps." *Annales de chimie* 66(1808):152–67, 290–317; 67(1808):5–25, 113–51.

———. "Versuche und Beobachtungen uber die Farben des Lichtes, etc.; c'est-à-dire, *Expériences et observations sur les couleurs de la lumière*, par Chrétien-Ernest Wunsch [1792]." *Annales de chimie*. 64(1807):135–63.

Haüy, René Just. *An Elementary Treatise on Natural Philosophy*. Translated by Olinthus Gregory. 2 vols. London: G. Kearsley, 1807.

———. *Traité de minéralogie*. 5 vols. Paris: Louis, 1801.

———. *Traité élémentaire de physique*. 2 vols. Paris: Delance et Lesueur, 1803.

———. *Traité élémentaire de physique*. 2 vols. 2nd ed. Paris: Courcier, 1806.

———. *Traité élémentaire de physique*. 2 vols. 3rd ed. Paris: Bachelier et Huzard, 1821.

Heath, Thomas. *Mathematics in Aristotle*. Oxford: Clarendon Press, 1949.

Hecht, Eugene, and Alfred Zajac. *Optics*. 2nd ed. Reading, Mass.: Addison-Wesley Publishing Co., 1987.

Heilbron, John L. *Electricity in the 17th and 18th Centuries: A Study of Early Modern Physics*. Berkeley: University of California Press, 1979.

———. *Physics at the Royal Society during Newton's Presidency*. Los Angeles: William Andrews Clark Memorial Library, University of California, 1983.

Heimann, P. M. "Ether and imponderables." In Geoffrey N. Cantor and M.J.S. Hodge, eds. *Conceptions of Ether: Studies in the History of Ether Theories, 1740–1900*. Cambridge University Press, 1981, pp. 61–83.

———. " 'Nature is a perpetual worker': Newton's aether and eighteenth-century natural philosophy." *Ambix* 20(1973):1–25.

Heimann, P. M., and J. E. McGuire. "Newtonian forces and Lockean powers: Concepts of matter in eighteenth-century thought." *Historical Studies in the Physical Sciences* 3(1971):233–306.

Heinrich, Placidus. "Ueber die Preisfrage: 'Kömmt das Newtonische, oder das Eulerische System vom Lichte mit dem neuesten Versuchen und Erfahrungen der Physik mehr überein?' Eine mit dem Preise belohnte Abhandlung." *Neue philosophische Abhandlungen der baierischen Akademie der Wissenschaften* 5 (1789):145–398.

———. *See also* Link and Heinrich.

Hellot, Jean. "Théorie chimique de la teinture des étoffes." *Mémoires de l'académie royale des sciences*, 1740 [Amsterdam, 1745]:176–208; 1741 [Amsterdam, 1747]:49–95.

Helmholtz, Hermann von. "On the theory of compound colours." *Philosophical Magazine*, ser. 4, 4(1852):519–34.

———. *Treatise on Physiological Optics*. Translated from the 3rd German ed. by James P. C. Southall, 3 vols. in two. Rochester, 1924–5; rpt. New York: Dover Publications, 1962.

———. "Ueber die Theorie der zusammengesetzten Farben." *Annalen der Physik*, ser. 2, 87(1852):45–66.

Helsham, Richard. *A Course of Lectures in Natural Philosophy*. Edited by Bryan Robinson. Dublin, 1739.

Henry, Thomas. "Considerations relative to the nature of wool, silk, and cotton, as objects of the art of dying; on the various preparations, and mordants, requisite for these different substances; and on the nature and properties of colouring matter. Together with some observations on the theory of dying in general, and particularly the Turkey red." *Memoirs of the Literary and Philosophical Society of Manchester* 3(1790):343–408.

Herschel, John Frederick William. "Analysis of scientific books and memoirs." *Edinburgh Journal of Science* 2(1825):344–8.

———. "Light." In *The Encyclopaedia of Mechanical Philosophy.* London: John Joseph Griffin & Co., 1848, pp. 341–586. [Reprinted from *The Encyclopaedia Metropolitana* (1828).]

———. "On the absorption of light by coloured media, and on the colours of the prismatic spectrum exhibited by certain flames; with an account of a ready mode of determining the absolute dispersive power of any medium, by direct experiment." *Transactions of the Royal Society of Edinburgh* 9(1823):445–60.

———. "On the absorption of light by coloured media, viewed in connexion with the undulatory theory." *Philosophical Magazine*, ser. 3, 3(1833):401–12.

———. "On the absorption of light by coloured media, viewed in connexion with the undulatory theory." *Report of the British Association* (1833) [1834]:373–4.

———. "On the action of crystallized bodies on homogeneous light, and on the causes of the deviation from Newton's scale in the tints which many of them develope on exposure to a polarized ray." *Philosophical Transactions* 110(1820):45–100.

———. *Traité de la lumière.* Translated by P. F. Werhulst and A. Quetelet. 2 vols. Paris: Mahler et Cie., L. Hachette, 1829–33.

Herschel, William. "Experiments for investigating the cause of the coloured concentric rings, discovered by Sir Isaac Newton, between two object-glasses laid upon one another." *Philosophical Transactions* 97(1807):180–233.

———. "Investigation of the powers of the prismatic colours to heat and illuminate objects; with remarks, that prove the different refrangibility of radiant heat." *Philosophical Transactions* 90(1800):255–83.

———. *The Scientific Papers of Sir William Herschel, Including Early Papers Hitherto Unpublished.* 2 vols. London: Royal Society and Royal Astronomical Society, 1912.

Holmes, Frederic L. "From elective affinities to chemical equilibria: Berthollet's law of mass action." *Chymia* 8(1962):105–45.

Home, R. W. "Newton on electricity and the aether." In Bechler, ed. *Contemporary Newtonian Research*, pp. 191–213.

———. "The notion of experimental physics in early eighteenth-century France." In Joseph C. Pitt, ed. *Change and Progress in Modern Science.* Dordrecht: Reidel, 1985, pp. 107–31.

Hooke, Robert. *Micrographia: Or Some Physiological Descriptions of Minute Bodies Made by Magnifying Glasses. With Observations and Inquiries Thereupon.* London, 1665; rpt. Brussels: Culture et Civilisation, 1966.

[Hutton, Charles.] "Chromatics." In *Encyclopaedia Britannica* 4:721–41.

———. *A Mathematical and Philosophical Dictionary.* 2 vols. London, 1795.

Hutton, James. *Dissertations on Different Subjects in Natural Philosophy.* Edinburgh, 1792.

Huygens, Christiaan. *Oeuvres complètes de Christiaan Huygens.* 22 vols. The Hague: 1888–1950.

———. *Treatise on Light. In Which Are Explained the Causes of That Which Occurs in Reflexion, and in Refraction. And Particularly in the Strange Refraction of Iceland Crystal.* Translated by Silvanus P. Thompson. London: 1912; rpt. New York: Dover Publications, 1962.

James, Frank A.J.L. "The creation of a Victorian myth: The historiography of spectroscopy." *History of Science* 23(1985):1–24.

———. "The debate on the nature of the absorption of light, 1830–1835: A core-set analysis." *History of Science* 21(1983):335–68.

———. "The discovery of line spectra." *Ambix* 32(1985):53–70.

———. "Spectro-chemistry and myth: A rejoinder." *History of Science* 24(1986):433–7.

Jardine, Nicholas. "Epistemology of the sciences." In Charles B. Schmitt et al., eds. *The Cambridge History of Renaissance Philosophy*. Cambridge University Press, 1988, pp. 685–711.

The Jewish National and University Library. *Catalog of the Sidney M. Edelstein Collection of the History of Chemistry, Dyeing and Technology*. Compiled by Moshe Ron. Jerusalem: Jewish National and University Library Press, 1981.

[Jordan, Gibbs Walker.] *An Account of the Irides or Coronae Which Appear Around, and Contiguous to, the Bodies of the Sun, Moon, and Other Luminous Objects*. London, 1799.

[———.] *New Observations concerning the Colours of Thin Transparent Bodies, Shewing Those Phaenomena to Be Inflections of Light, and That the Newtonian Fits of Easy Transmission and Reflection Derived from Them Have No Existence, but Fail Equally in Their Establishment and in Their Application by Newton to Account for the Colours of Natural Bodies*. London: T. Cadell, Jun. and W. Davies, 1800.

[———.] *The Observations of Newton concerning the Inflections of Light; Accompanied by Other Observations Differing from His; and Appearing to Lead to a Change of His Theory of Light and Colours*. London, 1799.

Kargon, Robert. "Newton, Barrow and the hypothetical physics." *Centaurus* 11(1965):46–56.

Kayser, Heinrich. *Handbuch der Spectroscopie*. Vol. 1. Leipzig: S. Hirzel, 1900.

Kepler, Johannes. *Gesammelte Werke*. Vol. 2. Edited by Franz Hammer. Munich: C. H. Beck, 1939.

Keyser, Barbara Whitney. "Between science and craft: The case of Berthollet and dyeing." *Annals of Science* 47(1990):213–60.

Kipnis, Naum S. *History of the Principle of Interference of Light*. Basel: Birkhäuser, 1991.

Kottler, Dorian Brooks. "Jean Senebier and the emergence of plant physiology, 1775–1802: From natural history to chemical science." Ph.D. diss. Johns Hopkins University, 1973.

Kronick, David A. *A History of Scientific Periodicals: The Origins and Development of the Scientific and Technical Press, 1665–1790*. 2nd ed. Metuchen, N.J.: Scarecrow Press, 1976.

Kuhn, Thomas S. *The Essential Tension: Selected Studies in Scientific Tradition and Change*. University of Chicago Press, 1977.

Koyré, Alexandre. *Newtonian Studies*. University of Chicago Press, 1968.

Koyré, Alexandre, and I. Bernard Cohen. "Newton's 'electric & elastic spirit.' " *Isis* 51(1960):337.

Laird, W. R. "The scope of Renaissance mechanics." *Osiris*, ser. 2, 2(1986):43–68.

Lamarck, Jean-Baptiste. *Mémoires de physique et d'histoire naturelle*. Paris, An V [1797].

———. *Recherches sur les causes des principaux faits physiques*. 2 vols. Paris, An II [1793/4].

Laplace, Pierre-Simon. *Exposition du système du monde*. 6th ed. Paris, 1835; rpt. Paris: Fayard, 1984.

———. *Oeuvres complètes de Laplace*. 14 vols. Paris: Gauthier-Villars, 1878–1912.

———. *Théorie du mouvement et de la figure elliptique des planètes*. Paris, 1784.

Lavoisier, Antoine-Laurent. *Elements of Chemistry, in a New Systematic Order, Containing All the Modern Discoveries*. Translated by Robert Kerr. Edinburgh, 1790; rpt. New York: Dover Publications, 1965.

———. *Oeuvres de Lavoisier*. 6 vols. Paris, 1862–93; rpt., New York: Johnson Reprint, 1965.

Lawrie, L. G. *A Bibliography of Dyeing and Textile Printing, Comprising a List of Books from the Sixteenth Century to the Present Time (1946)*. London: Chapman and Hall, 1949.

Le Blond, Jean-Baptiste. "Essais sur l'art de l'indigotier, pour servir à un ouvrage plus étendu;... Lus & approuvés par l'académie des sciences." *Journal de physique* 38(1791):141–50.

Leeuwen, Henry G. van. *The Problem of Certainty in English Thought, 1630–1690*. 2nd ed. The Hague: Martinus Nijhoff, 1970.

[Leslie, John.] "[Review of] *Experimental Researches Concerning the Philosophy of Permanent Colours*... by Edward Bancroft." *Monthly Review*, ser. 2, 17(1795):286–96, 376–89.

Levere, Trevor H. *Affinity and Matter: Elements of Chemical Philosophy 1800–1865*. Oxford: Clarendon Press, 1971.

Lewis, William. *Commercium Philosophico-Technicum: or, the Philosophical Commerce of Arts: Designed as an Attempt to Improve Arts, Trades, and Manufactures*. London, 1763–5.

Libes, Antoine. *Nouveau dictionnaire de physique, rédigé d'après les découvertes les plus modernes*. 3 vols. Paris: Giguet et Michaud, 1806.

———. *Traité complet et élémentaire de physique, présenté dans un ordre nouveau, d'après les découvertes modernes*. 3 vols. 2nd ed. Paris: Courcier, 1813.

Lindberg, David C. *Theories of Vision from al-Kindī to Kepler*. University of Chicago Press, 1976.

Link, Heinrich Friedrich, and Placidus Heinrich. *Ueber die Natur des Lichts. Zwey von der Kaiserl. Akademie der Wissenschaften zu St. Petersburg gekrönte Preisschriften*. St. Petersburg: Kaiserl. Akademie der Wissenschaften, 1808.

Lloyd, Humphrey. "Report on the progress and present state of physical optics." *Report of the British Association*, (1834)[1835]:295–413.

McClellan, James E. III. "The scientific press in transition: Rozier's Journal and the scientific societies in the 1770s." *Annals of Science* 36(1979):425–49.

McCluskey, Stephen C., Jr. "Nicole Oresme on light, color, and the rainbow: An edition and translation, with introduction and critical notes, of part of book three of his *Questiones super quatuor libros meteororum*." Ph.D. diss. University of Wisconsin, 1974.

McDonald, John F. "Properties and causes: An approach to the problem of hypothesis in the scientific methodology of Sir Isaac Newton." *Annals of Science* 28(1972):217–33.

McGucken, William. *Nineteenth-Century Spectroscopy: Development of the Understanding of Spectra 1802–1897*. Baltimore: Johns Hopkins Press, 1969.

McGuire, J. E. "Atoms and the 'analogy of nature': Newton's third rule of philosophizing." *Studies in History and Philosophy of Science* 1(1970):3–58.

———. "Force, active principles, and Newton's invisible realm." *Ambix* 15(1968):154–208.

———. "Neoplatonism and active principles: Newton and the *corpus hermeticum*." In Robert S. Westman and J. E. McGuire. *Hermeticism and the Scientific Revolution*. Los Angeles: William Andrews Clark Memorial Library, University of California, 1977, pp. 93–142.

———. "Newton's 'principles of philosophy': An intended preface for the 1704 *Opticks* and a related draft fragment." *British Journal for the History of Science* 5(1970):178–86.

———. "The origin of Newton's doctrine of essential qualities." *Centaurus* 12 (1968):233–60.

———. *See also* Heimann and McGuire.

McGuire, J. E., and P. M. Rattansi. "Newton and the 'Pipes of Pan.' " *Notes and Records of the Royal Society of London* 21(1966):108–43.

McGuire, J. E., and Martin Tammy. *Certain Philosophical Questions: Newton's Trinity Notebook*. Cambridge University Press, 1983.
McKie, Douglas. "Guillaume-François Rouelle (1703–70)." *Endeavour* 12(1953):130–3.
———. "The 'Observations' of the Abbé François Rozier (1734–93)." *Annals of Science* 13(1957):73–89.
———. *See also* Partington and McKie.
McMullin, Ernan. "The conception of science in Galileo's work." In Butts and Pitt, eds., *New Perspectives*, pp. 209–57.
———. "Conceptions of science in the scientific revolution." In David C. Lindberg and Robert S. Westman, eds. *Reappraisals of the Scientific Revolution*. Cambridge University Press, 1990, pp. 27–92.
———. "Newton and scientific realism." Forthcoming.
———. *Newton on Matter and Activity*. University of Notre Dame Press, 1978.
McNalty, Arthur S. "Edward Bancroft, M.D., F.R.S., and the war of American independence." *Proceedings of the Royal Society of Medicine* 38(1944):7–15.
Mach, Ernst. *The Principles of Physical Optics: An Historical and Philosophical Treatment*. Translated by John S. Anderson and A.F.A. Young. London: 1926; rpt. New York: Dover Publications, [1953].
Macquer, Pierre-Joseph. *Dictionnaire de chymie, contenant la théorie et la pratique de cette science*. 4 vols. 2nd ed. Paris, 1778.
———. *Elémens de chymie théorique*. Paris, 1749.
———. *Elémens de chymie-pratique*. Paris, 1751.
———. *Elements of the Theory and Practice of Chemistry*. Translated by Andrew Reid. 2 vols. London, 1758.
———. "Examen chymique du bleu de Prusse." *Mémoires de l'académie royale des sciences*, 1752 [Amsterdam, 1761]:87–113.
Magirus, Johannes. *Physiologiae peripateticae libri sex cum commentariis*. Cambridge, 1642.
Maistre, Xavier de. "Sur la cause des couleurs dans les corps naturels." *Bibliothèque universelle: Sciences et arts* 47(1831):17–39.
Mandelbaum, Maurice. *Philosophy, Science and Sense Perception: Historical and Critical Studies*. Baltimore: Johns Hopkins Press, 1964.
[Marivetz, Etienne Claude]. "Lettre adressée a l'auteur de ce recueil, par Madame T.E.S.A.V.L.M.O.R." *Journal de physique* 9(1777):330–7.
[———.] "Lettre de Madame de V***, à M. Senebier, Bibliothécaire de la République de Genève, sur les differénces qu'il établit entre la lumière et le phlogistique." *Journal de physique* 10(1777):206–13.
———. "Lettre de M. le Baron de Marivetz, à M. Senebier." *Journal de physique* 23(1783):270–8, 340–9.
Mauskopf, Seymour H. *Crystals and Compounds: Molecular Structure and Composition in Nineteenth-Century French Science. Transactions of the American Philosophical Society*, new ser., 66, part 3 (1976).
———. "Minerals, molecules, and species." *Archives internationales d'histoire des sciences* 23(1970):185–206.
Mayer, Johann Tobias. *Anfangsgründe der Naturlehre zum Behuf der Vorlesungen über die Experimental-Physik*. 2nd ed. Göttingen: Heinrich Dieterich, 1805.
Mazéas, Guillaume. "Observations sur des couleurs engendrées par le frottement des surfaces planes et transparentes." *Histoire de l'académie royale des sciences et belles-lettres* 8(1752) [Berlin, 1754]:248–61.
———. "Observations sur des couleurs engendrées par le frottement des surfaces planes et transparentes." *Mémoires de mathématique et de physique, présentés à l'académie royale des sciences, par divers savans, et lûs dans ses assemblées* 2(1755):26–43.

Meinel, Christoph. " 'Das letzte Blatt im Buch der Natur': Die Wirklichkeit der Atome und die Antinomie der Anschauung im den Korpuskulartheorien der frühen Neuzeit." *Studia Leibnitiana* 20(1988):1–18.

———. "Theory or practice? The eighteenth-century debate on the scientific status of chemistry." *Ambix* 30(1983):121–32.

Melhado, Evan M. "Chemistry, physics, and the chemical revolution." *Isis* 76(1985):195–211.

Melvill, Thomas. "Observations on light and colours." *Essays and Observations, Physical and Literary. Read before a Society in Edinburgh, and Published by Them* 2(1756):12–90.

Meteyard, Eliza. *The Life of Josiah Wedgwood.* 2 vols. London: Hurst and Blackett, 1866.

Metzger, Hélène. *Newton, Stahl, Boerhaave et la doctrine chimique.* Paris: Albert Blanchard, 1930.

Moigno, François Napoléon Marie. *Répertoire d'optique moderne.* 4 vols. Paris: A. Franck, 1847–50.

Morrison-Low, A. D., and J.R.R. Christie, eds. *'Martyr of Science': Sir David Brewster 1781–1868.* Edinburgh: Royal Scottish Museum, 1984.

Morse, Edgar W. "Young, Thomas." In *Dictionary of Scientific Biography* 14:562–72.

Musschenbroek, Pieter van. *Cours de physique expérimentale et mathématique.* Translated by Joseph Aignan Sigaud de la Fond. 3 vols. Paris, 1769.

———. *Introductio ad philosophiam naturalem.* 2 vols. Leyden, 1762.

Musson, Albert Edward, and Eric Robinson. *Science and Technology in the Scientific Revolution.* Manchester University Press, 1969.

Nangle, Benjamin Christie. *The Monthly Review, First Series, 1749–1789. Indexes of Contributors and Articles.* Oxford: Clarendon Press, 1934.

———. *The Monthly Review, Second Series, 1790–1815. Indexes of Contributors and Articles.* Oxford: Clarendon Press, 1955.

[Newton, Isaac.] "An account of the book entituled *Commercium epistolicum.*" *Philosophical Transactions* 29, no. 342 (January 1714/5):173–224.

———. *The Correspondence of Isaac Newton.* Edited by H. W. Turnbull, J. F. Scott, A. Rupert Hall, and Laura Tilling. 7 vols. Cambridge University Press, 1959–77.

———. *Correspondence of Sir Isaac Newton and Professor Cotes.* Edited by Joseph Edleston. London, 1850; rpt. London: Cass, 1969.

———. *Isaac Newton's Papers and Letters on Natural Philosophy and Related Documents.* Edited by I. Bernard Cohen. Cambridge, Mass.: Harvard University Press, 1958.

———. *Isaac Newton's 'Philosophiae naturalis principia mathematica.' The Third Edition (1726) with Variant Readings.* Edited by Alexandre Koyré and I. Bernard Cohen, with the assistance of Anne Whitman. 2 vols. Cambridge, Mass.: Harvard University Press, 1972.

———. *The Mathematical Papers of Isaac Newton.* Edited by D. T. Whiteside, with the assistance in publication of A. Prag. 8 vols. Cambridge University Press, 1967–81.

———. *The Optical Papers of Isaac Newton. Volume 1. The Optical Lectures, 1670–1672.* Edited by Alan E. Shapiro. 1 vol. to date. Cambridge University Press, 1984– .

———. *Opticks: Or, a Treatise of the Reflexions, Refractions, Inflexions and Colours of Light. Also Two Treatises of the Species and Magnitude of Curvilinear Figures.* London, 1704; rpt. Brussels: Culture et Civilisation, 1966.

———. *Opticks: Or a Treatise of the Reflections, Refractions, Inflections and Colours of Light.* Based on the 4th ed. London, 1730. London, 1931; rpt. New York: Dover Publications, 1952.

———. *Optice: Sive de reflexionibus, refractionibus, inflexionibus et coloribus lucis libri tres.* Translated by Samuel Clarke. London, 1706.

———. *Sir Isaac Newton's Mathematical Principles of Natural Philosophy and His System of the World.* Translated into English by Andrew Motte in 1729. Revised by Florian Cajori. Berkeley: University of California Press, 1934.

———. *Traité d'optique sur les reflexions, refractions, inflexions, et les couleurs, de la lumiere.* Translated by Pierre Coste. 2nd ed. Paris, 1722; rpt. Paris: Gauthier-Villars, 1955.

———. *Unpublished Scientific Papers of Isaac Newton. A Selection from the Portsmouth Collection in the University Library, Cambridge.* Edited by A. Rupert Hall and Marie Boas Hall. Cambridge University Press, 1962.

Nobili, Leopoldo. "Memoir on colours in general, and particularly on a new chromatic scale deduced from metallochromy for scientific and practical purposes." In *Scientific Memoirs* 1(1837):94–121.

Nollet, Jean-Antoine. *Leçons de physique expérimentale.* 6 vols. 6th ed. Paris, 1777.

Okruhlik, Kathleen. "The foundation of all philosophy: Newton's third rule." In James R. Brown and Jürgen Mittelstrass, eds. *An Intimate Relation: Studies in the History and Philosophy of Science Presented to Robert E. Butts on His 60th Birthday.* Boston Studies in the Philosophy of Science. Vol. 116. Dordrecht: Kluwer Academic Publishers, 1989, pp. 97–113.

Olson, Richard. *Scottish Philosophy and British Physics, 1750–1880: A Study in the Foundations of the Victorian Scientific Style.* Princeton University Press, 1975.

"On Dr. Parr's theory of light and heat. By a correspondent." *Journal of Natural Philosophy, Chemistry, and the Arts* 2(1799):547–8.

Opoix, Christophe. "Observations physico-chymiques sur les couleurs." *Journal de physique* 8(1776):100–116, 189–211.

———. *Théorie des couleurs et des corps inflammables, et de leurs principes constituants: La lumière et le feu; basée sur les faits, et sur les découvertes modernes.* Paris: Méquignon l'aîné et Gabon, 1808.

Palter, Robert. "Newton and the inductive method." *Texas Quarterly* 10(1967)160–73.

Pampusch, Anita M. " 'Experimental,' 'metaphysical,' and 'hypothetical' philosophy in Newtonian methodology." *Centaurus* 18(1974):289–300.

[Parr, Bartholomew.] [C. A.] "Observations on light, particularly on its combination and separation as a chemical principle." In *Essays by a Society of Gentlemen at Exeter.* Exeter, 1796, pp. 491–541.

Partington, J. R. *A History of Chemistry.* 4 vols. London: Macmillan, 1961–70.

Partington, J. R., and Douglas McKie. "Historical studies on the phlogiston theory. I. The levity of phlogiston. II. The negative weight of phlogiston. III. Light and heat in combustion. IV. Last phases of the theory." *Annals of Science* 2(1937):361–404; 3(1938):1–58, 337–71; 4(1939):113–49.

Pearson, George. "Experiments and observations made with the view of ascertaining the nature of the gaz produced by passing electric discharges through water; with a description of the apparatus for these experiments." *Journal of Natural Philosophy, Chemistry, and the Arts* 1(1797):241–8, 299–305, 349–55.

Péclet, Eugene. *Cours de physique.* Marseille: Antoine Ricard, 1823.

———. *Traité élémentaire de physique.* 2nd ed. Paris: L. Hachette, 1832.

Pedersen, Kurt Møller. "Roger Joseph Boscovich and John Robison on terrestrial aberration." *Centaurus* 24(1980):335–45.

Pemberton, Henry. "A letter to Dr. Jurin, Coll. Med. Lond. Soc. & Secr. R. S. concerning the abovementioned appearance in the rainbow, with some other reflections on the same subject." *Philosophical Transactions* 32(1722–3):245–61.

———. *A View of Sir Isaac Newton's Philosophy.* London, 1728; rpt. New York: Johnson Reprint, 1972.

Picard, Charles-Emile. "La vie et l'oeuvre de Jean-Baptiste Biot." *Mémoires de l'académie des sciences de l'Institut de France*, ser. 2, 59(1938):$_2$I–XXXIX.

Playfair, John. "Biographical account of the late John Robison." *Transactions of the Royal Society of Edinburgh* 7(1815):495–539.

Pouillet, Claude. *Elemens de physique expérimentale et de metéorologie*. 2 vols. Paris: Bechet, 1827–30.

Powell, Baden. *A Short Elementary Treatise on Experimental Optics Designed for the Use of Students in the University*. Oxford: D. A. Talboys, 1833.

Preston, Thomas. *The Theory of Light*. Edited by Charles Jasper Joly. 3rd ed. London: Macmillan, 1901.

Prevost, Isaac-Bénédict. "Sur le mode d'émission de la lumière qui part des corps colorés; moyen d'augmenter considérablement l'intensité de la couleur de ces corps." *Annales de chimie*, ser. 2, 4(1817):192–201.

———. "Nouvelles additions au mémoire sur le mode d'émission de la lumière qui part des corps colorés; moyens d'augmenter considérablement l'intensité de la couleur de ces corps." *Annales de chimie*, ser. 2, 4(1817):436–43.

Prevost, Pierre. "Quelques remarques d'optique, principalement rélatives à la réflexibilité des rayons de la lumière." *Philosophical Transactions* 88(1798):311–31.

Priestley, Joseph. *Disquisitions Relating to Matter and Spirit*. London, 1777.

———. *The History and Present State of Discoveries Relating to Vision, Light, and Colours*. London, 1772; rpt. Millwood, N.Y.: Kraus Reprint, 1978.

Prieur-Duvernois, Claude Antoine (Prieur de la Côte-d'Or). "Considérations sommaires sur les couleurs irisées des corps réduits en pellicules minces; suivies d'une explication des couleurs de l'acier recuit, et de celles des plumes de paon. Fragment d'un ouvrage sur la coloration." *Annales de chimie* 61(1807):154–79.

———. "De la décomposition de la lumière, en ses élémens les plus simples; fragment d'un ouvrage sur la coloration." *Annales de chimie* 59(1806):227–61.

———. "Extract from a memoir entitled 'Considerations on colours, and several of their singular appearances.' " *Philosophical Magazine* 22(1805):289–98.

———. "Extrait d'un mémoire ayant pour titre: *Considérations sur les couleurs, et sur plusieurs de leurs apparences singulières*." *Annales de chimie* 54(1805):5–27.

———. "On the decomposition of light into its most simple elements, being part of a work upon colours." *Philosophical Magazine* 28(1807):162–70, 210–19.

———. "Summary considerations upon variegated colours of bodies when reduced into thin pellicles; to which is added an explanation of the colours of tempered steel, and of those of peacocks' feathers. Extracted from a work on colours." *Philosophical Magazine* 28(1807):332–9, 29(1807):11–17.

Procès-verbaux des séances de l'académie tenues depuis la fondation de l'Institut jusqu'au mois d'août 1835. 10 vols. Hendaye: Imprimerie de l'Observatoire d'Abbadia, 1910–22.

"Propositions et demandes sur les couleurs des corps, au sujet du mémoire de M. Opoix, publié dans le *Journal de physique*, du mois d'août 1776; par un simple amateur de la physique." *Journal de physique* 10(1777):66–72.

Radicke, F.W.G. *Hanbuch der Optik, mit besonderer Rücksicht auf die neuesten Fortschritte der Wissenschaft*. 2 vols. in one. Berlin: Der Nicolaischen Buchhandlung, 1839.

Rappaport, Rhoda. "G.-F. Rouelle: An eighteenth-century chemist and teacher." *Chymia* 6(1960):68–101.

———. "Rouelle and Stahl – The phlogistic revolution in France." *Chymia* 7(1961):73–102.

"[Review of] *Experimental researches, &c.* . . . par Edward Bancroft." *Bibliothèque Britannique*. Sciences et Arts, 2(1796):265–86.

"[Review of] *Memoirs of the Literary and Philosophical Society of Manchester." Critical Review* 61(1786):343–53.
Robinson, Bryan. *A Dissertation on the Aether of Sir Isaac Newton*. Dublin, 1743.
Robison, John. "On the motion of light, as affected by refracting and reflecting substances, which are also in motion." *Transactions of the Royal Society of Edinburgh* 2(1790):83–111.
[———.] "Optics." In *Encyclopaedia Britannica* 13:231–364.
Rose, Paul Lawrence. *The Italian Renaissance of Mathematics: Studies on Humanists and Mathematicians from Petrarch to Galileo*. Geneva, Librairie Droz, 1975.
Rowning, John. *A Compendious System of Natural Philosophy: With Notes Containing the Mathematical Demonstrations, and Some Occasional Remarks. In Four Parts.* 2 vols. London, 1744.
———. *A Compendious System of Natural Philosophy*. 2 vols. London, 1744–5.
———. *A Compendious System of Natural Philosophy*. 2 vols. London, 1753.
Rutherforth, Thomas. *A System of Natural Philosophy, Being a Course of Lectures in Mechanics, Optics, Hydrostatics, and Astronomy: Which Are Read in St. John's College Cambridge*. 2 vols. Cambridge, 1748.
Sabra, A. I. *Theories of Light from Descartes to Newton*. London: Oldbourne, 1967.
Sadoun-Goupil, Michelle. *Le chimiste Claude-Louis Berthollet (1748–1822): Sa vie, son oeuvre*. Paris: J. Vrin, 1977.
———. "Science pure et science apliquée dans l'oeuvre de Claude-Louis Berthollet." *Revue d'histoire des sciences* 27(1974):127–45.
Salusbury, Thomas, ed. *Mathematical Collections and Translations*. Vol. 1. London, 1661; rpt. London: Dawsons, 1967.
Scheele, Carl Wilhelm. *The Collected Papers of Carl Wilhelm Scheele*. Translated by Leonard Dobbin. London: G. Bell & Sons, 1931.
Schofield, Robert E. "Josiah Wedgwood, industrial chemist." *Chymia* 5(1959):180–92.
———. *Mechanism and Materialism: British Natural Philosophy in an Age of Reason*. Princeton University Press, 1970.
Scientific Memoirs, Selected from the Transactions of Foreign Academies of Science and Learned Societies, and from Foreign Journals. Edited by Richard Taylor. 5 vols. London, 1837–52; rpt. New York: Johnson Reprint, 1966.
Senebier, Jean. *Mémoires physico-chymiques, sur l'influence de la lumière solaire pour modifier les êtres des trois règnes de la nature, & sur-tout ceux du règne végétal*. 3 vols. Geneva, 1782.
———. "Quatrième mémoire sur le phlogistique, ou réponse à la lettre de Madame de V***, contenue dans le *Journal de physique* pour le mois de septembre 1777, page 206, avec des remarques sur la nature du phlogistique." *Journal de physique* 11(1778):326–38.
———. *Recherches sur l'influence de la lumière solaire pour métamorphoser l'air fixe en air pur par la végétation*. Geneva, 1783.
———. "Réponse à la lettre de Madame de V*** contenue dans le supplément au *Journal de physique*, page 281; dans laquelle on trouvera, 1°. les raisons qui rendent probable le systême de l'émission de la lumière; 2°. des idées & des expériences nouvelles sur la nature de la lumière & de ses effets, & en particulier sur la décoloration des surfaces colorées qui sont exposées à la lumière, & sur l'étiollement des plantes." *Journal de physique* 14(1779):200–15.
———. "Seconde lettre à Madame de V***, ou mémoire sur la nature de la lumière & de ses effets, sur la décoloration des surfaces colorées exposées à son action, & sur l'étiolement des plantes." *Journal de physique* 14(1779):355–84.

———. "Sur le phlogistique, considéré comme la cause du développement, de la vie & de la destruction de tous les êtres dans les trois règnes." *Journal de physique* 8(1776):25–37, 9(1777):97–104, 366–76.

Shapin, Steven. "Robert Boyle and mathematics: Reality, representation, and experimental practice." *Science in Context* 2(1988):23–58.

Shapin, Steven, and Simon Schaffer. *Leviathan and the Air-Pump: Hobbes, Boyle, and the Experimental Life*. Princeton University Press, 1985.

Shapiro, Alan E. "Beyond the dating game: Watermark clusters and the composition of Newton's *Opticks*." In Harman and Shapiro, eds. *The Investigation of Difficult Things*, pp. 181–227.

———. "The evolving structure of Newton's theory of white light and color." *Isis* 71(1980):211–35.

———. "Experiment and mathematics in Newton's theory of color." *Physics Today* 37, no. 9 (September 1984):34–42.

———. "Huygens' *Traité de la lumière* and Newton's *Opticks*: Pursuing and eschewing hypotheses." *Notes and Records of the Royal Society of London* 43(1989):223–46.

———. "Kinematic optics: A study of the wave theory of light in the seventeenth century." *Archive for History of Exact Sciences* 11(1973):134–266.

———. "Newton's 'achromatic' dispersion law: Theoretical background and experimental evidence." *Archive for History of Exact Sciences* 21(1979):91–128.

Shapiro, Barbara J. *Probability and Certainty in Seventeenth-Century England: A Study of the Relationships between Natural Science, Religion, History, Law, and Literature*. Princeton University Press, 1983.

Shaw, Peter. *Chemical Lectures, Publickly Read at London, in the Years 1731, and 1732; and at Scarborough, in 1733; for the Improvement of Arts, Trades, and Natural Philosophy*. 2nd ed. London, 1755.

Sherman, Paul D. *Colour Vision in the Nineteenth Century: The Young-Helmholtz-Maxwell Theory*. Bristol: Adam Hilger, 1981.

Silliman, Robert H. "Fresnel and the emergence of physics as a discipline." *Historical Studies in the Physical Sciences* 4(1974):137–62.

Simpson, Allen. "The early development of the reflecting telescope in Britain." Ph.D. diss. University of Edinburgh, 1981.

Sivin, Nathan. "William Lewis (1708–1781) as a chemist." *Chymia* 8(1962):63–88.

Smeaton, W. A. *Fourcroy: Chemist and Revolutionary, 1755–1809*. Cambridge: W. Heffer and Sons, 1962.

Smith, Crosbie. " 'Mechanical philosophy' and the emergence of physics in Britain: 1800–1850." *Annals of Science* 33(1976):3–29.

Smith, Robert. *A Compleat System of Opticks in Four Books, viz. A Popular, a Mathematical, a Mechanical, and a Philosophical Treatise*. 2 vols. Cambridge, 1738.

———. *Traité d'optique . . . considérablement augmenté*. Translated by Juval le Roy. Brest, 1767.

Stahl, Georg Ernst. *Philosophical Principles of Universal Chemistry*. Translated by Peter Shaw. London, 1730.

Steffens, Henry John. *The Development of Newtonian Optics in England*. New York: Science History Publications/USA, 1977.

Stokes, George Gabriel. *Mathematical and Physical Papers*. Edited by G. G. Stokes and Joseph Larmor. 5 vols. Cambridge, 1880–1905; rpt. New York: Johnson Reprint, 1966.

Sutton, M. A. "Sir John Herschel and the development of spectroscopy in Britain." *British Journal for the History of Science* 7(1974):42–60.

———. "Spectroscopy and the chemists: A neglected opportunity?" *Ambix* 23(1976):16–26.

———. "Spectroscopy, historiography, and myth: The Victorians vindicated." *History of Science* 24(1986):425–32.

Talbot, W. H. Fox. "On the nature of light." *Philosophical Magazine,* ser. 3, 7(1835):113–18.

———. "Some experiments on coloured flames." *Edinburgh Journal of Science* 5(1826):77–81.

Taylor, Charles. "Biographical memoranda respecting Edward Hussey Delaval." *Philosophical Magazine* 45(1815):29–32.

Thackray, Arnold. *Atoms and Powers: An Essay on Newtonian Matter-Theory and the Development of Chemistry.* Cambridge, Mass.: Harvard University Press, 1970.

———. " 'Matter in a nut-shell': Newton's *Opticks* and eighteenth-century chemistry." *Ambix* 15(1968):29–53.

Thomson, Benjamin. *The Collected Works of Count Rumford.* Edited by Sanborn C. Brown. 5 vols. Cambridge, Mass.: Harvard University Press, 1968–70.

Thomson, Thomas. *A System of Chemistry.* 4 vols. 2nd ed. Edinburgh: Bell and Bradfute, and Balfour, 1804.

———. *A System of Chemistry.* From the 5th London ed. Notes by Thomas Cooper. 4 vols. Philadelphia: Abraham Small, 1818.

Tiraboschi, Girolamo, ed. *Notizie biografiche e litterarie in continuazione della biblioteca modenese.* 5 vols. Reggio: Torreggiani, 1833–7.

Vavilov, S. I. "Newton and the atomic theory." In the Royal Society, *Newton Tercentenary Celebrations, 15–19 July 1946.* Cambridge University Press, 1947, pp. 43–55.

Venel, Gabriel-François. "Chymie." In *Encyclopédie* 3(1753):408–37.

Venturi, Giovanni Battista. "Considerazioni ottiche." *Memorie di matematica e di fisica della società italiana* 3(1786):268–77.

———. "Indagine fisica sui colori." *Memorie di matematica e di fisica della società italiana* 8(1799):699–754.

———. *Indagine fisica sui colori.* 2nd ed. Modena: La Società Tipografica, 1801.

Verdet, Emile. *Leçons d'optique physique.* In *Oeuvres de E. Verdet.* Vols. 5, 6. Paris: Victor Masson et fils, 1869–70.

Voigt, Johann Gottfried. "Beobachtungen und Versuche über farbigtes Licht, Farben und ihre Mischung." *Neues Journal der Physik* 3(1796):235–98.

Wallace, William A. *The Scientific Methodology of Theodoric of Freiberg: A Case Study of the Relationship between Science and Philosophy.* Fribourg: The University Press, 1959.

Wescher, H. "The French dyeing industry and its reorganization by Colbert." *Ciba Review* 18(1939):643–6.

———. "Great masters of dyeing in 18th century France." *Ciba Review* 18(1939):626–41.

Westfall, Richard S. "The development of Newton's theory of color." *Isis* 53(1962):339–58.

———. *Force in Newton's Physics: The Science of Dynamics in the Seventeenth Century.* London: Macdonald, 1971.

———. "Isaac Newton's coloured circles twixt two contiguous glasses." *Archive for History of Exact Sciences* 3 (1965):181–96.

———. *Never at Rest: A Biography of Isaac Newton.* Cambridge University Press, 1980.

———. "Newton's reply to Hooke and the theory of colors." *Isis* 54(1963):82–96.

———. "Uneasily fitful reflections on fits of easy transmission." *Texas Quarterly* 10, no. 3 (1967):86–102.

Whewell, William. *History of the Inductive Sciences, from the Earliest to the Present Times.* 3 vols. London: J. W. Parker, 1837.

———. "Presidential address." *Report of the British Association* (1833) [1834]:xi–xxvi.

———. "Suggestions respecting Sir John Herschel's remarks on the theory of the absorption of light by coloured media." *Report of the British Association* (1834) [1835]:550–2.

Whiteside, D. T. "Newton the mathematician." In Bechler, ed. *Contemporary Newtonian Research*, pp. 109–27.

Willis, Thomas. *Diatribae duae medico-philosophicae: Quarum prior agit de fermentatione sive de motu intestino particularum in quovis corpore. Altera de febribus, sive de motu earundem in sanguine animalium.* London, 1659.

———. *Dr. Willis's Practice of Physick, Being the Whole Works of that Renowned and Famous Physician.* Translated by Samuel Pordage. London, 1684.

Wilson, David B. "The reception of the wave theory of light by Cambridge physicists (1820–1850): A case study in the nineteenth-century mechanical philosophy." Ph.D. diss. Johns Hopkins University, 1968.

Wisan, Winifred Lovell. "Galileo's scientific method: A reexamination." In Butts and Pitt, eds., *New Perspectives*, pp. 1–57.

Witte, A. J. de. "Interference in scattered light." *American Journal of Physics* 35(1967):301–13.

Wollaston, William Hyde. "A method of examining refractive and dispersive powers, by prismatic reflection." *Philosophical Transactions* 92(1802):365–80.

Wrede, Fabian Jakob von. "Attempt to explain the absorption of light according to the undulatory theory." In *Scientific Memoirs* 1(1837):477–502.

Wren, Christopher. *Parentalia: Or, Memoirs of the Family of the Wrens.* London, 1750; rpt. Farnborough, Hants.: Gregg Press, 1965.

Young, Matthew. "On the number of the primitive colorific rays in solar light." *Transactions of the Royal Irish Academy* 7(1800):119–37.

Young, Thomas. "An account of some cases of the production of colours, not hitherto described." *Philosophical Transactions* 92(1802):387–97.

———. *A Course of Lectures on Natural Philosophy and the Mechanical Arts.* 2 vols. London, 1807; rpt. New York: Johnson Reprint, 1971.

———. "Experiments and calculations relative to physical optics." *Philosophical Transactions* 94(1804):1–16.

———. "On the theory of light and colours." *Philosophical Transactions* 92(1802):12–48.

———. "Outlines of experiments and inquiries respecting sound and light." *Philosophical Transactions* 90(1800):106–50.

Ziemacki, Richard L. "Humphry Davy and the conflict of traditions in early nineteenth-century British chemistry." Ph.D. diss. University of Cambridge, 1974.

Zucchi, Niccolò. *Optica philosophia experimentis et ratione a fundamentis constituta.* Lyons, 1652.

INDEX

Aberdeen, University of, 273
absorption, selective: British debate on, 348–9, 351–4; caused by primordial particles, 122–8, 340; and chemical change, 298–9; explained in wave theory, 286–7, 349, 351–3, 360; introduced by Newton, 107–8; and mathematical description, 320, 325–6, 332–7; as nonmechanical process, 76, 123–4; *see also* affinities, elective: identified with selective absorption
absorption spectra: compared with Newton's theory of colored bodies, 285–7, 297–8, 310–14, 317–18, 345–7; described, 212–13; observed by Dutour, 238–41; predicted by nomograph, 94–6; representation of, 294–7, 301–2, 311–12, 318, 332–6, 347
absorption spectroscopy, method of, 213, 293–4, 310, 332; rejected by Biot, 324–5
Academy of Sciences: Bavarian, 257n; Berlin, 205, 234; Paris, 211, 218, 238n, 243, 251, 252, 260 (*see also* Institut National de France); St. Petersberg, 257n
active principles, 8, 74–6
Adet, Pierre-Auguste, 308
aether, 73–6, 222; as cause of optical phenomena, 76–85; and force, 197–9; after *Principia*, 141, 188, 197, 199
aethereal vibrations, 43, 57n, 117; as cause of optical phenomena, 54–5, 60–1, 69, 73, 76–85
affinities, elective, 245–7
——Berthollet's critique of, 322–3
——as bond of physics and chemistry, 219, 220, 246, 304–5
——identified with selective absorption, 275–6, 278, 280, 299, 310–12, 317, 321n, 349–51; unlawlike nature of, 307–8, 310–12, 320, 325–7, 355–6
——and Newton, 245–7
——*see also* chemistry of light; short-range forces
Airy, George Biddell, 344, 352, 356n
Albertus, Magnus, 5
alchemy, 74–6, 88
Alembert, Jean le Rond d', 204n, 218, 229
Algarotti, Francesco, 227n
analogy of nature, 43–4, 68, 84–5, 245
analysis and synthesis, 38–9, 200
Annales de chimie, 237n, 309, 328
Arago, Dominique-François-Jean, 221, 322, 328, 360–1, 363; and Biot, 319–20, 328
archetypal phenomena, 106–7, 109, 118, 239n
Arceuil, Society of, 221, 305
Archimedes, 27, 28
Aristotle, 5, 26–8, 243; Aristotelian conception of science, 12, 26–8, 34–5
Averroes, 5

Babbage, Charles, 338
Babinet, Jacques, 329n

Bacon, Roger, 6
Bailly, Charles, 329n
Bancroft, Edward, 272–8, 289, 303, 304; and Delaval, 232, 234, 274, 276–7; and dye chemistry, 212, 261, 262, 273–4
Barrow, Isaac, 11, 13; and mathematical science, 31–6, 37
Beccaria, Giacomo Battista, 255
Bechler, Zev, 19n, 85n, 145n
Beer, August, 334n
Beer's law, 295, 334n
Bentley, Thomas, 234n
Benvenuti, Carlo, 228; on theory of fits, 63n, 183, 184, 186, 203, 367
Bérard, Jacques-Etienne, 322n
Bergman, Torbern, 259, 260
Berthollet, Amédée-Barthelemy, 303n
Berthollet, Claude-Louis, 212, 220–1, 265–6, 278, 290, 309; and affinities, 247, 303, 304–5, 322–3; and Bancroft, 274–5, 275–6, 303, 304; and Delaval, 232, 303, 304; and dye chemistry, 259, 261, 262; and Hassenfratz, 309; and Haüy, on colored bodies, 302–3, 304, 305–8, 309, 310; and Prieur, 309, 316; and Venturi, 293
Berzelius, Jöns Jakob, 359
Beudant, François-Sulpice, 327–8
Bewley, William, 233n
Biot, Jean-Baptiste, 49, 221, 293, 321, 329n, 359; and Arago, 319–20, 328; and Brewster, 339, 343, 346; and defense of Newton's theory of colored bodies, 323–8; and Hassenfratz, 309, 322; and J. Herschel, 339, 341; and thick plates, 205n, 367–9
Black, Joseph, 221, 281
black bodies, 116–17
Blackburne, William, 280–1
Boerhaave, Hermann, 244
Bonnet, Charles, 228n, 255; and Senebier, 253, 256n, 258n
Bošković, Rudjer Josip, 126n, 183n, 203, 228; Boscovichean natural philosophy, 281, 282
Bouguer, Pierre, 218, 334
Bouguer's law, 334
Boyle, Robert, 6; on colors of bodies, 7, 8–9, 99–102; conception of science, 20, 24n, 31, 36; and mechanical chemistry, 88–9, 101–2, 119, 120; and transduction, 46–7, 102, 104; *see also* Newton, Isaac: and Boyle
Boyle's law, 20, 42
Brewster, David, 330n, 331–9, 341, 342–52, 363; and J. Herschel, 331–2, 341, 342, 345, 348, 352; and theory of fits, 136, 207n
British Association for the Advancement of Science, 341, 345n, 348, 351, 352
Brougham, Henry, 282–4; review of Venturi, 283–4, 292, 296–7, 298n, 301–2; and theory of fits, 202, 283
Buffon, Georges-Louis Leclerc, 247, 252
Bunsen, Robert, 350, 358
Burke, John G., 305n
Burlingame, Leslie J., 252n

C.A., *see* Parr, Bartholomew
caloric, 263–4, 280, 287n; *see also* heat; light: and heat
calxes, colors of, 231–2, 264
Cambridge, University of, 227, 230, 338; Lucasian Professor, 11, 31–2
Carnot, Lazare, 309
Cassini, Jean-Dominique, 146
Castillon, Johann, 234
Cavallo, Tiberius, 136, 363
certainty: derived from experiment, 21, 22, 35–6, 37–8, 39; and mathematical description, 13, 24–6, 31, 36; Newton's belief in, 12–14, 19, 21–2, 23n, 36, 39; *see also* probabilism
Chaptal, Jean-Antoine, 220, 221, 262, 266, 293
Charles, Jacques-Alexandre-César, 310, 314
Chaulnes, Michel-Ferdinand d'Albert d'Ailly, duc de, 154, 204–6
chemistry: and alliance with physics, 220, 246; development of, 219–25; as an independent science, 220, 222–5, 242–3, 258–9, 356–7; and optical instrumentation, 214, 255n, 267, 270, 357–9 (*see also* optical investigation of matter); vulgar or common, 75, 119, 222; *see also* mechanical chemistry; Stahlian chemistry
chemistry of light: and emission theory, 215–16, 257–8, 299; explained, 211, 246–7; and French physics, 319–20,

322–3; in Germany, 215, 257–8, 321; historiography of, x, 215, 254n; and Newton, 123, 142n, 245, 247; rejected by wave theory, 216, 286–7, 349, 352, 360–1
Cheyne, George, 226
Chisholm, Alexander, 277, 278
chlorine, 261, 265
chlorophyll, 11, 298, 359; *see also* green (of vegetation)
chromatic dispersion, 144; and chemical forces, 281–2; unlawlike nature of, 320
Clarke, Samuel, xii, 180n
Clavius, Christopher, 28, 32
Cohen, I. Bernard, 23n, 45
Colbert, Jean-Baptiste, 260
College Charlemagne, Paris, 322n
color, 5–7; compound and simple, denied, 10; harmonies, 43, 174; modification theory of, 7, 8, 9, 100–1; *see also* Newton's rings; thick plates, colors of; thin films, colors of; white light and color, Newton's theory of
color indicator test for acids and alkalies, 101, 120, 233, 248
color mixing, additive and subtractive, 299–301, 334
colored bodies, Newton's theory of
——analogical, 99, 117–22
——complementary reflected and transmitted colors in, 98–9, 106, 108–10, 324; denied, 239, 269–70, 297, 314, 317, 340
——and order and composition of colors, 94–6, 116n, 120–1, 274
——phenomenological, 98, 106–10, 134
——and size of corpuscles, 119–20, 227–8, 229, 286, 287
——weak version of, 228, 244
——*see also* absorption spectra: compared with Newton's theory of colored bodies; archetypal phenomena
colors: complementary, defined, 55; primary, 316, 343
colors of bodies: caused by a substance, 248, 255–6, 262–3, 272, 276; and chemical change, 100–1, 118–19, 248 (*see also* sequences of color changes; color indicator test for acids and alkalies); composition of, 94–6, 238–41, 249, 277, 291, 307–8; explained, 99, 347, 353–4; historiography of, 3, 5–8; *see also* Boyle, Robert: on colors of bodies; colored bodies, Newton's theory of; Hooke, Robert: on colors of bodies
Commercium epistolicum, 24n
confirmation: of calculation of thick plates, by measurements, 167, 169–70; of theory of fits, by thick plates, 169, 170, 175–7, 191–2; of theory of light, by Newton's rings, 26, 89, 96
corpuscles of matter: opaque or transparent, 101–2, 122, 125–8, 228–9; transparent, 101–2, 104, 112–13
corpuscular vibrations, 141–2, 171–9, 182
corpuscularity of matter, 114–16, 127–9
Coste, Pierre, 180n
Cotes, Roger, 15, 179n
Coulomb, Charles-Augustin, 218
crystallography, 305–6, 337–8, 339
Cudworth, Ralph, 74

Dalton, John, 221, 323, 328
Davy, Humphry, 221, 278
Déal (author of paper for Academy of Sciences, Paris), 361
Deane, Silas, 273
Defoe, Daniel, 180n
Delaval, Edward Hussey, 239, 297, 314, 326, 340; and Bancroft, 232, 234, 274, 276–7; and Berthollet, 232, 303, 304; and Brewster, 343, 345; Newton's theory of colored bodies, confirmed by, 230–4; refuted by, 110, 269–72, 277–8, 279n, 288, 354
Descartes, René, 6, 8, 16, 19; on transduction, 40–1; *see also* Newton, Isaac: and Descartes
Desmarets, Nicolas, 262n
Despretz, César-Mansuète, 329
dichroism, 299, 334
diffraction, 9, 141–2, 205, 317, 329n; in *Opticks*, 143, 148–9
"Discourse of Observations," *see* "Observations"
Dobbs, Betty Jo Teeter, 74n, 111n, 116n
Dollond, John, 282
Donovan, Arthur, 222n
double refraction, 319, 320, 322, 337; and Huygens, 147n, 203, 217

Draper, John William, 298, 344
Dufay, Charles-François de Cisternay, 255n, 259–60
Dutour, Etienne François, 205n, 206n, 237–41, 251, 293
dye chemistry, 212; development of, 259–63; mechanical theories of, 260, 262, 305; *see also* Prussian blue

eclipses, color of, 144–6
Ecole des Mines, 305
Ecole Polytechnique, 308, 309, 322, 327, 328, 361
Edinburgh, University of, 277–8, 280, 281, 282, 289
electric spirit, 16–17, 179n, 197–9
electroplating, 339n
emission spectra, 284–5, 330, 339
emission theory of light, 23; and absorption, 123; and colors of thin films, 60, 65, 67, 68, 76–82; and theory of fits, 186–9, 200–1; *see also* chemistry of light: and emission theory
Encyclopaedia Britannica, 205n, 279n, 281
Encyclopédie, 204n, 224, 229, 262
etiolation, 249, 254–6
Euler, Leonhard, 205, 234–7, 257, 287
Exeter, Society of Gentlemen, 280
experiment: ambiguity of results of, 318, 354; conclusions from, revisable, 19, 39; and conflicting results by different experimenters, 204–5, 239, 315–16, 345, 347; for discovering properties, 17–18, 200; few required, 35, 70–1; and mathematical law, 217–19, 220, 320–1, 334–7; and modern methodological canons, 71; Newton's erroneous conclusion from, 64–8; *see also* absorption spectra: compared with Newton's theory of colored bodies; absorption spectroscopy; certainty: derived from experiment; colored bodies: complementary reflected and transmitted colors in; confirmation; measurement; Newton's rings; reflection of light: not cause of colors of bodies; thick plates, colors of

fading, of colors, 249, 254–5, 260
Fatio de Duillier, Nicholas, 45
Figala, Karin, 88
filters, colored, 214, 255n, 299, 331–2, 340
fire, 244, 249, 252, 254, 279; *see also* phlogiston
Fischer, Carl, 215n
Fischer, Ernst Gottfried, 321
fits
———continuity of, 183, 190–1
———as disposition, 81–2, 160, 181–6
———etymology of, 180–1
———interval of, 181, 186n; and variation with different colors, 192; with direction, 191, 193–5; with index of refraction, 192; *see also* vibration length
———periodicity of, 181–6
———physical models of, 182, 196–9
———theory of: dated, 147; development of, 171–9; and emission theory of light, 186–9, 201; and force, 197–9, 202–3; intelligibility of, 136, 137, 195–6, 200–2; mathematical nature of, 137–8, 189–90, 192, 203, 206–7; reception of, 202–7, 236, 278, 283, 288–9, 314; and similarity to wave theory, 189–90, 201, 367–9
Flamsteed, John, 145–6
Florence, University of, 339n
fluorescence, 99, 107, 359
force, and Newton's optical theories, 74, 119, 196–99, 202–3; *see also* mechanical explanation, and Newton; short-range forces
Fouchy, Jean-Paul Grandjean de, 205n
Fourcroy, Antoine-François de, 220, 264, 265–6, 293, 309
"Fourth Book," xvii, 138, 141–3, 172
Franklin, Benjamin, 238, 273
Fraunhofer, Joseph von, 330, 347, 348
Freind, John, 221, 246
Fresnel, Augustin, 190, 322, 359–61
"Fundamentum Opticae," xvii, 139

Galilei, Galileo, 29–30, 217
Gehler, Johann Samuel Traugott, 205n, 215–16
Geneva, 253, 328, University of, 253
Geoffroy, Etienne-François, 246
Gilbert, William, 30
Gladstone, John Hall, 358
Glanvill, Joseph, 20
Glasgow, University of, 281, 289
glass, for prisms, 318n, 332, 347
Gobelins, tapestry works, 265

God, 74, 77, 127; regularity of actions of, 35, 36, 43, 71n
Goethe, Johann Wolfgang von, 216n
Good, Gregory, 338
Grassmann, Hermann Günther, 301
Gravesande, Willem Jacob's, 228–9, 244n
green (of vegetation): explained chemically, 255–6, 265 (*see also* photosynthesis); and order and composition of, 93, 121, 326, 345–6
Gregory, David, 145, 146, 226; memoranda, 124, 148, 149, 154n
Gregory, James, 31
Gregory, Joshua C., 86n, 89n
Gregory, Olinthus, 290
Gren, Friedrich Albrecht Carl, 257n
Grimaldi, Francesco Maria, 9, 28, 108
Grotthuss, Theodor von, 298
Grotthuss–Draper law, 298

Hacking, Ian, 19
Hakfoort, Casper, 257
Harman, P. M. (né Heimann), 222, 227n
Harrington, John, 43
Hartley, David, 230
Hassenfratz, Jean-Henri, 275, 321, 322, 361; on Newton's theory of colored bodies, 308–14, 318, 327–8, 355; relation to Prieur, 308–10
Hauksbee, Francis, 197
Haüy, René Just, 219, 275, 293, 309, 327, 337; and Berthollet, on Newton's theory of colored bodies, 302–3, 304, 305–8, 310; and defense of Newton's theory, 290, 322n, 329n
heat, 57n, 244, 287n; *see also* caloric; light: and heat
Heimann, P. M., *see* Harman, P. M.
Heinrich, Placidus, 257–8n
Hellot, Jean, 259, 260–1, 262, 305
Helmholtz, Hermann von, 301, 337n, 344
Helsham, Richard, 226
Henry, Thomas, 262n, 277
Herschel, John Frederick William, 151, 205n, 207n, 351–3; and Brewster, 331–2, 342, 344, 348, 352; on selective absorption and colors of bodies, 110, 212–13, 329n, 331–42
Herschel, William, 288–9, 338, 357
hierarchical structure of matter, 85–9, 102, 129, 131–3; Boyle on, 88–9, 102; matter in a nutshell, 226–7; porous nature of, 86–7, 128–9, 142; supported, 228–9, 306, 325
Hooke, Robert, 9, 20, 46; on colors of bodies, 7, 100, 102–5, 113–14; of thin films, 9, 50–9; *see also* Newton, Isaac: and Hooke
Hutton, Charles, 206n, 279n
Hutton, James, 279–80
Huygens, Christiaan, 203, 217; *see also* Newton, Isaac: and Huygens
hypotheses: contrasted with properties, 13, 17–18, 22–3; demarcated from certain principles, 12, 16, 17, 24, 72, 85, 97, 195; experimental, 13, 17, 18, 57, 176; imaginary, 13, 15–16, 18–19; and intelligibility, 23–4 (*see also* fits: theory of: intelligiblity of); role of, in Newton's science, 17–19, 178–9, 199–201
"Hypothesis," xvi, 60, 73–85
hypothesis-free science, 13, 18; theory of fits as, 187–9, 195–6, 202
hypothesis of vibrations, transformed into theory of fits, 171–9, 182
hypothetico-deductive method: and deductive sciences, 34; and Newton, 39, 191, 200–2

indigo (dye), 259, 260
inflection, *see* diffraction
infrared rays, 214, 289, 322n, 357
Ingenhousz, Jan, 255
Institut National de France, 309–10, 313, 315, 328, 361
interval, *see* fits: interval of; vibration length
Italian Society of Science, 292

James, Frank A.J.L., 350, 358
Jardin du Roi, 219
Jesuits, 17, 28
Jordan, Gibbs Walker, 289; critique of transduction, 112, 113–14n, 117–20n, 135, 363–6
Journal de physique, 237, 241, 249, 253; *see also* Rozier, François

Kargon, Robert, 32n
Kayser, Heinrich, 354–5
Keill, John, 221, 246
Kepler, Johannes, 29, 108
Kirchoff, Gustav, 350, 358
Klügel, Georg Simon, 206n
Kottler, Dorian Brooks, 211, 254n, 258n
Koyré, Alexandre, 15n

Lamarck, Jean-Baptiste, 251–2, 307n
Laplace, Pierre-Simon, 218–19, 247, 293, 309, 316, 328; Laplacian physics, 219, 304, 318–21, 327–8; Laplacians, 206n, 313, 314; and Lavoisier, 220, 246
Lavoisier, Antoine-Laurent, 219, 225n, 246, 252–3, 308; and chemical revolution, 220, 263–5
Le Blond, Jean-Baptiste, 260
Le Roy, Juval (translator), 203n, 205n
Leeuwen, Henry G. van, 19
Leibniz, Gottfried Wilhelm, 16, 24n, 140n, 217
Leslie, John, 277–8, 356–7
Lewis, William, 224–5, 277
Libes, Antoine, 322n
Libri, Guglielmo, 328
light: as active principle, 76, 124, 245; convertible into matter, 44n, 123, 256; and heat, 76, 78–9, 172, 263–4, 279, 280; speed of, 77; speed of, for different colors, 144–6; *see also* chemistry of light; emission theory of light; monochromatic light; periodicity: of light; wave theory of light
lignum nephriticum, 106–7, 109; *see also* archetypal phenomena
Link, Heinrich Friedrich, 258n
Lloyd, Humphrey, 352

McGuire, J.E., 40n, 58n, 74n, 111n
McMullin, Ernan, 23n, 44n
Macquer, Pierre-Joseph, 219, 225, 246; on dyeing, 255–6, 259, 261, 262; and Opoix, 211, 243, 251
Magirus, Johannes, 5n
Mairan, Dortous de, 238
Maistre, Xavier de, 328n
Malus, Etienne-Louis, 219, 221, 322
Mandelbaum, Maurice, 16n, 40n, 102n
Marivetz, Etienne Claude de, 249, 251, 254
Marseilles, University of, 327
mathematical laws, and French physics, 309, 320–1, 361
mathematics, mixed, 25–7, 28, 32–6, 216–17
mathematization of nature, 8, 12–13, 24–6, 26–36, 37, 216–18; and Newton, 13–14, 21, 30, 32, 36, 39–40; *see also* fits: theory of; fits: mathematical nature of; white light and color, Newton's theory of, as mathematical science
matter theory, *see* corpuscularity of matter; hierarchical structure of matter; light: convertible into matter
Maxwell, James Clerk, 301
Mayer, Johann Tobias, 257n
Mazéas, Guillaume, 204–6, 235
measurement: and Newton's rings, 54, 61, 63–8; and thick plates, 166–70; precision, doubted by Mazeas, 204; of thin film, Hooke's failure, 52, 57, *see also* monochromatic light: insufficiently intense for measurement; physics: quantitative
mechanical chemistry, 75–6, 101, 119, 120, 260; attacked, 222–5, 232–4, 262–3, 274–5, 304–5, 306–7
mechanical explanation, and Newton, 8, 73–6, 77, 85, 119, 246–7
mechanical philosophy, 6, 73–6, 77; and mathematics, 33, 36; *see also* transduction: and mechanical philosophy
Melloni, Macedonio, 344
Melvill, Thomas, 125n, 229–30
metallic oxides, *see* calxes
microscope, and visibility of corpuscles of matter, 46, 104, 113, 121–2
Miller, William Allen, 358
mineralogy, 305, 337–8
mirror, telescope, 155–6
Mitscherlich, Eilhard, 337
Modena, University of, 292–3
Moigno, François Napoleon Marie, 328n, 348n
Monge, Gaspard, 220, 309, 310, 314
monochromatic light, 68–70, 330, 331–2, 339; insufficiently intense for measurements, 61, 70, 155; *see also* vibration length
Montpellier, University of, 253n
More, Henry, 74
Muséum d'Histoire Naturelle, 305
musical scales, 91–2, 155n, 174, 192–3; *see also* sound, analogy with light
Musschenbroek, Pieter van, 205n, 228n, 244n

natural bodies, 98n
natural philosophy, 204, 218, 238, 239–41, 249, 315, 317
"New theory," 10, 12, 21, 38
Newton, Isaac
——and Barrow, 31–2, 36, 37

———and Boyle, 20, 88–9, 118–19; on colors of bodies, 8, 9, 99–101, 107
———and Descartes, 8, 16, 41–2, 73
———and Galileo, 30n
———and Hooke, 9–10, 12, 21, 36–7, 62, 122; on colors of bodies, 100, 103–5, 112–15; on thin films, 50–9, 72n; notes on *Micrographia*, 9, 50, 56n, 58, 65
———and Huygens, 21, 23, 71, 129, 140, 148n
Newtonianism, 218–19, 225–9, 268–9; *see also* Laplace, Pierre-Simon: Laplacian physics
Newton's rings: complementary colors of, 55, 69; composition of colors of, 55, 93; discovered, 52–4; experiment criticized, 204, 288; experimental arrangement of, 54; explanation of, 79–83; periodicity of, 54, 63–4, 80–3; sequence of colors of, 55–6, 62–3; variation of, with index of refraction, 56, 58, 68, 70–71, 93–4; with obliquity, 58, 64–8, 83–5
Newton's table of colors, 94, 95; and color sequences, 121, 233, 326 (*see also* sequences of color changes); and size of corpuscles, 119, 133, 227–8
Nobili, Leopoldo, 339n
Nollet, Jean-Antoine, 218, 229, 237–8, 240, 251, 266
nomograph, 89–96, 287n, 310–12

"Observations," xvi, 59–60, 62–73; incorporated in *Opticks*, 139–40
"Of Colours" (Add. 3975), 54–6, 58, 87, 107; described, xvi, 11
"Of Colours" (Add. 3996), xvi, 8–9
"Of ye coloured circles," 59–62, 64n, 65–7, 68, 69–70; described, xvi, 11, 65n
Oldenburg, Henry, 21, 22, 24, 38, 60, 72, 187
"On vegetation," 74–6
106-mean-proportionals, law of, 68, 164–5, 193–5
opacity of bodies, 112, 114–15, 116–17, 123, 125; Hooke on, 105; *see also* corpuscles of matter; transparency of bodies
Opoix, Christophe, 211, 242–3, 248–53, 255, 264, 361
optical investigation of matter, 337–9, 350–1, 359; *see also* chemistry: and optical instrumentation
Optical Lectures, 11, 24–5, 37, 38, 59n; and Venturi, 293
Opticks, xvii, 229; Advertisement, 140–1; Book IV, Part I, xvii, 138–9, 147, 151; composition of, 138–50; draft preface, 127–8, 200; manuscript of, 138; renumbering of books, 139, 148, 149–50; *see also* queries
Oresme, Nicole, 5, 6
Orval, Hecquet d' (chemist), 260n
oxygen, as cause of color, 263–6, 275–6, 278, 280, 303

Palladino, Paolo S., 294
Pampusch, Anita M., 17n
Pardies, Igace Gaston, 17–19, 21–2
paroxysms, 148, 180
Parr, Bartholomew (pseud. C.A.), 280
partial reflection, 79, 184–5, 196; *see also* reflection of light
Pasteur, Louis, 358
path difference, 84, 159–60, 162–4, 190, 367
Pavia, University of, 293
Pearson, George, 280n
Péclet, Eugene, 329n, 348n
Pemberton, Henry, 203n, 226
periodicity: of colors of thin films, 51, 55–6; in light, 182, 185–6, 189, 196; of light, 49, 64, 69, 167, 184–6; proposed by Hooke, 51; *see also* Newton's rings: periodicity of
phenomenological theories, 13, 17–18, 22–4
Philosophical Transactions, 21
phlogiston, 243–5; as cause of color, 243–4, 248–9, 254–6, 272; in light, 248–9, 251n, 254, 272; and oxygen, 244, 263–5, 272n; and refractive force, 247; *see also* calxes
phosphorescence, 212, 230
photochemistry, 254–6, 278n; and chemistry of light, 212, 253–4, 257–8; *see also* chemistry of light; photosynthesis
photography, 254, 278n
photosynthesis, 255, 265
physics, 217–19, 221, 237n; Aristotelian, 26–8; mathematical, 26, 29–30, 33, 217 (*see also* mathematics, mixed; mathematization of nature); quantitative, 205, 206, 217–19, 221,

321, 348, 355–6; *see also* chemistry; natural philosophy
Pitcairne, Archibald, 226
Playfair, John, 282
polarimeter, 358
polarization, 147n, 319, 322, 337, 361; and J. Herschel, 332, 339
Pordage, Samuel, 180
Pörner, Carl Wilhelm, 262n
Pouillet, Claude, 329n
Powell, Baden, 353n
Prevost, Isaac-Bénédict, 328
Prevost, Pierre, 207n
Priestley, Joseph, 151n, 221, 226–7, 255, 273, 289n; on Newton, 205–6, 228
Prieur-Duvernois, Claude Antoine (pseud. Prieur de la Côte-d'Or), 314–18, 321, 345; and relation to Hassenfratz, 308–10
Principia mathematica, xiv, xvii, 38, 217; General Scholium, 14–17, 43, 198–9; planned appendix, 179n, 197; Rules of Reasoning, 19, 42–6, 47–8, 118, 126–7
probabilism, 12–13, 19–20, 35; and Newton, 38–9; *see also* certainty
properties, 13, 17–18, 22–3
Prussian blue, 255, 260, 261, 265
Ptolemy, 27, 28

Quatremère-Disjonval, 260n
quercitron, 273
queries: as hypothesis, 24n, 60, 143; Latin, xiii; numbering of, xiii, xvii
"Questiones quaedam philosophicae," *see* "Of Colours" (Add. 3996)

Radicke, F.W.G., 353n
rainbow, 5, 7
reflection of light: explained with aethereal surface, 77–91; explained with difference of density, 112–13; explained with theory of fits, 183–6; explained without collision with parts of bodies, 77–8, 122–5; not cause of colors of bodies, 105, 108, 235–6, 269–72, 278, 314, 327n, 340–1; *see also* colored bodies: complementary reflected and transmitted colors in; partial reflection
refractive power: and affinities, 319–20; dependent on density, 231, 247; dependent on phlogiston, 247; distinct from absorption, 250, 258, 261, 275; *see also* chromatic dispersion
refutation, *see* absorption spectra: compared with Newton's theory of colored bodies; colored bodies: complementary reflected and transmitted colors in, denied; reflection of light: not cause of colors of bodies
Ribaucour, de (chemist), 260n
Ritter, Johann Wilhelm, 357
Robinson, Bryan, 228n
Robison, John, 281–2
Rømer, Ole, 77
Rouelle, Guillaume-François, 219, 244, 253, 264
Rowning, John, 203n, 204n; and matter in a nutshell, 226–7
Royal Military Academy, Woolwich, 279, 290
Royal Society of London, 11, 37, 60, 72, 232, 273, 284; controversies at, 12, 24n, 57, 288–9; and probabilism, 14, 19, 20, 21, 31
Rozier, François, 237n, 251n
Rutherforth, Thomas, 203n, 227–8

St. Bartholomew's Hospital, London, 273
Salusbury, Thomas, 30n
Saury (or Sauri), Jean, 252n
Scheele, Carl Wilhelm, 255n, 261, 265, 358
Schulze, Johann Heinrich, 254–5
Senebier, Jean, 212, 253–9, 265, 275, 298, 358
sequences of color changes, 230, 233, 298n, 326, 340, 346; predicted by Newton, 120–1, 230
Shapiro, Barbara, 19
Shaw, Peter, 223n, 224, 244n
short-range forces, 203, 220, 221–2, 318–19; and Laplacian physics, 219, 318–21 (*see also* Laplace, Pierre-Simon); proposed by Newton, 43–4, 141, 143n, 197, 219, 245–7; *see also* affinities, elective
silver salts, sensitivity to light, 254–5
simple amateur of physics, 249–50
Simpson, Allen, 154–5
skepticism, 19, 28–9, 34–5
Smith, Robert, 203n, 205n, 227
Sorbonne, 327
sound, analogy with light, 43, 174, 236–7, 352–3; *see also* musical scales

spectral lines, 284, 330, 347, 350–1
spectrochemical analysis, 350–1
spectroscopy, historiography of, 354–9
spectrum: analysis by absorption, 316, 343–5
Stahl, Georg Ernst, 220, 223, 243–4
Stahlian chemistry, 220, 222–5, 243–4, 264; and anti-physics polemics, 220, 223–4, 242–3, 251, 258–9
Stokes, George Gabriel, 151, 205n, 358, 359
Sutton, M. A., 358
Sydenham, Thomas, 180n

Talbot, W. H. Fox, 347n, 350
Tamny, Martin, 58n
Tartaglia, Niccolò, 29
Taylor, Charles, 231n
telescope, reflecting, 11, 154–5
Tessier, Henri Alexandre, 298
Thackray, Arnold, 226–7
Theodoric of Freiberg, 5, 6n
thick plates, colors of: analogous to thin films, 150, 153, 155–7, 159–64, 167, 173; calculation of, 164–9, 367–9; caused by scattering, 152, 154, 205; discovered, 143, 145–6, 154–5; experiments on, by duc de Chaulnes, 204–5; explained by Newton, 157–60; explained by wave theory, 151–3, 367
thin films, colors of: explained by ordinary refraction, 240, 251; investigated by Hooke, 50–9; investigated by Mazéas, 204; of uniform thickness, 71–2, 84, 150, 162–4; *see also* Newton's rings
Thomson, Benjamin (Count Rumford), 287n
Thomson, Thomas, 289–90, 360
Tingry, Pierre-François, 253
transduction, 40–8; in eighteenth century, 126n, 134–5, 230, 271, 298n; explained, 4–5, 40; and mechanical philosophy, 40–1, 46–7, 102, 104–5; and Newton's theory of colored bodies, 113–18, 130, 134–5; problems with, 5, 45, 111–12, 115n, 125–7, 134–5 (*see also* Jordan, Gibbs Walker: critique of transduction)
transmittance curves, spectral, 332–4
transparency of bodies, 113, 114, 115–17; *see also* corpuscles of matter; opacity of bodies
Trinity College, Dublin, 226

ultraviolet rays, 322n, 357

Vavilov, S. I., 86n
vegetative processes, 74–6
Venel, Gabriel-François, 219, 224, 262–3
Venturi, Giovanni Battista, 292–302, 335–6; and Dutour, 240, 293, 315; and theory of fits, 207n; *see also* Brougham, Henry: review of Venturi
Verdet, Emile, 353n
vibration length, 61, 63–4, 80–1; first measurement of, 54; final measurement of, 168–9; interim measurement of, 64; ratio of extreme values, 61, 91–2, 155–6, 174; relation to wavelength, 64n; variation with different colors, 55, 61, 69–70, 90–3; with direction, 67–8, 84–5; with index of refraction, 56, 68, 70–1, 93–4; of yellow light, 64, 164; *see also* fits: interval of
vibrations, *see* aethereal vibrations; corpuscular vibrations; waves, overtaking
vision, 43, 73; color, 284, 301
Voigt, Johann Gottfried, 258n

Wallis, John, 31, 149
wave theory of light, 216, 340; and absorption, 286–7, 349, 351–3, 360; debated in England, 331, 352; and Euler, 234–7; and Newton's attack on, 188; and theory of fits, 64n, 186–90, 201; and thick plates, 151–3, 367–9; and thin films, 84–5, 162–4; *see also* chemistry of light: rejected by wave theory; path difference
waves, overtaking, 79–1, 182, 196, 197–9; difficulty of mathematizing, 81n, 173–4, 189–90
Wedgwood, Josiah, 277
Wedgwood, Tom, 278n
Westfall, Richard S., 49n, 178–9, 180n
Whewell, William, 215, 344, 352
white light, composed of three primaries, 310, 316–17, 343–4
white light and color, Newton's theory of, 6–7, 9–12; as mathematical science, 21, 23n, 25–6, 36–8, 96–7
Willis, Thomas, 180
Wilson, Benjamin, 230
Wollaston, William Hyde, 284–5, 337, 358
Wrede, Fabian Jakob von, 353

Wren, Christopher, 13, 31
Wünsch, Christian Ernst, 310n

Young, Matthew, 316, 345
Young, Thomas, 52, 204n, 221, 302, 332; on absorption and Newton's theory of colored bodies, 284–8; on colors of thick plates, 151; and the duc de Chaulnes, 154, 205n; identification of Dutour, 237; identification of Jordan, 363; on Priestley, 206; and theory of fits, 136, 190

Zucchi, Niccolò, 108